ALLE · ZEIT · WACH · 1842

G. Obe A.T. Natarajan (Eds.)

Chromosomal Aberrations

Basic and Applied Aspects

With 100 Figures

Springer-Verlag
Berlin Heidelberg New York London
Paris Tokyo Hong Kong Barcelona

Professor Dr. GÜNTER OBE
FB9 der Universität
Gesamthochschule Essen
Universitätsstraße 5
Postfach 10 37 64
4300 Essen 1, FRG

Professor Dr. A.T. NATARAJAN
State University of Leiden
Department of Radiation Genetics
and Chemical Mutagenesis
P.O. Box 95 03
2333 AL Leiden, The Netherlands

ISBN-13: 978-3-642-75684-9 e-ISBN-13: 978-3-642-75682-5
DOI: 10.1007/978-3-642-75682-5

Library of Congress Cataloging-in-Publication Data. Chromosomal aberrations: basic and applied aspects / G. Obe, A.T. Natarajan (eds). p. cm. Includes index. ISBN-13: 978-3-642-75684-9 (U.S.: alk.paper) 1. Human chromosome abnormalities. 2. Medical genetics. I. Obe, G. II. Natarajan, A. T. [DNLM: 1. Chromosome Aberrations. 2. Chromosome Abnormalities. WH 462.A1 C557] RB 155.5.C47 1990 616'.042-dc20 DNLM/DLC for Library of Congress 90-9804

Softcover reprint of the hardcover 1st edition 1990

2131/3145-543210 – Printed on acid-free paper

Preface

Eukaryotic chromosomes are complex structures containing very long DNA molecules, histones, and nonhistone proteins. The structural features of the association of DNA with histones are relatively well understood. The impact of nonhistone proteins on the structure of chromosomes is still a mystery. Chromosomes are dependent on the cellular environment in which they exist and their activities are part of a complex cellular network. The present volume deals mainly with chromosomal aberrations. Understanding the mechanisms of the origin of such aberrations would give us a better insight in the structure and function of the chromosomes and this is one aspect of the present volume, namely, the basic one. Chromosomal aberrations are indicators of mutagenic activity and are widely used as end points in testing for mutagens; some articles of the volume deal with this applied aspect. The following topics are discussed: chromosome structure, repair of genetic damage and chromosomal aberrations (Chaps. 1–6), induction of chromosomal aberrations with restriction endonucleases (Chaps. 7–9), chromosomes and cancer (Chaps. 10–12), human disorders with chromosomal instabilities (Chap. 13), the phenomenon of adaptive response (Chaps. 14–17), the use of chromosomal aberration frequencies as biological dosimeters of radiation exposure (Chaps. 18–22), and chromosomal aberrations as indicators of mutagenic activities of environmental chemicals and life-style factors (Chaps. 23–27).

We thank the authors for their contributions and the staff of Springer-Verlag, especially Dr. Dieter Czeschlik and Mrs. Antonella Cerri for their help.

We dedicate this volume to Professor Rigomar Rieger on the occasion of his 60th birthday, in recognition of his classical contributions in the area of cytogenetics.

Essen and Leiden, 1990 G. Obe and A.T. Natarajan

Prof. Dr. Rigomar Rieger
(Photo: Peter Wieler)

Contents

List of Contributors

Quantitative Detection of Chromosome Structures by Computerized Microphotometric Scanning

M. E. DRETS, G. A. FOLLE, and F. J. MONTEVERDE[1]

Contents

1 Introduction

Mammalian chromosomes are highly complex structures as revealed with different banding techniques obtained using fluorescent dyes, proteolytic treatments or differential denaturating methods (Arrighi and Hsu 1971; Caspersson et al. 1969; Drets and Shaw 1971; Dutrillaux and Lejeune 1971; Dutrillaux 1973; Sumner et al. 1971; Yunis 1976).

Specific banding patterns are induced by restriction endonucleases (RE) in fixed chromosomes (Bianchi and Bianchi 1987). Exposure of living mammalian cells to REs leads to the production of structural chromosomal aberrations (Obe et al. 1987) and to sister chromatid exchanges (Natarajan and Mullenders 1987).

Different banding patterns reflect the DNA base composition, Giemsa (G) and quinacrine (Q) bands are relatively rich in adenine and thymine (AT) and the reverse (R) bands are relatively rich in guanine and cytosine (GC) and thought to be chromosome segments with concentrated active genes (Korenberg and Engels 1978; Korenberg and Rykowski 1988; Weisblum and de Haseth 1972; review by Therman 1986). A precise localization of bands is critical at the organizational level of chromosomes, e.g. R-G/Q band junctions are believed to be sites of exchanges and rearrangements induced by clastogenic agents (Morgan and Crossen 1977) and also "hot spots" for the occurrence of mitotic chiasmata (Korenberg et al. 1978; Kuhn and Therman 1986).

Chromosome bands were the subject of several international meetings on chromosome nomenclature (ISCN 1985). Complete maps of banding patterns resulted from these conferences but no quantitative data on band localization, band size and

[1] Division of Human Cytogenetics and Quantitative Microscopy, Instituto de Investigaciones Biologicas "Clemente Estable", Avda, Italia 3318, Montevideo, Uruguay

band-interband junctions were reported. Published maps were thus largely based on direct microscope observations and not on quantitative estimations on the position and size of bands.

The problem of the quantitative band localization has not been completely solved as yet. Numerous changing parameters found in usual metaphase spreads such as division stage, degree of chromosome condensation, chromosome bending or overlapping pose serious difficulties in developing a reliable method for quantitative chromosome image analyses.

We studied the problem of quantitative band localization using centromeric (C) banded human Y-chromosomes and G-banded human chromosomes No. 1. The chromosomes were scanned using a semi-automatic analogue recording microphotometer. Densitometer tracings thus obtained were measured and quantitative maps of the relative band localization of C- and G-bands were drawn (Drets and Seuanez 1974).

Based on this method a computer program was written (Bandscan Program) (Drets 1978). The Bandscan program allowed the detection of the relative position of several characteristic bands and landmarks of human chromosome No. 1, confirming data previously obtained from analogue densitometer tracings. The Bandscan program was subsequently rewritten and extensively reviewed for our present instrumentation and now it allows the detection of the relative positions of band densitometer peaks and of band-interband junctions, thus quantitative information on the chromosome structure can be obtained (Drets and Monteverde 1987).

Since the induction of chromosome aberrations is closely related to the organization of chromosomes and with the banding patterns, this chapter reports briefly on the quantitative analytical methodology developed in our laboratory concerning the densitometric analysis of chromosomes using microscope photometric chromosome scanning and graphics computer diagrammatic imaging, which can be useful in cytogenetics research.

2 Materials and Methods

Human lymphocyte cultures were prepared according to Edwards (1962). Chinese hamster ovary (CHO) cells were cultured in Petri dishes or, alternatively, in flasks containing McCoy's 5 A Medium (Gibco) supplemented with 200 mM glutamine (Sigma). Cells were exposed to colchicine (Merck) prior to harvesting, fixed in methanol-acetic acid (3:1) and the preparations stained with Giemsa (Merck) stain. C-banding was obtained following the procedure of Arrighi and Hsu (1971). G-banding was induced by treating the chromosomes with trypsin (Seabright 1971), R-banding was produced following a modification of the fluorescence plus Giemsa procedure reported by Perry and Wolff (1974), T-banding was obtained by Dutrillaux's procedure (1973). A Zeiss Photomicroscope II and a 63X Plan Apochromatic phase immersion objective was used. Reflected light microscope observations were performed with a Zeiss vertical illuminator III C/45 mm system and reflector H-Pl-Pol. Microscope photographs were taken with High Contrast Copy Film (Kodak, Rochester) exposed at DIN 8 and developed with Microdol (Kodak) at 20 °C for 9 min. Negatives were enlarged for scannings on Fine Grain Positive Film (Kodak) and developed in Dektol (Kodak) developer for 1–2 min. Chromosomes were scanned using a Zeiss microscope photometer MP01 with a 10-μm step scanning stage and Zeiss Luminar lenses (40 mm 1:4/A. 0.13; 25 mm 1:3,5/A. 0.15). Electronic instrumentation associated with the MP01 system was described previously (Drets 1978) except that a Digital PDP 11/23 computer and a graphics color terminal from Tektronix model 4107 were associated on-line for image analysis. Quantitative localization of densitometric band peaks was based on Bandscan, an interactive program developed for the Wang programmable calculator 720C (Drets 1978). Software was developed by one of us (FJM) for band-interband junction localization, graphics band quantitative analyses, pixel image and pseudo-third dimension diagrammatic computer displays and sister chromatid exchange detection (SCE-SCAN program). A description of algorithms and computer programs developed will be reported elsewhere.

3 Results and Discussion

3.1 Improved Mapping of Bands

The variable density and staining intensity of chromosome bands makes a complete detection of all bands in densitometric analogue curves obtained after scanning chromosome arms difficult. Figure 1a illustrates a computer diagram generated from a single scanning of the long arm of chromosome No. 1 from a CHO cell. The left chromatid shows the relative position of band densitometer peaks as detected by the Bandscan program. Values obtained on the relative positions of band-interband junctions are seen on the right chromatid. In both chromatids, bands and lines were displayed according to the relative positions and size detected as reported previously (Drets and Monteverde 1987). A number of densitometric peaks were detected (left chromatid) but only seven bands were displayed, including the centromeric one.

To overcome this problem, an algorithm to transform densitometric curves was developed. The sequence of analysis was as follows: (1) chromosomes were scanned and computer files generated; (2) the detected curves were transformed and modified data saved in a new file which was subsequently used for graphics banding analysis.

Figure 1b shows the result obtained after transforming the analogue curve corresponding to the chromosome arm illustrated in Fig. 1a. In the new diagram the number of bands detected increased to 12 and, on the whole, a better quantitative definition was obtained as indicated by the number of band-interband junction localizations detected. This method of curve transformation allows band analyses independently of stain intensity, degree of chromosome contraction or banding pattern procedure used.

We consider that this kind of computerized analysis will increase the precision of the localization of break points along the chromatids produced by mutagenic agents, and will relate them to specific bands or interbands and, by this to the general organization of the chromosomes.

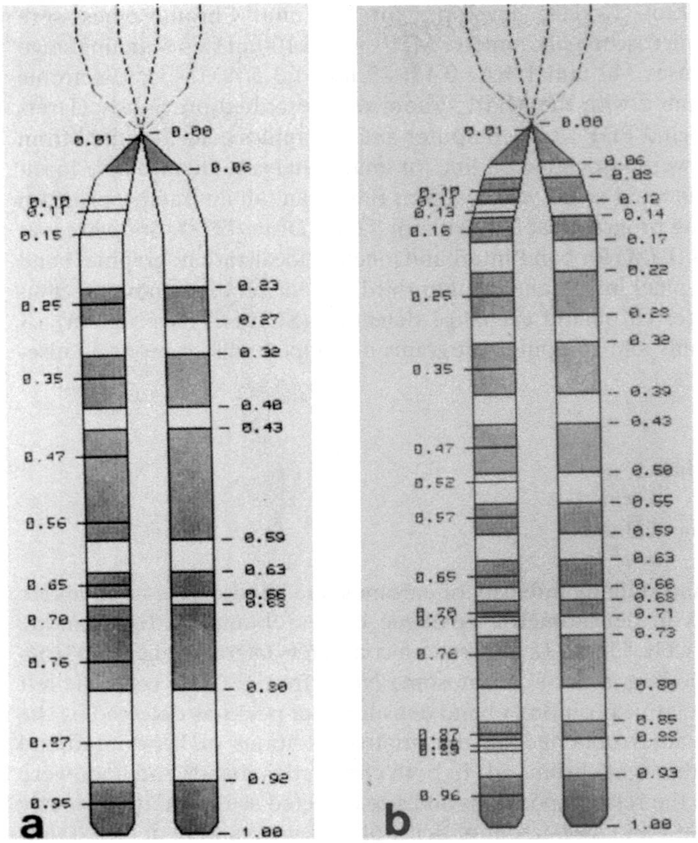

Fig. 1.a–b. Computer graphics diagrams obtained by scanning the long arm of chromosome No. 1 of a CHO cell. **a** Graphics diagram without transformation of the analogue densitometric curve. Lines and values appearing on the *left* chromatid represent densitometric peaks and on the *right* chromatid diagrams of bands and band-interband junctions values are displayed according to their relative positions. **b** Computer graphics diagram of the same chromosome after curve transformation. Note that in **b** the number of bands is higher than in **a**

3.2 Detection of Image Density Distribution

Appropiate data manipulation and the use of modern color graphics terminals have added a new dimension to quantitative cytogenetic analysis of banding patterns and related analytical problems.

Special computer graphics programs for displaying chromosomes and nuclei as pseudo-third dimension graphics diagrams, color pixels or numerical displays, representing the different densities measured, were also developed.

The application of this methodology to banded chromosomes generated interesting images. In particular, we detected that densities of individual bands can be differentially distributed between sister chromatids. This observation seems to be rather general and independent of the banding procedure followed.

Figure 2a–d shows four examples of this differential density distribution as observed using different banding procedures. Figure 2a illustrates an R-banded human chromosome No. 9 in a pixel graphics image showing differential staining density distribution between the chromatids of the long arm. Figure 2b is a C-banded human Y chromosome where the centromeric heterochromatin as well as the terminal heterochromatic block show asymmetric distribution between the chromatids. Figure 2c illustrates a G-banded dicentric chromosome from a CHO cell with irregular distribution of band material between the sister chromatids; and Fig. 2d presents a CHO Giemsa-stained chromosome presenting numerous sister chromatid exchanges showing that the highest chromatin densities were limited to two chromosome exchanged segments.

The scanning of T-banded chromosomes showed that T-material was more concentrated in the telomere region of one chromatid as compared to the other (Fig. 3). Graphics windowing allowed a comparison on screen images obtained simultaneously with differnt dwell scanning times and averaged measurements. This technique is exemplified in Fig. 3a, b for one- and ten-step dwell times showing that T-material was found to be denser in one chromatid of the long arm of human chromosome No. 1.

The differential distribution of banded material could result from different reactivities of sister chromatids to the treatments or from real differences between homologous chromatids which could result from unequal crossing over in regions of highly repeated sequences.

3.3 Detection of Intercalary Heterochromatin

We developed graphics programs for the analysis of C-banded chromosomes to measure band size and localization and band-interband junctions of heterochromatic segments. These programs are particularly useful for studying the variability of intercalary C-segments (Pätau 1973) as observed in CHO chromosomes. Typically, there are intercalary C-segments in chromosome No 1. and in the X-chromosome located close to secondary constrictions. Figure 4a shows intercalary dots of heterochromatin as seen with Giemsa staining in chromosome No. 1 and the X-chromosome (arrows). The use of reflection optics combined with Giemsa staining confirmed these findings and showed that the whole long arms of the X-chromosomes were heterochromatic (Fig. 4b). Pixel imaging of the X-chromosome enhanced the centromeric heterochromatin and the two intercalary segments located close to a secondary constriction (Fig. 5a). In some cells, two or three C-segments were spread along,the chromatids, particularly in the X-chromosome. An example of three intercalary heterochromatic segments observed in one X-chromosome from a CHO cell is shown in Fig. 5b (inset). A pixel image of this chromosome (Fig. 5b) showed that the machine detected these intercalary C-segments but also heterochromatic material spread in the two distal thirds of only one chromatid (arrows). With our system it was possible to locate these segments quantitatively. Computer graphics diagrams of these chromosomes are presented in Fig. 6a–b. Relative values of densitometric peaks and band-interband junctions showed that the extra intercalary C-segments were located at positions different to the ones found in most of the normal X-

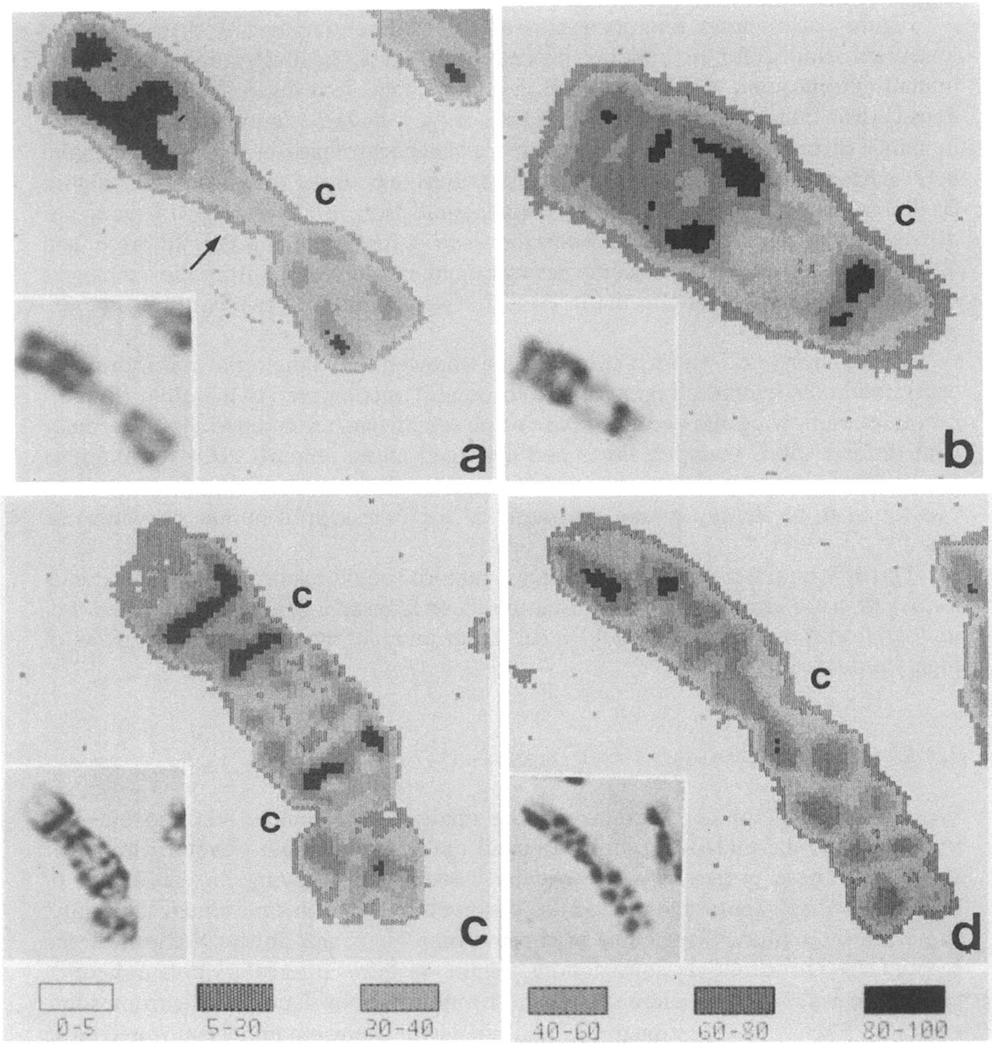

Fig. 2.a–d. Differential distribution of chromatin material as detected by pixel imaging of banded chromosomes. **a** R-banded human chromosome No. 9 showing denser R-segments in one chromatid. Chromosome arms, chromatids, secondary constriction (*arrow*), centromere region (*c*) and several bands are seen as different density structures. **b** Human C-banded Y-chromosome with a higher concentration of heterochromatic material in one chromatid. Centromeric heterochromatic segments are seen in both chromatids. The terminal C-segment is composed of three sub-bands (see *inset*). **c** G-banded dicentric chromosome of a CHO cell with irregular distribution of Giemsa-stained band densities. **d** CHO chromosome presenting multiple sister chromatid exchanges. Distal region of the long arm shows two SCE segments with increased density. All *insets* illustrate scanned chromosomes. Absorbances detected by the system and displayed as predefined dither color patterns by the computer terminal were only limited to six sorting values ranging from 0 to 100. This arbitrary scaling appears *below* as a row of rectangles filled with shades of gray patterns ranging from white (absorbance: 0–5) to black (absorbance: 80–100), which indicated the highest densities detected

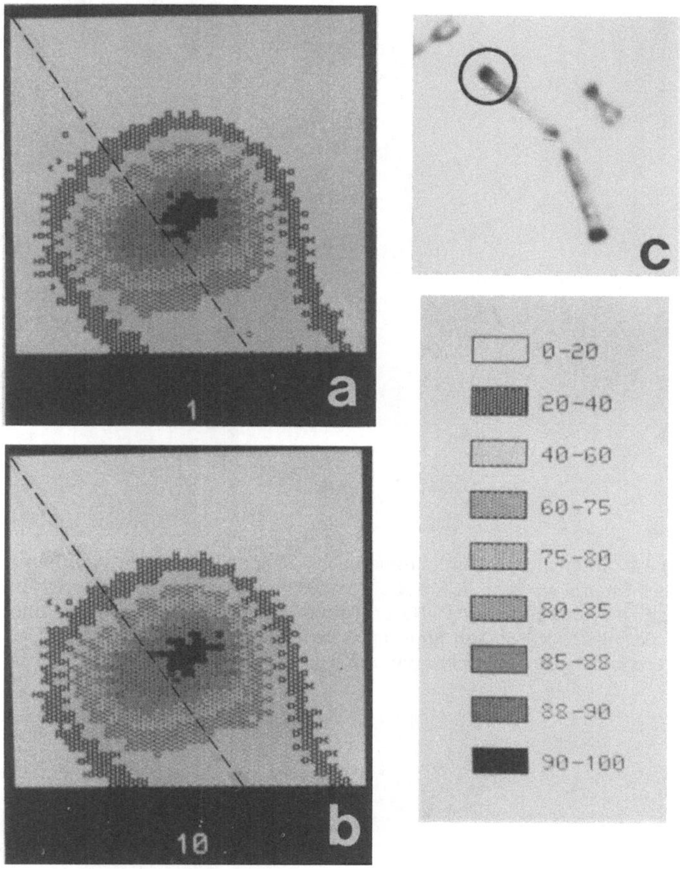

Fig. 3.a–c. Telomeric T-band in the long arm of a human chromosome No. 1 scanned at 1- and 10-step scanning stage dwell times (**a, b**). T-bands are more intense in one chromatid than in the other. *Broken lines* denote separation between chromatids. Dithered gray pixels surrounding the chromosome tips correspond to diffraction images produced by the microscope objects. **c** The chromosome and telomeric region (*encircled*) used for scanning. The vertical row of *rectangles* represents different detected absorbances sorted and transformed into a series of nine arbitrary gray shades following the procedure mentioned in Fig. 2

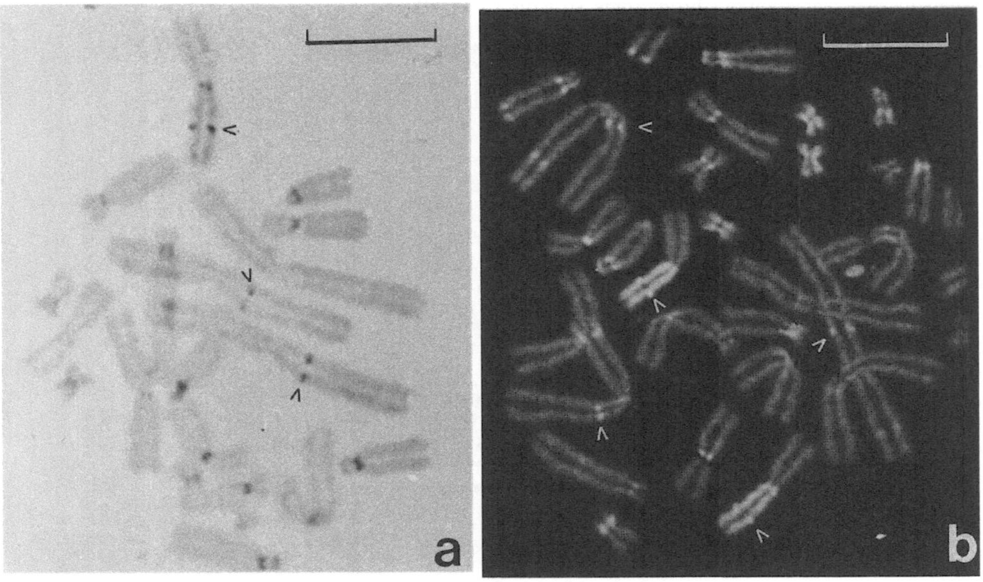

Fig. 4.a,b. Intercalary heterochromatin in C-banded CHO chromosomes. **a** Dots of intercalary heterochromatin are seen in the X-chromosome and in chromosome No. 1 (*arrow heads*); Giemsa staining. **b** Giemsa-stained CHO polyploid metaphase as seen with reflected light microscopy. Intercalary heterochromatin appears as brilliant spots (*arrow heads*). The long arms of the two X-chromosomes are heterochromatic. *Bars* indicate 10 μm

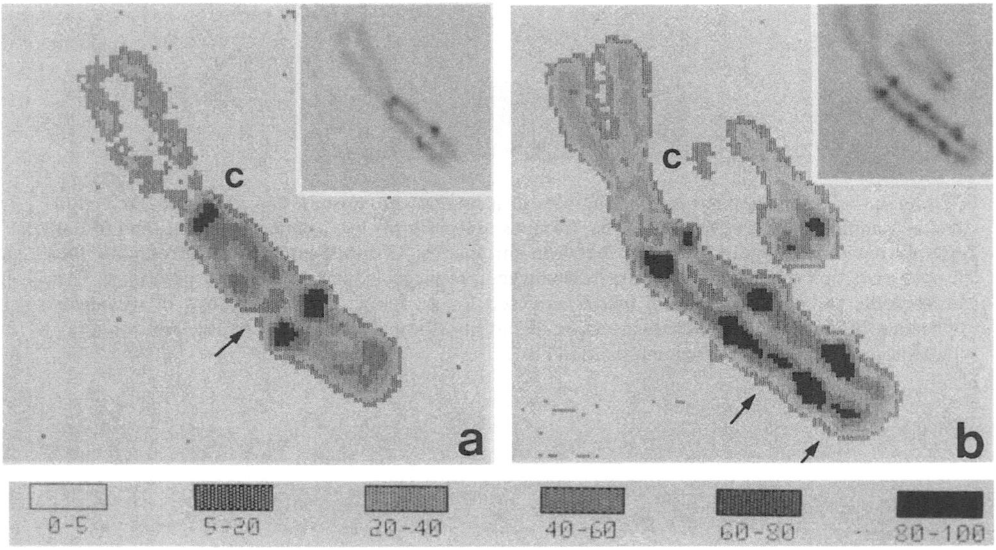

Fig. 5.a,b. Computer graphics image of scanned X-chromosomes from CHO cells with centromeric and intercalary heterochromatin. **a** A normal X-chromosome presenting centromeric (*c*) and intercalary heterochromatin in the long arm. A secondary constriction close to the intercalary heterochromatic segments can be seen (*arrow*). **b** Graphics image resulting from the scanning of the long arm of one X-chromosome with three intercalary heterochromatic segments. *Arrows* point out extra amounts of heterochromatin present in one chromatid. The series of *rectangles* appearing *below* represents absorbance values as is indicated in Fig. 2

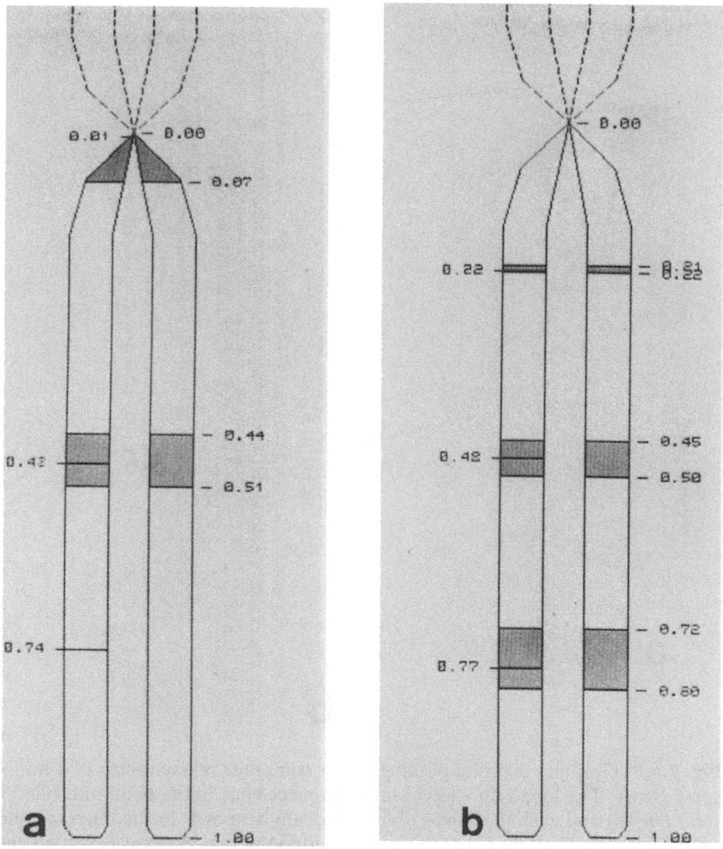

Fig. 6.a,b. Computer diagrams of centromeric and intercalary C-band localizations. **a** This diagram corresponds to the X-chromosome appearing in the inset of Fig. 5a; **b** corresponds to the X-chromosome shown in Fig. 5b (inset). Comparison of both sets of relative values shows that the extra heterochromatic segments are located at different sites than in the normal chromosome

chromosomes suggesting that complex rearrangements involving heterochromatic segments occurred.

3.4 Localization of Chromosome Breaks

Another area of application, which is under development in our laboratory, is the possibility of precisely detecting chromosome break regions and relating them to specific bands.

A reverse-banded human chromosome No. 1 was taken as a model for developing this analytical tool. A photograph of this chromosome was cut at a predefined location of the long arm and both pieces scanned separately (Fig. 7a). This procedure generated two separated data files in the computer system. Figure 7b illustrates the

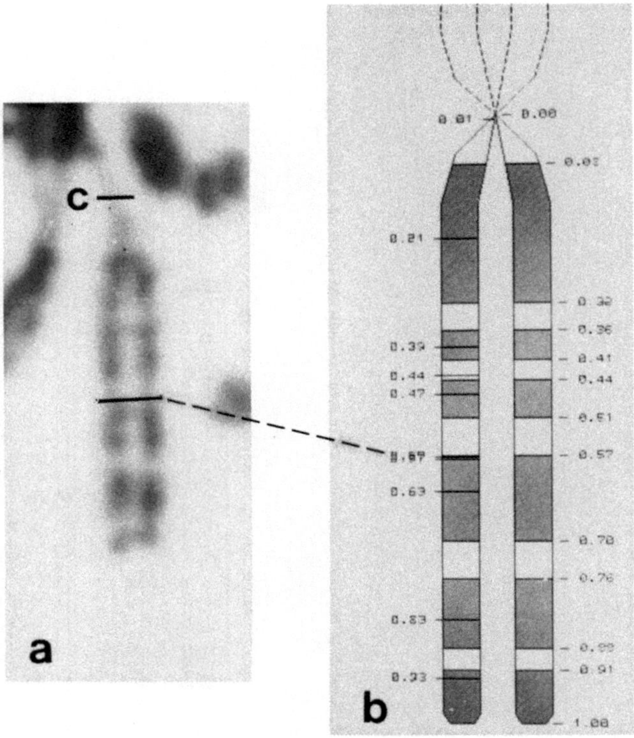

Fig. 7.a,b. Graphics diagram obtained after computer reassembling of a human R-banded chromosome No. 1. The long arm was cut into two pieces just in the band-interband region, indicated by a *solid line* (**a**) and each piece was photometrically scanned. In the diagram, the computer-generated extra lines at 0.57 relative units corresponded to the region cut as expected and overlapped with the localization of the chromosome band-interband region. *Broken lines* interconnect the chromosome arm region cut with the diagrammatic extra line

graphics diagram generated after reassembling both chromosome pieces with the computer. Two extra lines closely located at 0.568 and 0.573 relative units were displayed in an overlapped way and values rounded by the computer to 0.57, which corresponded precisely to the predetermined cut region (Fig. 7a, solid line). This system is considered potentially useful for localizing breaks along the chromosome arms.

3.5 Localization of Sister Chromatid Exchanges

Computer localization of sister chromatid exchanges (SCE) has been another area of application of microphotometric methodology. As discussed in a previous report, we prefer Giemsa staining for SCEs which results in a higher contrast, thus facilitating scanning and relative localization. Although an obvious cytogenetic interest to precisely locate the exchange points exists, in order to relate them to specific bands, no quantitative method for detecting them has been reported as yet. To study this problem, a computer program (SCE-SCAN) for the quantitative localization of SCEs in chromosomes was developed. Tests on the program with our system and the results obtained were reported previously (Drets and Monteverde 1987). A CHO chromosome with several SCEs is shown in Fig. 2d. Manipulation of acquired data after scanning this type of chromosome allowed the localization of exchange points. Figure 8 presents data obtained after scanning both chromatids separately; SCEs are displayed and distributed on a computer-generated vector diagram. Diagrammatic analysis of SCE was also used as reported previously together with the application of quantitative vectorial SCE diagrams (Drets and Monteverde 1987). This system of SCE computer analysis facilitated quantitative comparisons of data obtained on exchange regions and related them to banding patterns as well.

Fig. 8. Computer graphics diagram representing both chromatids of the long arm of a CHO chromosome No. 1 with exchange points detected after scanning both chromatids separately. Several relative exchange values detected by the system in both chromatids showed good quantitative agreement

Acknowledgements. We thank to G. A. Drets for developing a curve transformation algorithm for this research. This work was supported in part by the Cámara de Industrias del Uruguay (FJM), the Commission of the European Economic Communities (Contract CI*.0433.UY) and the Consejo Nacional de Investigaciones Científicas y Tecnológicas (CONICYT, Uruguay) Project No. 025/87 (GAF).

References

Arrighi FE, Hsu TC (1971) Localization of heterochromatin in human chromosomes. Cytogenetics 10: 81–86

Bianchi NO, Bianchi MS (1987) Analysis of the eukaryotic chromosome organization with restriction endonucleases. In: Obe G, Basler A (eds) Cytogenetics: basic and applied aspects. Springer, Berlin Heidelberg New York, pp 280–299

Caspersson T, Zech L, Modest EJ, Foley GE, Wagh U (1969) Chemical differentiation with fluorescent alkylating agents in *Vicia faba* metaphase chromosomes. Exp Cell Res 60: 315–319

Drets ME (1978) Bandscan. A computer program for on-line linear scanning of human banded chromosomes. Comp Prog Biomed 8: 283–294

Drets ME, Monteverde FJ (1987) Automated cytogenetics with modern computerized scanning microscope photometer systems. In: Obe G, Basler A (eds) Cytogenetics: basic and applied aspects. Springer, Berlin Heidelberg New York, pp 48–64

Drets ME, Seuanez H (1974) Quantitation of heterogeneous human heterochromatin: microdensitometric analysis of C- and G-bands. In: Coutinho EM, Fuchs F (eds) Physiology and genetics of reproduction, part A. Plenum, New York, pp 29–52

Drets ME, Shaw MW (1971) Specific banding patterns of human chromosomes. Proc Natl Acad Sci USA 68: 2073–2077

Dutrillaux B (1973) Nouveau système de marquage chromosomique: les bands T. Chromosoma (Berl) 41: 395–402

Dutrillaux B, Lejeune J (1971) Sur une nouvelle technique d'analyse du caryotype humain. CR Acad Sci Paris 272: 2638–2640

Edwards JH (1962) Chromosome analysis from capillary blood. Cytogenetics 1: 90–96

ISCN (1985) An international system for human cytogenetic nomenclature. Report of the Standing Committee on Human Cytogenetic Nomenclature. In: Hardnden, DG, Klinger HP (eds) Karger, Basel, p 118

Korenberg JR, Engels WR (1978) Base ratio, DNA content, and quinacrine-brightness of human chromosomes. Proc Natl Acad Sci USA 75: 3382–3386

Korenberg JR, Rykowski MC (1988) Human genome organization: Alu, Lines and the molecular structure of metaphase chromosome bands. Cell 53: 391–400

Korenberg JR, Therman E, Denniston C (1978) Hot spots and functional organization of human chromosomes. Hum Genet 43: 13–22

Kuhn EM, Therman E (1986) Cytogenetics of Bloom's syndrome. Can Genet Cytogenet 22, 1–18

Morgan WF, Crossen PE (1977) The frequency and distribution of sister chromatid exchanges in human chromosomes. Hum Genet 38: 271–278

Natarajan AT, Mullenders LHF (1987) Sister chromatid exchanges. In: Obe G, Basler A (eds) Cytogenetics: basic and applied aspects. Springer, Berlin Heidelberg New York, pp 338–344

Obe G, Vasudev V, Johannes C (1987) Chromosome aberrations induced by restriction endonucleases. In: Obe G, Basler A (eds) Cytogenetics: basic and applied aspects. Springer, Berlin Heidelberg New York, pp 300–314

Pätau K (1973) Three main classes of constitutive heterochromatin in man: intercalary, Y-type and centric. In: Wahrman J, Lewis KR (eds) Chromosomes today, vol 4. Wiley, New York, p 430

Perry P, Wolff S (1974) New Giemsa method for the differential staining of sister chromatids. Nature 251: 156–158

Seabright M (1971) Rapid banding technique for human chromosomes. Lancet 2: 971–972

Sumner AT, Evans HJ, Buckland RA (1971) A new technique for distinguishing between human chromosomes. Nature New Biol 232: 31–32

Therman E (1986) Human chromosomes: structure, behaviour, effects. Springer, Berlin Heidelberg New York, p 380

Weisblum B, Haseth PL de (1972) Quinacrine, a chromosome stain specific for deoxyadenylate-deoxythymidylate-rich regions in DNA. Proc Natl Acad Sci USA 69: 629–632

Yunis JJ (1976) High resolution of human chromosomes. Science 191: 1268–1270

Heterogeneity of DNA Repair in Relation to Chromatin Structure

L. H. F. MULLENDERS[1,2], J. VENEMA[1], A. VAN HOFFEN[1], A.T. NATARAJAN[1,2], A. A. VAN ZEELAND[3], and L.V. MAYNE[3]

Contents

1 Introduction

Analysis of repair processes in mammalian cells has largely been focused on induction and repair of DNA damage in the genome overall. In particular the repair of ultraviolet light-induced photoproducts has been intensively studied in a variety of mammalian cells and in most cases UV-induced cytotoxicity can be correlated to the extent of unscheduled DNA synthesis or removal of pyrimidine dimers from the nuclear DNA. For example variation in UV-induced cytotoxicity found both within and between the various complementation groups of the human UV-sensitive disorder xeroderma pigmentosum (XP) generally correlates with the extent of defective excision repair (Kantor and Hull 1984). However, a notable exception to this is found in nondividing XP-cells belonging to complementation group C, which are relatively resistant to the lethal effects of UV (Kantor and Hull 1984; Mayne and Lehmann 1982). Also, in a number of other cases the removal of pyrimidine dimers from the genome overall turned out to be an invalid parameter to predict UV-induced cytotoxicity. Cockayne's syndrome (CS) is a human disorder characterized at the cellular level by an increased sensitivity to the killing effects of UV-light, but with an apparently normal capacity to perform unscheduled DNA synthesis or to remove pyrimidine dimers (Mayne and Lehmann 1982). The various rodent cell lines consistently exhibit low levels of pyrimidine dimer removal for the genome overall (Van Zeeland et al. 1981) but are equally resistant to lethal effects of UV-light as human cells which are capable of performing fast and efficient repair of pyrimidine dimers.

---

[1] Department of Radiation Genetics and Chemical Mutagenesis, State University of Leiden, Sylvius Laboratory, Wassenaarseweg 72, 2300 RA Leiden, The Netherlands
[2] J.A. Cohen Institute, Interuniversity Research Institute for Radiopathology and Radiation Protection, Leiden, The Netherlands
[3] Biological Sciences Division, Centre for Medical Research, University of Sussex, Falmer BN1 9Qg, UK

An obvious explanation for these observations is that not all parts of the genome are equally important for lethal effects of DNA damaging agents and that DNA repair processes may operate heterogeneously with a strong preference for removal of damage from functionally important domains (Keyse and Tyrrell 1987). The various conformational states of chromatin being the proper substrates for DNA repair systems might play a role in determining the efficiencies as well as kinetics of DNA repair.

We have studied the role of chromatin structure in DNA repair. In interphase nuclei and metaphase chromosomes the chromatin is folded into loop domains by anchorage to a skeletal structure termed nuclear matrix or scaffold (Dijkwel et al. 1979; Paulson and Laemmli 1977). These loops are 50–150 kb long and are equivalent to the size of replicons.There are approximately 50000 to 100000 chromatin loops per nucleus. The folding of chromatin into loops is not a random phenomenon and numerous studies have demonstrated that transcriptionally active genes (Small et al. 1985) and origins of replication (Van der Velden et al. 1984) are in close proximity to the nuclear matrix. Consistent with the compartimentalization of templates and regulatory elements at the nuclear matrix fundamental processes such as replication and transcription are intimately associated with this structure. A model in which attachment to the nuclear matrix is a necessary precondition for replication and transcription can be readily extended to include repair. Repair can occur when chromatin loops are reeled through or when lesions become attached to repair complexes fixed at the nuclear matrix. Repair of transcriptionally active genes might proceed by such a mechanism as RNA synthesis occurs at the nuclear matrix (Jackson et al. 1981) and the transcription process itself may be directly involved in the removal of bulky adducts (Mellon et al. 1987).

We have analyzed UV-induced repair at the level of DNA loops, and at the level of single copy genes employing the method described by Bohr et al. (Bohr et al. 1985). Both approaches differ not only with respect to the genomic regions under study but also with respect to measurement of excision repair. Repaired sites studied at the level of DNA-nuclear matrix complexes were detected by radioactive labelling and will originate from all types of damage being subject to repair, whereas repair at the gene level was quantified for a single lesion, i.e. the pyrimidine dimer. Experiments were performed with normal human fibroblasts and UV-sensitive cells having normal (Cockayne's syndrome, CS) or reduced repair capacities (xeroderma pigmentosum complementation groups C and D, XP-C, XP-D). The various cell lines exhibit striking differences in overall repair capacity, UV-induced cytotoxicity and cellular responses such as recovery of UV-inhibited DNA and RNA synthesis.

2 Results and Discussion

2.1 Is Repair Synthesis Confined to the Nuclear Matrix?

Intact supercoiled DNA loops attached to the nuclear matrix can be isolated by extraction of nuclei with either high salt (2 M NaCl) or with low salt (25 mM lithium diodosalicylate). The relative position of repair events within DNA loops can be examined by producing random breaks in DNA loops employing the enzyme DNase

1 or in case of unligated repair patches employing the single-strand specific enzyme nuclease S_1 (Fig. 1). The probability of a DNA fragment being released from the nuclear matrix will decrease the closer the fragment is situated to an attachment site at the nuclear matrix (Fig. 1). Matrix attached DNA and loop DNA can be separated in neutral sucrose gradients by virtue of fast sedimentation of the nuclear matrix in sucrose gradients (Dijkwel et al. 1979).

In order to determine the intranuclear localization of DNA repair confluent normal human fibroblasts were UV-irradiated (5 and 30 J/m^2) and pulse labelled for 5–10 min immediately or after various time intervals following irradiation. Pulse labelling of the cells was performed in the presence of hydroxyurea to reduce incorporation by replicative synthesis to a sufficiently low level. Under these experimental conditions the majority of the repair events is not completed during the pulse (Mullenders et al. 1987) virtually ruling out the possibility that the repair process is too fast to be trapped at the nuclear matrix. In cells exposed to 5 J/m^2 a distinct enrichment of repaired sites at the nuclear matrix was observed, although much less than the very profound enrichment of newly replicated DNA at the nuclear matrix (Fig. 2). However, at a higher UV-dose repair approached a random distribution. These observations as well as the inability to chase the repair label from the nuclear matrix into loop DNA (Mullenders et al. 1988) do not support a general model of DNA repair confined to the nuclear matrix compartment as found for replication and transcription. Instead the preferential occurrence of repaired sites at the nuclear matrix following 5 J/m^2 UV-irradiation reflects the preferential repair of DNA permanently bound to the nuclear matrix. The majority of DNA damage is repaired at the sites of the lesions and does not require attachment to the nuclear matrix prior to repair.

The preferential repair of nuclear matrix associated DNA following 5 J/m^2 UV-irradiation turned out to be restricted to the first hour following treatment. Sites of repair pulse labelled 2 h after UV-exposure tended to be distributed randomly along the DNA loops. The preferential repair of nuclear matrix associated DNA during a

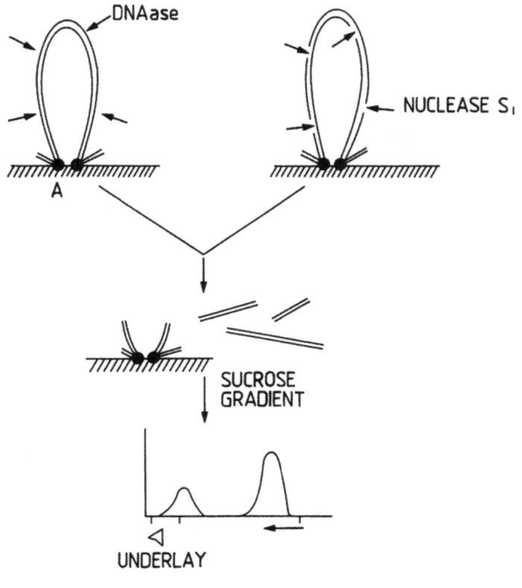

Fig. 1. Schematic view of release of DNA from the nuclear matrix by nucleases. The induction of DNA double-strand breaks in intact DNA loops by either DNase 1 directly or at single-strand regions by nuclease S_1 will result in a detached DNA loop fraction and a nuclear matrix associated DNA fraction. Both fractions can be separated in neutral sucrose gradients. The amount of DNA at the nuclear matrix is directly related to the induced number of DNA breaks. Single-strand DNA regions can be generated at the vicinity of repair sites by inhibition of the repair process

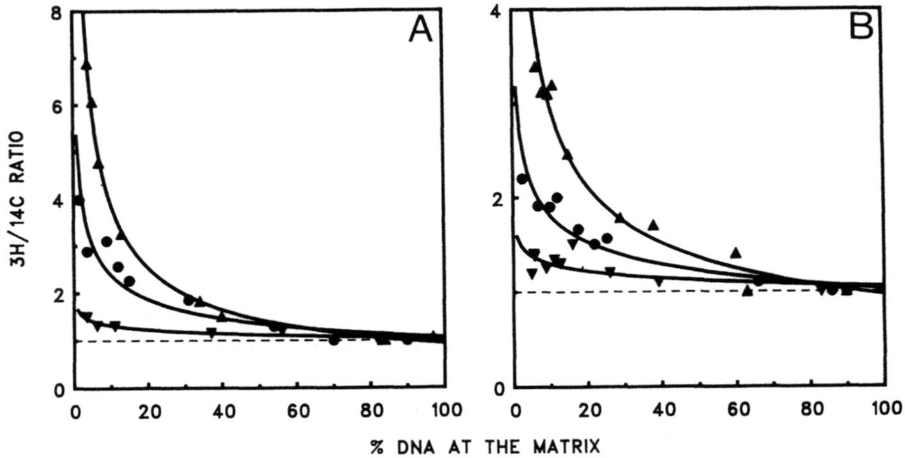

Fig. 2A,B. Dose- and time-dependent preferential repair of nuclear matrix associated DNA in UV-irradiated human fibroblasts. ^{14}C-prelabelled confluent normal human fibroblasts were UV-irradiated (5 or 30 J/m^2) and subsequently pulse labelled with ^{3}H-TdR for 10 min. DNA-nuclear matrix complexes were isolated and digested with DNase 1 as described in Fig. 1. *A* 5 J/m^2 (●); 30 J/m^2 (▼). Pulse labelling was achieved immediately after irradiation. For comparison the localization of newly replicated (10-min pulse) DNA (▲)at the nuclear matrix is shown. *B* 5 J/m^2 and 10-min pulse (▲); 5 J/m^2, 1-h post-UV incubation and 10-min pulse (●); 5 J/m^2, 2-h post-UV incubation and 10-min pluse (▼)

short period directly after irradiation fits in with the concept that in human fibroblasts exposed to a low dose of UV-light the repair of functionally important domains in the genome occurs quickly during a short period after treatment. Repair of potentially lethal damage (Keyse and Tyrrell 1987) and UV-inhibited RNA synthesis (Mayne and Lehmann 1982) is virtually completed within 2 h after treatment, and pyrimidine dimer removal from the transcribed strand of the human DHFR gene is most pronounced during the early period after UV-irradiation (Mellon et al. 1987). Taken together these domains, located proximal to the nuclear matrix, are likely to be identical to transcriptionally active DNA, and the preferential localization of repaired sites at the nuclear matrix might reflect the preferential repair of 6–4 photoproducts or pyrimidine dimers or both types of lesions. We note here that preferential repair of nuclear matrix associated DNA is observed at the biological relevant dose of 5 J/m^2, but not at the high dose of 30 J/m^2. This phenomenon is possibly due to different saturation levels of nuclear matrix and loop DNA repair as discussed previously (Mullenders et al. 1988).

2.2 Distribution of Repaired Sites in Normal Human and UV-Sensitive Fibroblasts

The ultimate distribution of repaired sites accomplished during the first 2 h following 5 J/m^2 UV-irradiation of normal human fibroblasts is shown in Fig. 3. The frequency of repaired sites in nuclear matrix associated DNA was approximately twofold higher than in loop DNA. Profound differences in distribution of repaired sites were found among the UV-sensitive cells. In XP-C the residual repair was highly specific for nuclear matrix associated DNA (about four fold enrichment), whereas in XP-D the distribution of the limited number of repaired sites was comparable to normal cells. CS fibroblasts with normal overall repair capacity showed the opposite results: nuclear matrix associated DNA contained less repaired sites than loop DNA. These differences in distribution indicate that the residual repair in XP-C cells is confined to transcriptionally active DNA, and that CS cells are defective in performing this type of repair. The observation that XP-C and CS cells are proficient and deficient respectively in the recovery of UV-inhibited RNA synthesis (Mayne and Lehmann 1982) is consistent with the above mentioned hypothesis. The heterogeneous distribution of repaired sites found in XP-C cells did not change into a random distribution even after an extended repair period of 24 h, suggesting that these cells are fully defective in repair of loop DNA, i.e. transciptionally inactive DNA. We note here that not all loops in XP-C cells are repaired efficiently at their bases during the first 2 h after treatment. Nuclease S_1 analysis (Fig. 1) revealed that in about 30% of the DNA loops repair occurred at both attachment sites, and that certain regions of the chromatin were excluded from the repair process (Mullenders et al. 1986).

The observation that the distribution of repaired sites in XP-D mimicked normal cells suggests that repair of nuclear matrix associated DNA and loop DNA are both impaired to the same extent, implying the involvement of common factors in both pathways.

2.3 Nonrandom Distribution of Repaired Sites in DNA Loops and Its Relationship to Heterogeneity in Removal of Pyrimidine Dimers from Defined DNA Sequences

To obtain direct experimental support for the hypothesis that XP-C and CS cells are proficient and deficient respectively in repair of transcriptionally active DNA, we studied the removal of pyrimidine dimers from active and inactive sequences using the dimer-specific enzyme T4 endonuclease V as described by Bohr et al. (1985).

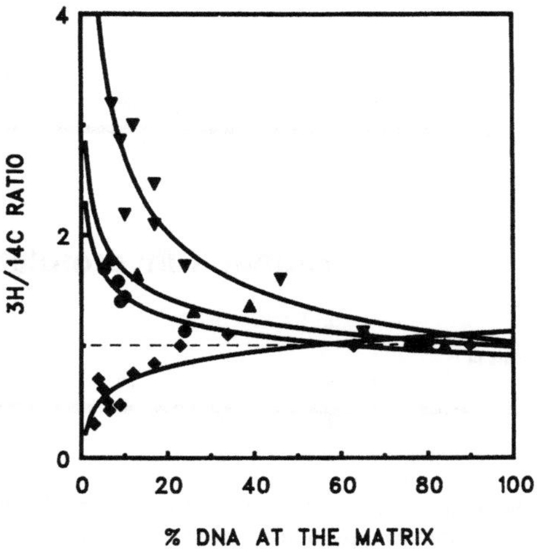

Fig. 3. Different distributions of repaired sites in DNA-nuclear matrix complexes prepared from normal and UV-sensitive human cells. ^{14}C-prelabelled confluent fibroblasts were UV-irradiated (5 J/m^2) and pulse labelled for 120 min starting directly after irradiation. DNA-nuclear matrix complexes were digested and analyzed as described in Fig. 1 Normal (●), XP-D (▲), XP-C (▼) and CS (◆)

Pyrimidine dimer removal from active genes was determined in 18.5 and 20-kb fragments generated from the 3‘ and 5‘ sites of the adenosine deaminase (ADA) gene by restriction of genomic DNA with $EcoR_1$ and Bcl1 respectively. Alternatively, repair was measured in a 20-kb fragment from the active dihydrofolate reductase (DHFR) gene (Fig. 4). A 14–kb fragment of the X-chromosomal 754 locus was used to determine repair of pyrimidine dimers from a transcriptionally inactive DNA sequence. Briefly, equivalent amounts of restricted DNA were digested or mock digested with T4 endonuclease V, subjected to alkaline gel electrophoresis and Southern transferred. After hybridization and exposure to X-ray films, band intensities were quantified and used to calculate the average number of pyrimidine dimers per fragment using the Poisson expression. Figure 5 shows the result of a representative experiment. Table 1 summarizes the results obtained with stationary normal, XP-C and CS fibroblasts exposed to 10 J/m^2 of UV-light. In the case of normal human cells it is very evident that the active ADA and DHFR genes were repaired faster and more efficiently than the inactive 754 gene. The latter was comparable to the rate and extent of pyrimidine dimer removal from the genome overall in normal cells, being 31% and 69% in 8 and 24 h respectively (Mayne et al. 1988).

Twenty-four hours following 10 J/m^2 UV-irradiation, repair of pyrimidine dimers in active genes was almost complete, whereas only about 50% of the pyrimidine dimers was removed from an inactive gene. There appeared to be no substantial differences in rate of repair of the two active genes. Consistent with the preferential repair of nuclear matrix associated DNA XP-C cells turned out to be proficient in repair of active genes, although there were marked differences in efficiency of repair among the various fragments generated from active genes. Removal of pyrimidine dimers from the inactive 754 locus was virtually absent, as would be expected from the low overall repair capacity. Within the ADA gene only the 3′ located $EcoR_1$ fragment was repaired with a similar rate and to the same extent as in normal cells. Following 24-h post-UV incubation both the Bcl1 fragment of the ADA gene as well as the Hind III fragment of the DHFR gene were repaired to a lesser extent in XP-

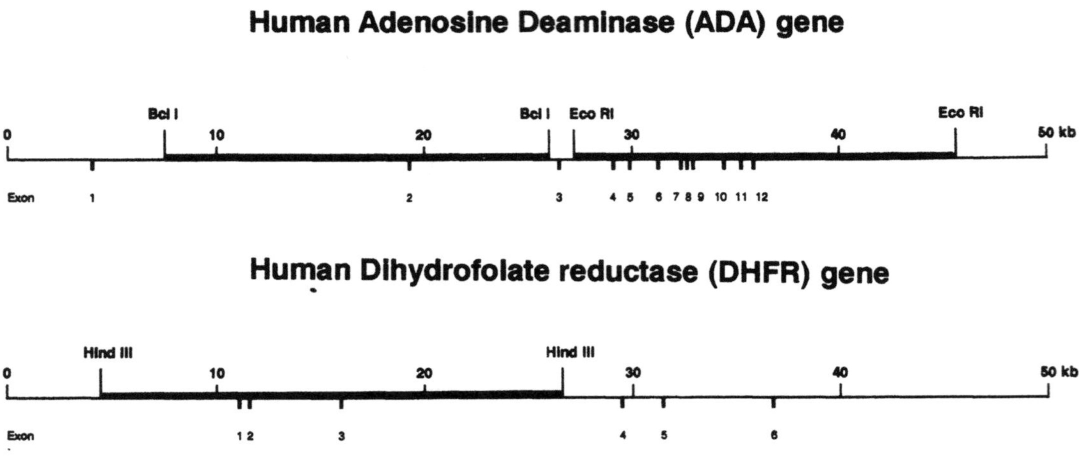

Fig. 4. Genomic organization of the human adenosine deaminase and dihydrofolate reductase genes. The genomic maps indicate positions of exons and relevant restriction sites

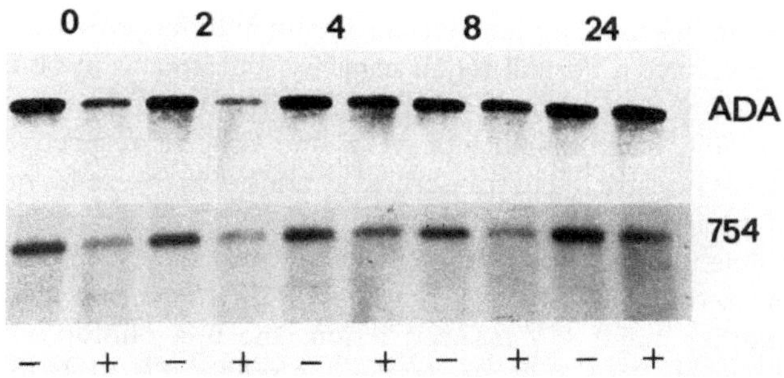

Fig. 5. Removal of pyrimidine dimers from the $EcoR_1$ fragment of the ADA and 754 genes. Confluent normal human cells were UV-irradiated (10 J/m^2) and incubated for the indicated time periods. DNA was isolated and analyzed for the presence of pyrimidine dimers by digesting (+) or mock-digesting (−) equivalent amounts of $EcoR_1$ restricted DNA. After Southern transfer the membrane was first hybridized with an ADA probe, washed and hybridized again using a 754 probe

Table 1. Percentages of removal of pyrimidine dimers from defined DNA fragments in normal, xeroderma pigmentosum group C and Cockayne's syndrome fibroblasts

Fragment	Repair time (h)	Normal	XP-C	CS
ADA ($EcoR_1$)	8	74	62	30
	24	93	93	56
ADA (Bcl1)	8	72	49	17
	24	92	69	36
DHFR	8	62	50	32
	24	87	57	52
754	8	34	8	23
	24	52	6	40

C than in normal cells. A possible explanation for these differences may be that efficient repair of active genes in XP-C cells is restricted to transcribed strands only as has been described for hamster cells (Mellon et al. 1987). Recently, antisense transcription at the 3' end of the ADA has been reported (Lattier et al. 1989) and this could account for the very proficient repair of the ADA $EcoR_1$ fragment in XP-C cells.

The repair capabilities of XP-C cells appear to be the reverse of those in CS. In contrast to XP-C, CS cells were unable to repair active genes preferentially. The rates of repair of the ADA and DHFR genes were similar to that found for the inactive 754 gene, suggesting that CS cells are able to repair active genes, but lost their ability to process damage in active DNA in preference to the bulk of damage. The reduced ability to perform preferential repair of active genes provides an explanation for the lack of recovery of transcription, and is consistent with the increased levels of cell killing and mutagenesis seen in CS cells after UV-irradiation. These data and the results described in Section 2 show that repair pathways operating in active and inactive chromatin are at least partially independent. XP-C cells have lost the ability to repair inactive chromatin and the reverse situation is found in CS cells, which have a normal repair capacity, but appear to be unable to perform efficient repair of active genes. Further experiments will be performed to find out whether the ability and inability of XP-C and CS-cells respectively to perform preferential repair is restricted to transcribed strands of active genes only. With respect to kinetics of pyrimidine dimer removal XP-C cells resemble Chinese hamster cells, which efficiently remove pyrimidine dimers from active genes but inefficiently from nonexpressed DNA and the genome overall (Bohr et al. 1985). The repair capacity for another major UV-induced lesion, the 6–4 photoproduct, however, is strikingly different. XP-C cells are deficient in repair of 6–4 photoproducts, whereas hamster cells are very efficient in repair of these lesions (Mitchell et al. 1984). Proliferating XP-C cells are very sensitive to lethal effects of UV-light in contrast to hamster cells, which suggests that 6–4 photoproducts are the main cytotoxic lesions in UV-irradiated growing cells.

Acknowledgements. This work was supported by the association of the University of Leiden with Euratom (contract No. B16-E-166) and Medigon (contract No. 900-501-074), MRC (UK) and the Wellcome Trust.

References

Bohr VA, Smith CA, Okumoto DS, Hanawalt PC (1985) DNA repair in an active gene: removal of pyrimidine dimers from the DHFR gene of CHO cells is much more efficient than in the genome overall. Cell 40: 359–369

Dijkwel PA, Mullenders LHF, Wanka F (1979) Analysis of the attachment of replicating DNA to a nuclear matrix in mammalian interphase cell nuclei. Nuc Acids Res 6: 219–230

Jackson DA, McCready ST, Cook PR (1981) RNA is synthesized at the nuclear cage. Nature 292: 522–525

Kantor GJ, Hull DR (1984) The rate of removal of pyrimidine dimers in quiescent cultures of normal human and xeroderma pigmentosum cells. Mutat Res 132: 21–31

Keyse SM, Tyrrell RM, (1987) Rapidly occurring DNA excision repair events determine the biological expression of UV-induced damage in human cells, Carcinogenesis 8: 1251–1256

Lattier DL. States IC, Hutton II, Wiginton DA (1989) Cell type-specific transcriptional regulation of the human adenosine deaminase gene. Nucl Acids Res 17: 1061–1076

Mayne LV, Lehmann AR (1982) Failure of RNA synthesis to recover after UV irradiation: an early defect in cells from individuals with Cockayne's syndrome and xeroderma pigmentosum. Cancer Res 42: 1473–1478

Mayne LV, Mullenders LHF, Zeeland AA van (1988) Cockayne's syndrome: a UV sensitive disorder with a defect in the repair of transcribing DNA but normal overall excision repair. In: Friedberg EC, Hanawalt PC (eds) Mechanisms and consequences of DNA damage processing. Liss New York, pp. 349–353

Mellon I, Spivak G, Hanawalt PC (1987) Selective removal of transcription blocking DNA damage from the transcribed strand of the mammalian DHFR gene. Cell 51: 241–249

Mitchell DL, Haipek CA, Clarkson IM (1984) (6–4) Photoproducts are removed from the DNA of UV-irradiated mammalian cells more efficiently than cyclobutane pyrimidine dimers. Mutat Res 143: 109–112

Mullenders LHF, Kesteren AC van, Bussmann CIM, Zeeland AA van, Natarajan AT (1986) Distribution of UV-induced repair events in higher order chromatin loops in human and hamster fibroblasts. Carcinogenesis 7: 995–1002

Mullenders, LHF, Zeeland AA van, Natarajan AT (1987) The localization of ultraviolet-induced excision repair in the nucleus and the distribution of repair events in higher order chromatin loops in mammalian cells. J Cell Sci Suppl 6: 243–262

Mullenders LHF, Kesteren-van Leeuwen AC van, Zeeland AA van, Natarajan AT (1988) Nuclear matrix associated DNA is preferentially repaired in normal human fibroblasts exposed to a low dose of ultraviolet light but not in Cockayne's syndrome fibroblasts. Nucl Acid Res 16: 10607–10622

Paulson JR, Laemmli UK (1977) The structure of histone-depleated chromosomes. Cell 12: 817–828

Small D, Nelkin B, Vogelstein P (1985) The association of transcribed genes with the nùclear matrix of Drosophila cells during heat shock, Nucl Acids Res 13: 2413—2431

Velden HMW van der, Willigen G van, Wetzels RHW, Wanka F (1984) Attachment of origin of replication to the nuclear matrix and chromosomal scaffold. Febs Lett 161: 13–16

Zeeland AA van, Smith CA, Hanawalt PC (1981) Sensitive determination of pyrimidine dimers in DNA of UV irradiated mammalian cells: introduction of T4 endonuclease V into frozen and thawed cells. Mutat Res 82: 173–189

The Poly-ADP-Ribosylation System of Higher Eukaryotes: How Can It Do What?

F. R. ALTHAUS, M. COLLINGE, P. LOETSCHER, G. MATHIS, H. NAEGELI, P. PANZETER, and C. REALINI[1]

Contents

1 Introduction

The poly-ADP-ribosylation system of higher eukaryotes is thought to modulate chromatin functions. The molecular mechanism of this presumed action is unknown (for review see Althaus and Richter 1987). In 1985 Ueda and Hayaishi stated that "... a major difficulty in obtaining direct and definitive evidence for the role(s) of poly-ADP-ribosylation has stemmed from the complexity of systems used. A major breakthrough would therefore be expected by reconstituting a model system in vitro from well defined components ...". A few years later, the first full length cDNA of the gene of poly(ADP-ribose)polymerase (e.g., Alkhatib et al. 1987; Cherney et al. 1987; Herzog et al. 1989; Kurosaki et al. 1987; Schneider et al. 1987; Suzuki et al. 1987; Uchida et al. 1988), a key component of the poly-ADP-ribosylation system, has become available. It is likely that molecular genetic approaches will soon provide us a clue to the biological role of this posttranslational protein modification in chromatin. However, the phenotype of a genetically engineered cell under- or overexpressing components of the poly-ADP-ribosylation system may be difficult to interpret, since these cells, if viable at all, may exhibit a complex pattern of imbalanced cellular functions. In fact, this has already been observed in the early 1980s when numerous studies involving inhibitors of poly(ADP-ribose)polymerase were performed (Althaus und Richter 1987). For these reasons, we have decided to complement molecular genetic approaches with in vitro reconstitution experiments to study the molecular function of poly-ADP-ribosylation in chromatin.

We have recently shown that protein-bound-ADP-ribosyl polymers may cause the release of core DNA fragments from nucleosomal core particles (Mathis and Althaus 1987). In the present study, we have analyzed the molecular properties of

[1] University of Zürich-Tierspital, Institute of Pharmacology and Biochemistry, Winterthurerstrasse 260, CH-8057 Zürich, Switzerland

ADP-ribosyl polymers responsible for this phenomenon. In addition, we have reconstituted an in vitro poly-ADP-ribosylation system with the goal to study the role of the enzyme poly(ADP-ribose)polymerase as a possible modulator of DNA-protein interactions under defined experimental conditions. This in vitro system has also provided valuable insights into the mode of polymer biosynthesis and the molecular factors which determine a given pattern of posttranslational protein modification with these polymers. Based on these results we propose a model for the biological function of poly-ADP-ribosylation in chromatin. The key feature of this model is derived from the observation that the noncovalent association of DNA-binding proteins with ADP-ribosyl polymers may be part of a protein shuttle mechanism on DNA templates (vide infra).

2 Enzymatic Components of the Poly-ADP-Ribosylation System

The metabolism of poly(ADP-ribose) involves three enzymes: (1) the DNA-dependent enzyme poly(ADP-ribose)polymerase (EC 2.4.2.30), which utilizes the respiratory coenzyme NAD^+ as the substrate for the biosynthesis of a homopolymer composed of ADP-ribosyl residues. We and others (Leduc et al. 1986; Mathis and Althaus 1987) have found this enzyme to be closely associated with the nucleosomal core. It is important to note that this enzyme serves as a major acceptor for ADP-ribose polymers, i.e., it catalyzes its automodification. (2) The enzyme poly(ADP-ribose) glycohydrolase, which degrades ADP-ribosyl polymers in an exoglycosidic reaction mode; and (3) the enzyme ADP-ribosyl protein lyase, which removes the protein-proximal ADP-ribosyl residue. The total ADP-ribose processing capacity of this system in mammalian cells is quite impressive, i.e., a total of 10 million ADP-ribosyl residues per min per cell for poly(ADP-ribose)polymerase and likewise for poly(ADP-ribose)-glycohydrolase, and 72 million residues for ADP-ribosyl protein lyase (for review, see Althaus and Richter 1987).

3 Mode of Polymer Addition to Proteins

We have investigated the properties of ADP-ribose polymer molecules which constitute the molecular signal for the biological function of these polynucleotides. For this purpose, the polymer pattern produced by poly(ADP-ribose)polymerase under in vitro conditions was analyzed as shown in Fig. 1. Polymers were radiolabeled with $[^{32}P]NAD^+$ as the substrate, detached from the acceptor proteins, and purified by boronate affinity chromatography (Jacobson et al. 1984). The different polymer size classes were separated on sequencing gels, and the relative frequency of each polymer size class was plotted using a 3-D computer graphics program (Naegeli et al. 1989). The results revealed a discontinuous polymer size pattern which was strictly maintained throughout the reaction (Fig. 2). This pattern indicates that polymer elongation is completed at early reaction times and that subsequent polymer synthesis produces larger numbers of polymers rather than longer polymers. These results then suggest that poly(ADP-ribose)polymerase modifies acceptor proteins in a processive mode. An obvious prediction of this concept is that the proportion of $(ADP\text{-}ribose)_n$-

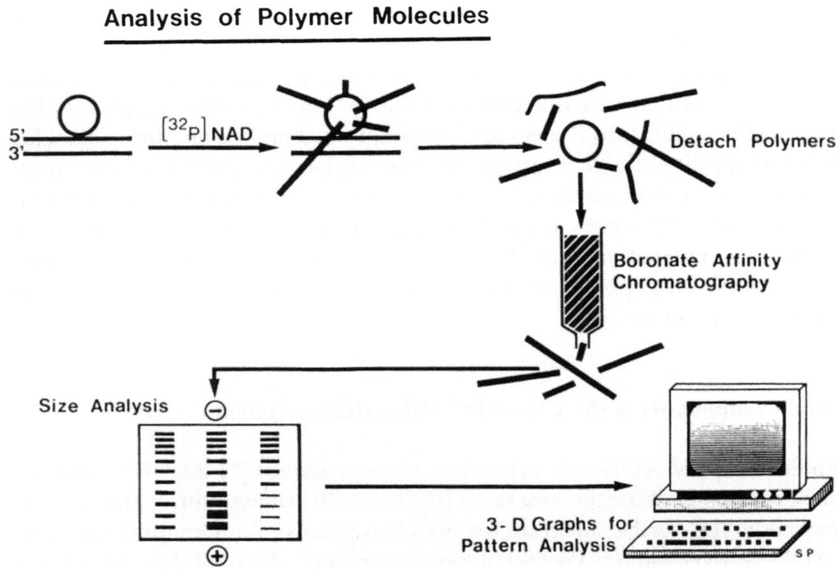

Fig. 1. Schematic illustration of the experimental approach. Purified poly(ADP-ribose)polymerase was incubated for various intervals with a 146-bp core DNA fragment in the presence of [^{32}P]NAD$^+$. The radiolabeled polymers were chemically detached from the enzyme molecules, purified by boronate affinity chromatography, and separated into individual polymer size classes. The amount of radioactivity in each polymer class was quantified. The relative size distribution patterns at different stages of the reaction were analyzed using three-dimensional graphics microcomputer software

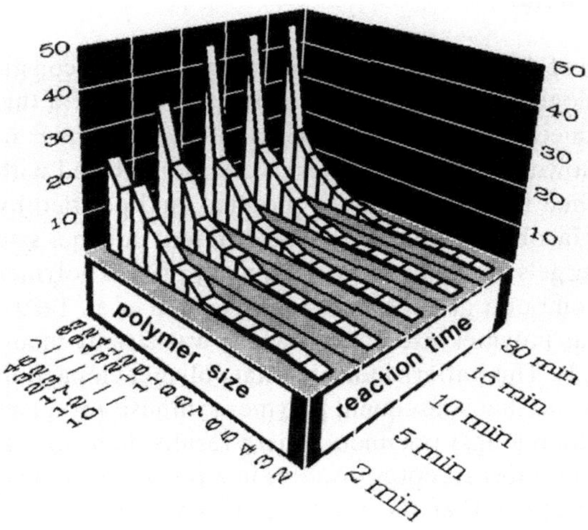

Fig. 2. Results of an experiment, in which [^{32}P]-labeled polymers were isolated from automodified poly (ADP-ribose)polymerase and then separated by polyacrylamide gel electrophoresis as described (Naegeli et al. 1989). The relative distribution of ADP-ribose residues in each size class was determined by quantification of the radioactivity contained in each band

modified proteins should increase gradually as larger numbers of otherwise identical polymers appear in the reaction. This prediction was tested in the experiment shown in Fig. 3. The results again fit the processivity model schematically outlined in Fig. 4. Similar results were obtained with endogenous enzyme which copurifies with nucleosomal core particles (Naegeli et al. 1989). In summary, we have shown that the poly-ADP-ribosylation of proteins involves a strictly processive reaction mechanism. In order to study the consequences of this mechanism on DNA-protein reactions, we set up a reconstituted in vitro system involving highly purified components.

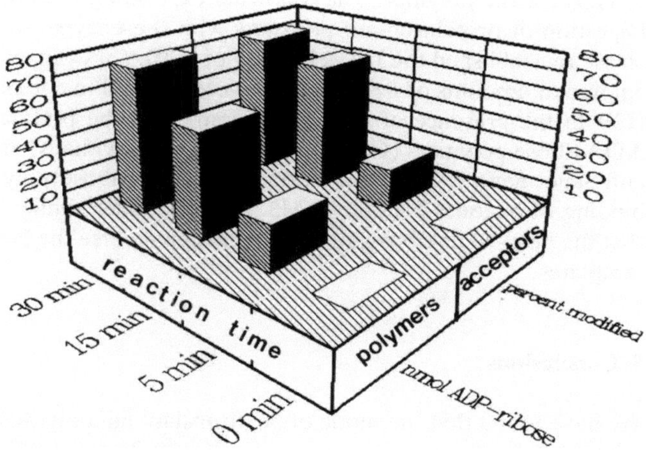

Fig. 3. Comparison between ADP-ribose polymer quantity and proportion of automodified poly(ADP-ribose)polymerase as a function of reaction time. The reaction conditions were as in Fig. 2

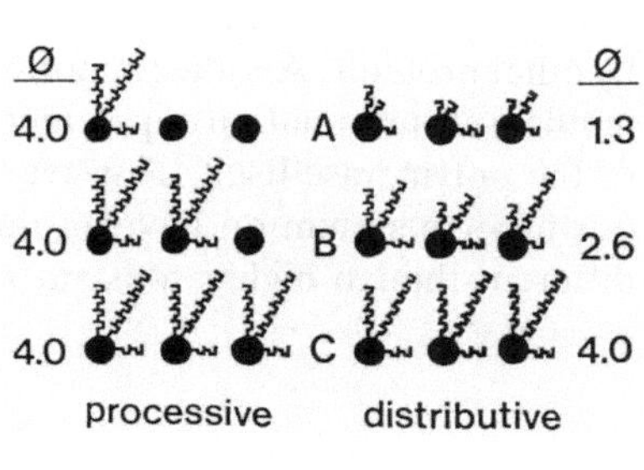

Fig. 4. The reaction intermediates produced in a processive or a distributive mode of protein modification by poly(ADP-ribose)polymerase. The letters *A–C* denote three hypothetical stages of the reaction, each of them involving the production of 12 ADP-ribose residues. The results shown in the present study are in agreement with the processivity model shown on the *left*

4 In Vitro System to Study Protein Shuttling by Poly-ADP-Ribosylation

We have reconstituted the following components in vitro: 5'-[^{32}-P]-end-labeled 146-bp core DNA fragments, an electrophoretically pure preparation of poly(ADP-ribose)polymerase, and histone H2B, purified to homogeneity by reverse-phase HPLC (Loetscher 1988). Figure 5 shows that incubation of a saturated histone H2B-DNA complex with the enzyme poly(ADP-ribose)polymerase in the presence of

NAD causes the release of the DNA fragment as detectable by mobility shift gel electrophoresis. This dissociation of the DNA-protein complex was again dependent on the formation of ADP-ribosyl polymers. In addition, we found a linear dose-response relationship between the quantity of DNA released and the concentration of polymer formed (Loetscher 1988). Thus, the DNA binding of a specific protein can be reversed by the action of poly(ADP-ribose)polymerase. Furthermore, addition of histone H2B to an incubation with poly(ADP-ribose)polymerase, which had been automodified in the presence of core DNA alone, produced identical results (not shown). This indicated that ADP-ribosyl polymers successfully compete for the binding of histone H2B in the presence of DNA, suggesting a higher binding affinity of H2B for the polymer. The following experiments confirmed this conclusion. (1) Digestion of protein-bound polymers with the enzyme snake venom phosphodiesterase reestablished the DNA binding of H2B. (2) Competition experiments, where equimolar amounts of *free* polymers were allowed to compete for binding of histone H2B in the presence of core DNA confirmed the preferential binding of H2B to ADP-ribose polymers (C. Realini, C. and F.R. Althaus, unpubl., Wesierska-Gadek and Sauermann 1988). In addition, poly(ADP-ribose)polymerase also reversed the binding of histones H1, H2A, H3, and H4 under similar conditions. We conclude that the poly-ADP-ribosylation reaction can reverse the binding of histones to DNA templates.

5 Conclusions

We have found that the mode of posttranslational poly(ADP-ribose)modification of proteins follows a processive reaction mode. These results were obtained with two experimental models of different complexities, i.e., in the presence or absence of various other proteins, including known acceptors of poly(ADP-ribose). Thus, the processive mode of operation of the enzyme poly(ADP-ribose)polymerase is an inherent property of the enzyme protein itself and apparently is not further regulated by other proteins. Another important aspect of this observation is that this mechanism implies self-termination of polymer elongation, which is also attributable to the action of the polymerase itself. However, we already know that other proteins present in our in vitro system do have an impact on this termination mechanism, generating different though highly constant polymer size distributions. We speculate that the

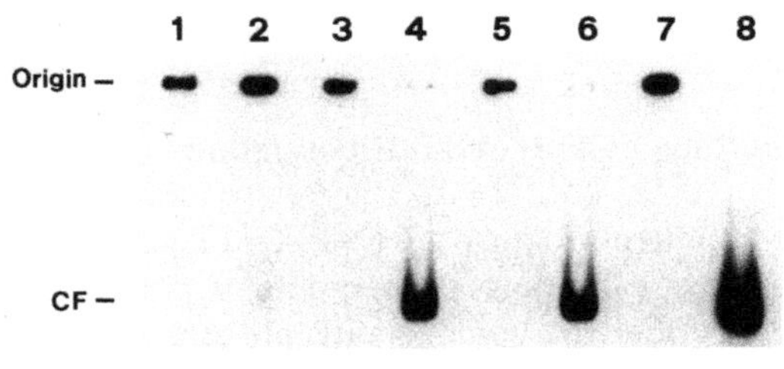

Fig. 5. Release of a 146-bp core DNA fragment for its association with histone H2B following incubation of the DNA-histone complex in the presence of poly(ADP-ribose)polymerase and 100 μM NAD. The incubation times were 0 min (lane *1*), 2 min (landes *2, 3*), 5 min (lanes *4, 5*), and 10 min (lanes *6, 7*). Benzamide (10 mM) was present in the incubations run on lanes *3, 5* and *7*. On lane *8*, 5′-end labeled 146-bp core DNA was run as a marker

distinct polymer size patterns observed in the presence of various DNA-binding proteins may reflect an adaptive response of this polymer-generating mechanism to some as yet unrecognized molecular properties of the shuttle target (e.g., charge distribution within the protein, hydrophobic versus hydrophilic domains etc.). This initial finding and further information amenable in this model system should greatly enhance our understanding of the complex polymer patterns found in intact cells following DNA damage, and in other active processes on chromatin. Another important aspect of these findings is that we have for the first time been able to define in molecular terms the complex pattern of protein-bound polymers, which is responsible for a specific function of the polymerase in a clearly defined in vitro system, i.e., the modulation of DNA-histone interactions.

6 Poly-ADP-Ribosylation System: A Protein Shuttle Mechanism in Chromatin?

The starting point for our studies on the biological function of poly-ADP-ribosylation has been the observation that poly-ADP-ribosylation is involved in several chromatin functions (Althaus et al. 1982a, b; Loetscher et al. 1987) including the repair of various types of DNA damages (Althaus et al. 1982a). Subsequent analyses showed that the nucleosomal unfolding of damaged DNA domains is deficient in poly(ADP-ribose)-depleted cells (Mathis and Althaus 1986; Althaus et al. 1989). This deficiency was coupled with the lack of excision of bulky DNA adducts. This suggested an involvement of the poly(ADP-ribose) generating system in nucleosomal unfolding, and indirectly, in DNA excision repair. These studies also revealed that the unfolding process generates DNA domains which are indistinguishable from linker DNA with respect to their accessibility to chemical or enzymatic probes (Mathis and Althaus 1986; Althaus et al. 1989). Whether this unfolding requires a complete stripping of proteins from DNA is currently not known. However, it is likely that the increased accessibility of these domains in vivo reflects a reduced binding of associated proteins. In accordance with this concept, we observed a significant reduction of histone binding to nucleosomal core DNA following in vitro ADP-ribosylation of nucleosomal core particles (Mathis and Althaus 1987). In fact, the poly-ADP-ribosylation of nucleosomal core particles reduced the separating forces required to release core DNA from histones by a factor of 2.5, an effect which is likely to be underestimated in this model system. Likewise, this phenomenon could be reproduced with electrophoretically pure preparations of histones and poly(ADP-ribose)polymerase in the presence of core DNA and NAD^+. In addition, when histones were given the choice to associate either with DNA or preexisting polymerase-bound ADP-ribosyl polymers, they exhibited a clear preference for the polymers. This binding of histones to polymers was saturable (Realini and Althaus, unpubl. obs.). Taken together, these results suggest that poly(ADP-ribose)polymerase may act as a histone shuttle mechanism in chromatin, in which the catabolic counterpart, poly(ADP-ribose)glycohydrolase could assume the role of reestablishing DNA-binding of histones. The scheme shown in Fig. 6 provides a summary of the mechanistic details of the protein shuttle mechanism that emerge from our in vitro studies. First, the enzyme poly(ADP-ribose)polymerase operates in a processive mode on DNA templates. The presence of DNA single- or double-strand breaks causes binding and activation of this enzyme (Benjamin and Gill 1980; Berger and Petzold 1985; Sastry et al. 1989; Mazen et al. 1989). A Zinc finger motif in the DNA binding domain of the polymerase apparently is involved in this function (Mazen et al. 1989). Following activation of the polymerase, a distinct pattern of polymers is sequentially added to the acceptor proteins. These polymers represent dynamic sites for protein interactions and reduce protein interactions with DNA. Enzymatic degradation of polymers releases bound proteins which are now able to interact with DNA again. This reversible shuttling of proteins causes temporary exposure of the DNA to other proteins which may interact with a damaged site. Preliminary experiments from our laboratory suggest that this mechanism may be involved in the unfolding of nucleosomes during DNA excision repair (Mathis and Althaus 1987; Realini and Althaus, unpubl. results). In view of our findings on the unfolding of chromatin domains in DNA excision repair in vivo (Mathis and Althaus 1986; Althaus et al. 1989), the rapid turnover of ADP-ribosyl residues on chromatin proteins, which varies with the level of DNA damage (Alvarez-Gonzales and Althaus 1989), could reflect adaptation of the shuttle mechanism on damaged templates to the level of DNA repair activity.

The concept of poly-ADP-ribosylation as part of a protein shuttle mechanism raises a number of questions. For example, it will be very important to understand the specificity, capacity, and potency of ADP-ribosyl polymers in reducing DNA binding of a larger spectrum of proteins, which may also act on DNA in DNA excision repair or in other active processes on chromatin such as replication and transcription. For example, it will be important to study the interaction of DNA repair enzymes with their substrate in the presence of histones and poly(ADP-ribose)polymerase. Also, more work is need to precisely define the cooperation of poly(ADP-ribose)glycohydrolase with poly(ADP-ribose)polymerase. These studies are now under way in our laboratory.

Acknowledgements. This study was supported by grant No. 3.161.0.88 from the Swiss National Foundation for Scientific Research, awarded to F.R.A.

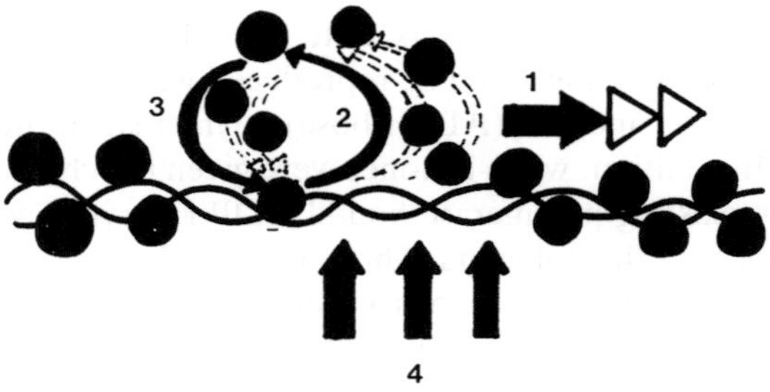

Fig. 6. Scheme summarizing essential features of the protein shuttling mechanism. The numbers *1* to *4* denote individual reaction steps of the shuttle mechanism. *1* Processive mode of the poly(ADP-ribose) polymerase reaction; *2* binding of histones to polymerase-bound polymers; *3* reestablishment of histone binding to DNA following digestion of polymers by poly(ADP-ribose) glycohydrolase; *4* increased accessibility of DNA template to other proteins. For further discussion, see text

References

Althaus FR, Richter C (1987) ADP-ribosylation of proteins: enzymology and biological significance. Molecular biology, biochemistry and biophysics, Vol 37. Springer, Berlin Heidelberg New York

Althaus FR, Lawrence SD, Sattler GL, Pitot HC (1982a) ADP-ribosyl-transferase activity in cultured hepatocytes: interactions with DNA repair. J Biol Chem 257: 5528–5535

Althaus FR, Lawrence SD, He YZ, Sattler GL, Tsukada Y, Pitot HC (1982b) Effects of altered $(ADP\text{-ribose})_n$-metabolism on expression of fetal functions by adult hepatocytes. Nature 300: 366–368

Althaus FR, Mathis G (1989) Alvarez-Gonzales R, Loetscher P, Mattenberger M, ADP-ribosylation and chromatin function. In: Jacobson MK, Jacobson EL (eds) Springer, Berlin Heidelberg New York, pp 151–157

Alkhatib HM, Chen D, Cherney D et al. (1987) Cloning and expression of cDNA for human poly(ADP-ribose)polymerase. Proc Natl Acad Sci USA 84: 1224–1228

Alvarez-Gonzalez R, Althaus FR (1989) Poly(ADP-ribose) catabolism in mammalian cells exposed to DNA damaging agents. Mutat Res 218: 67–74

Benjamin RC, Gill DM (1980) Poly(ADP-ribose) biosynthesis in vitro programmed by damaged DNA. J Biol Chem 255: 10502–10508

Berger NA, Petzold SJ (1985) Identification of the requirements of DNA for activation of poly(ADP-ribose)polymerase. Biochemistry 24: 4352–4355

Cherney BW, McBride OW, Chen D, Alkhatib H, Bhatia K, Hensley P, Smulson ME (1987) cDNA sequence, protein structure, and chromosomal location of the human gene for poly(ADP-ribose)polymerase. Proc Natl Acad Sci USA 84: 8370–8374

Herzog H, Zabel BU, Schneider R, Auer B, Hirsch-Kaufmann M, Schweiger M (1989) Human nuclear NAD^+ ADP-ribosyltransferase: localization of the gene on chromosome 1q41-q42 and expression of an active human enzyme in *Escherichia coli*. Proc Natl Acad Sci USA 86: 3514–3518

Jacobson MK, Payne DM, Alvarez-Gonzalez R, Juarez-Salinas H, Sims JL, Jacobson EL (1984) Determination of in vivo levels of polymeric and monomeric ADP-ribose by fluorescence methods. Meth Enzymol 106: 483–494

Kurosaki T, Ushiro H, Mitsuuchi Y et al. (1987) Primary structure of human poly(ADP-ribose)synthetase as deduced from cDNA sequence. J Biol Chem 262: 5990–5997

Leduc Y, Murcia G de, Lamarre D, Poirier GG (1986) Visualization of poly(ADP-ribose)synthetase associated with polynucleosomes by immunoelectron microscopy. Biochim Biophys Acta 375: 243–255

Loetscher P (1988) Funktionelle Bedeutung der Poly(ADP-ribosylierung) von Chromatinproteinen. Ph D Thesis, ETH Zürich.

Loetscher P, Alvarez-Gonzalez R, Althaus FR (1987) Poly(ADP-ribose) may signal changing metabolic conditions to the chromatin of mammalian cells. Proc Natl Acad Sci USA 84: 1286–1289

Mathis G, Althaus FR (1986) Periodic changes of chromatin organization associated with rearrangement of repair patches accompany DNA excision repair of mammalian cells. J Biol Chem 261: 5758–5765

Mathis G, Althaus FR (1987) Release of core DNA from nucleosomal core particles following $(ADP\text{-ribose})_n$-modification in vitro. Biochem Biophys Res Commun 143: 1049–1054

Mazen A, Menissier-De Murcia G, Molinete M, Simonin F, Gradwohl G, Poirier G, Murcia G De (1989) Poly(ADP-ribose)polymerase – a novel finger protein. Nucl Acids Res 17: 4689–4698

Naegeli H, Loetscher P, Althaus FR (1989) Poly ADP-ribosylation of proteins: processivity of a posttranslational modification. J Biol Chem 264: 14382–14385

Sastry SS, Buki KG, Kun E (1989) Binding of adenosine diphosphoribosyltransferase to the termini and internal regions of linear DNAs. Biochemistry 28: 5670–5680

Schneider R, Auer B, Hirsch-Kauffmann M, Wintersberger U, Schweiger M (1987) Isolation of a cDNA clone for human NAD^+:protein ADP-ribosyltransferase. Eur J Cell Biol 44: 302–307

Suzuki H, Uchida K, Shima H, Sato T, Okamoto T, Kimura T, Miwa M (1987) Molecular cloning of cDNA for human poly(ADP-ribose)polymerase and expression of its gene during HL-60 cell differentiation. Biochem Biophys Res Commun 146: 403–409

Uchida K, Morita T, Sato et al. (1988) Nucleotide sequence of a full-length cDNA for human fibroblast poly(ADP-ribose)polymerase. Biochem Biophys Res Commun 148: 617–622
Ueda K, Hayaishi O (1985) ADP-ribosylation. Annu Rev Biochem 54: 73–100
Wesierska-Gadek J, Sauermann G (1988) The effect of poly(ADP-ribose) on DNA-core histone interaction. Biol Chem Hoppe-Seyler 369–945

DNA Lesions, DNA Repair, and Chromosomal Aberrations

A. T. NATARAJAN, R. C. VYAS, F. DARROUDI, L. H. F. MULLENDERS, and M. Z. ZDZIENICKA[1]

Contents

1 Introduction

Among the DNA lesions induced by ionizing radiations, DNA double-strand breaks (DSBs) appear to be the most important, leading to chromosomal aberrations and cell death. We have presented biochemical and cytological evidence for this conclusion. These include the increase in the frequeny of chromosomal aberrations and DSBs in CHO cells X-irradiated and posttreated with *Neurospora* endonuclease, an enzyme which specifically cuts single-stranded DNA, thus generating DSBs (Natarajan and Obe 1978; Natarajan et al. 1980). In addition, restriction endonucleases, which induce only one type of lesion, namely DSBs, induce chromosomal aberrations in a pattern similar to that induced by ionizing radiations, i.e., chromosome type of aberrations in G1 and chromatid type of aberrations following G2 treatment (Natarajan and Obe 1984, Obe and Winkel 1985).

In this chapter, we discuss the data obtained from radiosensitive mutants from Chinese hamster cells (CHO and V79) which throw some light on the mechanisms in the formation of chromosomal aberrations. In addition, we have utilized the technique of premature chromosome condensation (PCC) to monitor the chromosomal repair kinetics in human lymphocytes and compared these data with the conventional analysis of metaphases following X-irradiation.

2 X-Ray Sensitive Mutants

Several mutants deficient in DNA DSB repair have been isolated in CHO cells (Jeggo and Kemp 1983). We have studied two of these mutants, namely, xrs 5 and xrs 6 in detail and tried to correlate the degree in the deficiency in rejoining of DSBs to the frequencies of chromosomal aberrations induced by X-rays. Though both the cell

[1] Department Radiation Genetics and Chemical Mutagenesis, State University of Leiden, Leiden, The Netherlands

lines belong to the same complementation group they exhibit different characteristics with regard to biological response to X-rays.

When the cells were irradiated in G2, the frequencies of induced aberrations correlated well with the defect in rejoining of DSBs(Kemp and Jeggo 1986). *Xrs* 5 cells which are 90% deficient had more aberrations than xrs 6 cells which are 60% deficient in repair of DSBs (Table 1); (Darroudi and Natarajan 1987a). Both the cell lines had a G2 block following X-irradiation which could be reversed by caffeine in xrs 5 but not in *xrs* 6 cells (Darroudi and Natarajan 1987a).

When G1 cells were irradiated, these mutants responded with higher, but similar frequencies of aberrations than the wild type (Table 2). The probable reason for the lack of difference between xrs 5 and xrs 6 may be that the proportion of heavily damaged cells reaching mitosis may vary between the mutants. This assumption is supported by the observation that when G1 cells are irradiated and the frequencies of breaks assessed by the premature chromosome condensation technique, xrs 5 cells have more breaks than xrs 6 (Darroudi and Natarajan 1989a). One of the character-

Table 1. Degree of DNA DSB repair and the frequency of chromosomal aberrations induced in wild-type CHO cells and the X-ray sensitive mutants xrs 5 and xrs 6

Cell type	Repair[a] (%)	Aberrations from 100 cells[b]		
		0.7 Gy	1.0 Gy	1.4 Gy
CHO	73	47	81	121
xrs 6	42	202	305	455
xrs 5	10	399	576	874

[a] Extent of repair at 120 min following irradiation (Kemp et al. 1984).
[b] Total aberrations after G2 irradiation (Darroudi and Natarajan 1987a).

Table 2. Frequencies of chromosomal aberrations induced in G1 cells by X-rays (data from Darroudi and Natarajan 1987a)

Cell type	Dose (Gy)	Aberrations/100 cells				
		Chromatid		Chromosome		Total
		Breaks	Exchanges	Breaks	Exchanges	
CHO	0	4	1	1	1	7
	0.7	5	1	3	6	15
	1.0	5	2	7	9	23
	1.4	9	2	8	13	32
xrs 5	0	5	0	1	1	7
	0.7	35	32	45	33	145
	1.0	44	43	68	42	189
	1.4	75	69	163	74	381
xrs 6	0	6	1	1	1	9
	0.7	32	53	44	38	167
	1.0	52	74	85	50	261
	1.4	88	120	160	152	420

istic features of G1 irradiated xrs mutant cells is the yield of both the chromosome and chromatid types of aberrations (Darroudi and Natarajan 1987a). This is very similar to the observation made in ataxia telangiectasia cells (Taylor 1978, Natarajan and Meijers 1978). The increase in the chromatid type of aberrations following G1 Irradiation could be due to a proportion of unrepaired DSBs reaching the S-phase. If this assumption is correct, one would expect that the frequency of induced chromatid aberrations in xrs 5 should be greater than the one obtained for xrs 6 since the former has a larger defect in DSB repair than the latter. However, it was found that the reverse was true, namely xrs 6 had more aberrations of the chromatid type than xrs 5. This led us to search for alternative lesions for the induction of the chromatid-type aberrations. Since both these mutant cells are known to be proficient in repair of DNA single-strand breaks (SSBs) (Kemp et al. 1984), X-ray-induced base damages appeared to be a possible candidate. Ionizing radiations are known to be a poor inducer of sister chromatid exchanges (SCEs). It is known that lesions like radiation-induced base damage can lead to SCEs (Uggla and Natarajan 1983). If the persisting lesions in xrs 6 are radiation-induced base damages, then one would expect an increase in the frequency of SCEs in this cell line following X-rays. When xrs 6 cells were irradiated with different doses of X-rays, a dose-dependent increase in the frequency of SCEs was obtained (Fig. 1; Darroudi and Natarajan 1987b). There was a small increase in xrs 5 cells and no increase in wild-type cells (Fig. 1).

It is possible to alter the proportion of radiation-induced DNA strand breaks to base damage by altering the irradiation conditions. Irradiation in air leads to more strand breaks due to indirect action by free radicals, in comparison to irradiation in nitrogen, while the frequency of base damage remains relatively unaltered (Paterson and Setlow 1972). In earlier experiments we have demonstrated that when wild-type CHO cells are irradiated under O_2 and N_2 conditions, the frequency of chromosomal

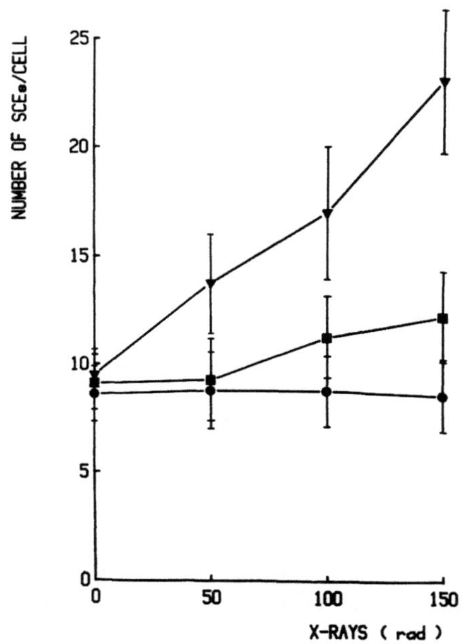

Fig. 1. Frequency of SCEs ± SD by X-rays in CHO K1 and xrs mutant cell lines. ● CHO K1; ■ xrs 5; ▼ xrs 6

aberrations increased under O_2 conditions and the frequency of SCEs increased under N_2 conditions, and if one compares the relative induction of SCEs to chromosomal aberrations, the ratio varies very much, a higher ratio being obtained for irradiation under nitrogen conditions in comparison to oxygen conditions (Uggla and Natarajan 1983). When similar experiments were done with xrs mutants, the increase in the frequency of SCEs was higher in the cells which were irradiated under nitrogen in comparison to irradiation under air. The ratio between SCEs and aberrations was highest for xrs 6 cells in comparison to xrs 5 and wild-type CHO cells (Fig. 2; Darroudi et al. 1990a). These results support the conclusion that the xrs 6 mutant is defective in the repair of both DSB and base damage.

Inhibitors of DNA replication such as cytosine arabinoside (ara C) inhibit also repair of DNA strand breaks and base damage (Hiss and Preston 1977; Iliakis and Bryant 1983). In view of the complexity in the repair defect of the xrs 6 mutant, we investigated the influence of this analog (along with hydroxyurea, HU) on the frequency of UV (to mimic base damage) induced chromosomal aberrations. The frequency of aberrations increased in both xrs 5 and wild-type cells, whereas there was no increase in xrs 6 cells (Darroudi et al. 1990b). This response is not due to the uptake of ara C into the cells as a similar response was observed following treatment with aphidicolin (Darroudi and Natarajan 1987b). Posttreatment with ara C and HU of UV-irradiated cells leads to the accumulation of SSBs and reflects the incision capacity of the cells. We measured the accumulation of SSBs at 20 and 120 min in CHO cells and xrs mutant cells following UV irradiation and posttreatment with ara C and HU (Fig. 3; Darroudi et al. 1990a). While the frequency of SSBs increased with time in the wild-type and xrs 5 cells, the frequency remained at the same level in xrs 6 at both times of determination, namely around 35% of the wild-type CHO cells. These results indicate that *xrs* 6 cells, in addition to the defect in the repair of DSBs, are deficient in incision of other types of damages (base damage, dimers) as

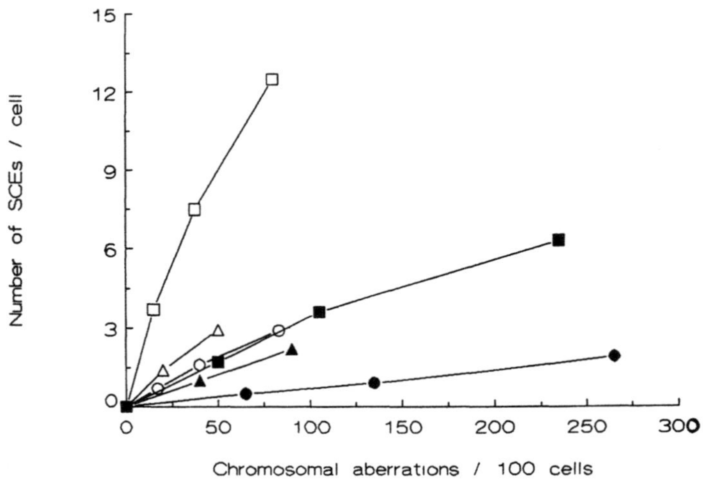

Fig. 2. Induced frequencies of SCEs plotted against the percentage of chromosomal aberrations induced by X-rays (similar doses) in air or in nitrogen in different cell lines. CHO K1 – air (▲); nitrogen (△); xrs 5 – air (●); nitrogen (○); xrs 6 – air (■); nitrogen (□)

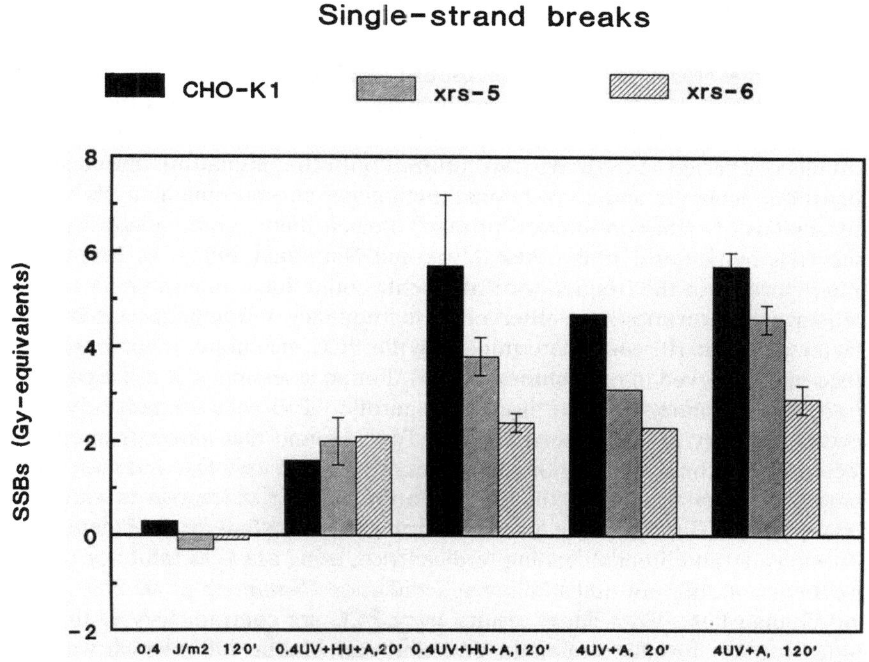

Fig. 3. Accumulation of SSBs (Gy equivalent) in the presence of ara C (A) and HU (20 or 120 min posttreatment) after UV exposure (dose of 0.4 or 4 J/m2) in CHO K1, xrs 5 and xrs 6 cells. Each point is the mean of 4 determinations with SEM from two separate experiments

well. This additional defect is reflected also by increased sensitivity of xrs 6 mutants to bleomycin and monofunctional alkylating agents in comparison to xrs 5 cells (Darroudi and Natarajan 1989b). While xrs 5 cells respond with a higher frequency of mutations (HPRT-) following X-irradiation than xrs 6 and wild-type cells, xrs 6 cells respond with higher frequency of mutations (HPRT-, Oua R) following treatment with MMS (Darroudi and Natarajan 1989a).

We have recently examined another group of radiation-sensitive mutants derived from V79 cells. These mutants (VC-4 and VG-8, VE-5, belonging to the same complementation group) resemble ataxia telangiectasia cells in the following respects. They are sensitive to cell killing by X-rays and bleomycin, exhibit no detectable defect in rejoining of single- or double-strand breaks and are resistant to radiation-induced inhibition of DNA sythesis (Zdzienicka et al. 1989). Like A-T cells, following irradiation of cells at G1 stage, these mutants respond with increased frequencies of both chromosome and chromatid types of aberrations (Zdzienicka et al. 1989). Since these mutants are not defective in DSB repair, some X-ray induced lesion(s) other than DSBs must be responsible for the increased frequency of aberrations. We have checked the response of these mutants to restriction endonucleases. Unlike xrs mutants, the frequency of aberrations induced in these mutant cells was very similar to the one obtained for wild-type cells (Table 3; Natarajan et al., unpubl.). These results indicate that under suboptimal levels of repair, X-ray-induced lesions, other than DSBs can lead to chromosomal aberrations.

3 Studies with Human Lymphocytes

Earlier studies (Holmberg and Gumauskus 1986; Natarajan et al. 1986) demonstrated that in human peripheral blood lymphocytes most of the lesions that give rise to dicentrics are repaired or misrepaired within the first 30 min, in contrast to earlier findings of Preston (1980). We have utilized both the premature chromosome condensation technique and conventional metaphase chromosome analysis to examine the kinetics of radiation-induced primary breaks, their rejoining and formation of dicentrics in human lymphocytes (Vyas and Natarajan 1989). A dose-dependent linear increase in the frequency of fragments and a linear-quadratic increase in the frequency of dicentrics were observed. The frequency of fragments was much higher (factor of about 10) when determined by the PCC technique in comparison to the frequency observed in metaphases (Fig. 4). For an assessment of the repair kinetics, irradiated lymphocytes were fused with mitotic CHO cells immediately and after several recovery periods after radiation. It was found that chromosome fragments decrease with time, whereas the dicentrics are formed very fast and their frequency remained the same, despite the decline in the number of fragments with increased recovery time (Fig. 5). These results confirm earlier findings on the kinetics of repair of lesions (strand breaks) leading to dicentrics, using ara C as inhibitor of repair of the breaks at different times following irradiation (Natarajan et al. 1986; Holmberg and Gumauskus 1986). These results using PCC are contradictory to the ones obtained for the formation of rings (Pantelias and Maillie 1985) which was found to increase with time of postincubation before fusion. This discrepancy has to be resolved and we are at present doing such experiments.

We have also investigated the role of chromosome condensation at the time of irradiation on the frequency of chromosomal aberrations (Vyas and Natarajan 1989). Human lymphocytes were fused with mitotic CHO cells and irradiated with 2 Gy immediately or 30 or 60 min after fusion. The frequency of induced breaks fell dramatically with increased condensation. It has been suggested that treatment with hypertonic salt solution suppresses the fast component of repair of radiation-induced strand breaks (Hittelman 1987). This should increase the frequency of X-ray-induced breaks. Hypertonic solution is also expected to condense the chromatin which could prevent the accessibility to the repair enzyme, thus increasing the frequency of breaks. Two types of experiments were carried out to study the role of hypertonic salt solution

Table 3. Frequency of chromosomal aberrations induced by restriction enzyme Bam H1 (50 units) in V79 CHO cells and their mutants

Cell type	Abnormal cells (%)	Aberrations per 100 cells		Total	Aberrations per aberrant cell
		B', B'', Ex	CR, DC		
V79	23	59	5	64	2.8
VC-4	14	29	3	32	2.3
VG-8	16	32	1	33	2.1
CHO-K1	23	53	14	67	2.9
xrs 5	46	176	14	190	4.1

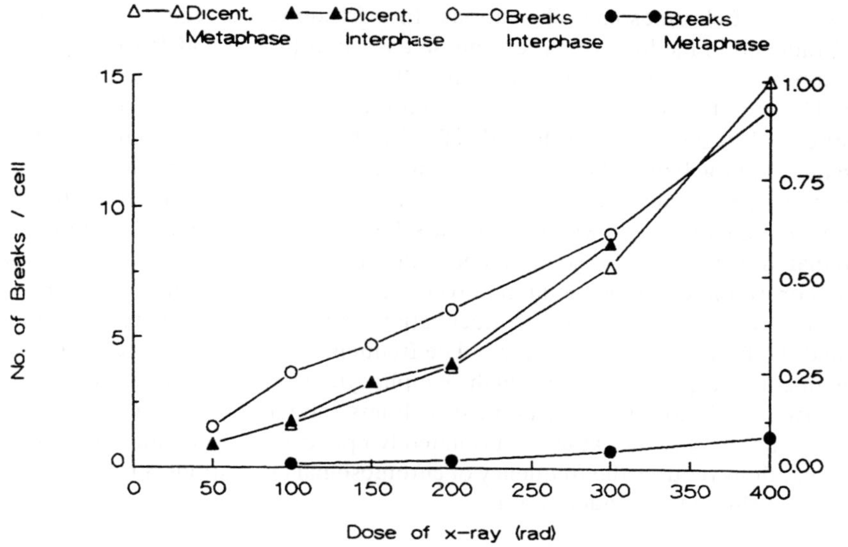

Fig. 4. Frequency of dicentrics and fragments detected following different doses of X-rays, either in interphase (premature chromosome condensation) or in metaphases

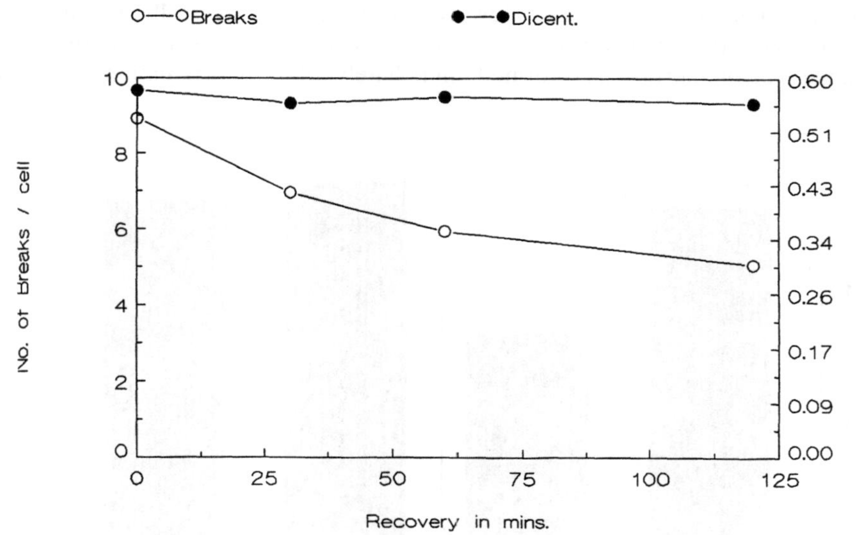

Fig. 5. The frequencies of chromosome breaks and dicentrics detected in PCCs following irradiation of lymphocytes with 3 Gy of X-rays at different recovery times. Note that the frequency of dicentrics remains constant while the frequency of breaks declines with time

on the yield of chromosomal aberrations. In a fractionation regime, lymphocytes were irradiated in medium or hypertonic salt solution (0.3 M NaCl) with 1 Gy and allowed to recover in the same medium for 30 or 90 min and irradiated further with 1 Gy. The lymphocytes were fused with mitotic CHO cells and the frequency of dicentrics and fragments was evaluated in PCCs. The yield of dicentrics and fragments was reduced in the lymphocytes irradiated in the presence of hypertonic solution, in contrast to the expectation. In another experiment, lymphocytes were irradiated with 3 Gy in medium, or treated with hypertonic salt solution before and after irradiation, or irradiated in cold isotonic solution and transferred immediately to hypertonic salt solution or irradiated in the medium at room temperature and transferred to hypertonic salt solution. After 30 min postincubation, the lymphocytes were cultured for 48 h and the frequency was evaluated. The frequencies of dicentrics wère similar in all treatments except the one in which the lymphocytes were irradiated in the presence of hypertonic solution which responded with less (about 40%) dicentrics (Fig. 6). These results indicate that at least in isolated lymphocytes, irradiation under hypertonic conditions reduces the frequency of chromosomal aberrations, probably due to the induced chromatin condensation.

4 Conclusions

DNA DSB appears to be the most important lesion leading to chromosomal aberrations in repair-competent cells. However, in repair defective cells, radiation-induced lesions other than DSBs may also lead to chromosomal aberrations. The exchange type of aberrations is formed immediately following irradiation, as visual-

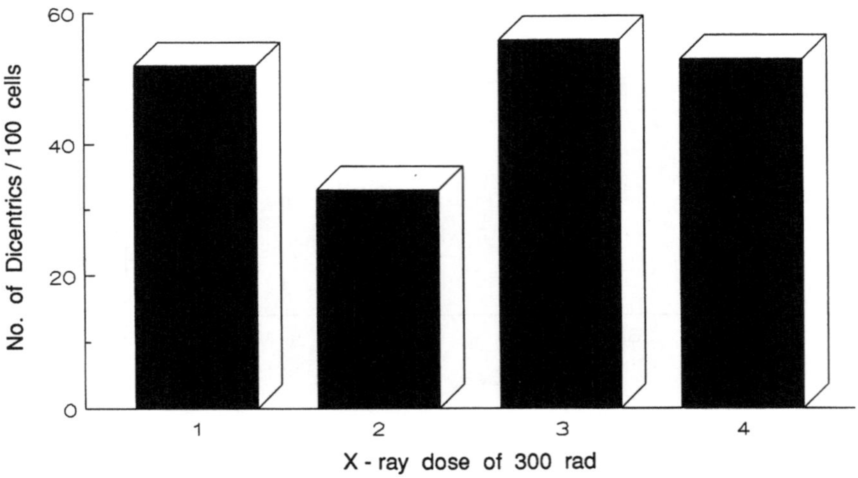

Fig. 6. Yield of dicentrics following irradiation of lymphocytes with 3 Gy and scored at mitosis. *1* Irradiation of lymphocytes in the medium; *2* irradiation of lymphocytes incubated in 0.3 M Na Cl for 30 min and irradiated in the presence of this solution; *3* irradiation of lymphocytes at 4 °C and immediately posttreated with 0.3 M Na Cl for 30 min before changing to the medium; *4* irradiation of lymphocytes at 4 °C, followed by culturing in normal medium

ized by the PCC technique, whereas the frequency of breaks decay with time in G0 lymphocytes. When the chromatin in human lymphocytes is condensed either by PCC or by hypertonic treatment before X-irradiation, the initial DNA damage is reduced leading to lower frequencies of induced aberrations.

Acknowledgments. These studies were financialy supported by J. A. Cohen Inter-University Research Institute, Leiden and an EEC Radiation Protection Programme (B-16-166 NL).

References

Darroudi F, Natarajan AT (1987a) Cytological characterization of Chinese hamster ovary X-ray-sensitive mutant cells xrs 5 and xrs 6. I. Induction of chromosomal aberrations by X-irradiation and its modulation with 3-aminobenzamide and caffeine. Mutat Res 177: 133–148

Darroudi F, Natarajan AT (1987b) Cytological characterization of Chinese hamster ovary X-ray sensitive mutant cells xrs 5 and xrs 6. II. Induction of sister-chromatid exchanges and chromosomal aberrations by X-rays and UV irradiation and their modulation by inhibitors poly (ADP-ribose) synthetase and α polymerase. Mutat Res 177: 149–160

Darroudi F, Natarajan AT (1989a) Cytological characterization of Chinese hamster ovary X-ray sensitive mutant cells xrs 5 and xrs 6. VII. Complementation analysis of X-irradiated wild type CHO-K1 and xrs mutants using premature chromosome condensation technique. Mutat Res 213: 249–255

Darroudi F, Natarajan AT (1989b) Cytogenetical characterization of Chinese hamster ovary X-ray sensitive mutant cells xrs 5 and xrs 6. III. Induction of cell killing, chromosomal aberrations and sister chromatid exchanges by bleomycin, mono- and bi-functional alkylating agents. Mutat Res 212: 123–136

Darroudi F, Natarajan AT, Schans GP van der, Loon AAWM van (1990a) Biochemical and cytogenetical characterization of X-ray sensitive Chinese hamster ovary mutant cells xrs 5 and xrs 6. V. The correlation between DNA strand breaks and base damage to chromosomal aberrations and sister chromatid exchanges induced by X-irradiation. Mutat Res 235: 119–127

Darroudi F, Natarajan AT, Schans GP van der (1990b) Biochemical and cytogenetical characterization of Chinese hamster ovary X-ray sensitive mutant cells xrs 5 and xrs 6. VI. The correlation between UV-induced DNA lesions and chromosomal aberrations and their modulations with inhibitors of DNA repair synthesis. Mutat Res 235:129–135

Hiss EA, Preston RJ (1977) The effect of cytosine arabinoside on the frequency of single strand breaks in mammalian cells following irradiation or chemical treatment. Biochim Biophys Acta 478: 1–8

Hittelman WN (1987) Inhibition of fast and slow components of chromosome repair. Proc 8th Int Conf Radiation Research, Edinbourgh, (Abstracts), p 212

Holmberg M, Gumauskus E (1986) The role of short-lived DNA lesions in the production of chromosome exchange aberrations. Mutat Res 160: 221–229

Iliakis G, Bryant PE (1983) Effect of nucleoside analogues ara A, ara A and ara C on cell growth and repair of potentially lethal damage and DNA double strand breaks in mammalian cell cultures. Cancer Res 3: 143–150

Jeggo PA, Kemp LM (1983) X-ray-sensitive mutants of Chinese hamster ovary cell line. Isolation and cross sensitivity to other DNA damaging agents. Mutat Res 112: 313–327

Kemp LM, Jeggo PA (1986) Radiation induced chromosome damage in X-ray sensitive mutant (xrs) of the Chinese hamster cell line. Mutat Res 166: 255–263

Kemp LM, Sedgwick SG, Jeggo PA (1984). X-ray sensitive mutants of Chinese hamster ovary cells defective in double strand break rejoining, Mutat Res 132: 189–196

Natarajan AT, Meijers M (1978) Chromosomal radiosensitivity of ataxia-telangiectasia cells at different cell cycle stages. Hum Genet 52: 127–132

Natarajan AT, Obe G (1978) Molecular mechanisms involved in the production of chromosome aberrations. I. Utilization of *Neurospora* endonuclease for the study of aberration production in G2 stage of the cell cycle. Mutat Res 52: 137–149.

Natarajan AT, Obe G (1984) Molecular mechanisms involved in the production of chromosomal aberrations. III. Restriction endonucleases. Chromosoma (Berl) 90: 120–127

Natarajan AT, Obe G, Zeeland AA van, Palitti F, Meijers M, Verdegaal-Immerzeel EAM (1980) Molecular mechanisms involved in the production of chromosomal aberrations. II. Utilization of *Neurospora* endonuclease for the study of aberration production by X-rays in G1 and G2 stages of the cell cycle. Mutat Res 69: 293–305

Natarajan AT, Darroudi F, Mullenders LHF, Meijers M (1986) The nature and repair of DNA lesions that lead to chromosomal aberrations induced by ionizing radiations. Mutat Res 160: 231–236

Obe G, Winkel EU (1985) The chromosome breaking activity of restriction endonuclease Alu I in CHO cells is independent of the S phase of the cell cycle. Mutat Res 69: 71–79

Pantelias GE, Maillie HD (1985) Direct analysis of radiation indued chromosome fragments and rings in unstimulated human peripheral blood lymphocytes by means of the premature chromosome condensation technique. Mutat Res 139: 67–72

Paterson MC, Setlow RB (1972) Endonucleolytic activity from *Micrococcus luteus* that acts on X-ray-induced damage in plasmid DNA of *Escherichia coli*. Proc Natl Acad Sci USA 69: 2927–2931

Preston RJ (1980) The effect of cytosine arabinoside on the frequency of X-ray induced chromosome aberrations in normal leucocytes. Mutat Res 69: 71–79

Taylor AMR (1978) Unrepaired DNA strand breaks in irradiated ataxia telangiectasia lymphocytes suggested from cytogenetic observations. Mutat Res 50: 407–418

Uggla AH, Natarajan AT (1983) X-ray induced SCEs and chromosomal aberrations in CHO cells. Influence of nitrogen and air during irradiation at different stages of the cell cycle. Mutat Res 122: 193–200

Vyas RC, Natarajan AT (1989) Radiation induced chromosomal breakage and rejoining evaluated in interphase and mitotic chromosomes of human lymphocytes (in prep)

Zdzienicka MZ, Jaspers NGJ, Schans GP van der, Natarajan AT, Simons JWIM (1989) Ataxia-telanngiectasia-like Chinese hamster V79 cell mutants with radioresistant DNA synthesis, chromosomal instability and normal DNA strand break repair. Cancer Res 49: 1481–1485

Is It Misrepair or Lack of Repair Which Kills Cells Irradiated in G_2?

R. C. Moore, L. Barber, and C. G. Bingham[1]

Contents

1 Introduction

A number of studies have been carried out to test the hypothesis that chromosome aberrations are the chief cause of cell death after irradiation, and the presence of chromosome aberrations has been found to correlate with the loss of clonogenic ability (Dewey et al. 1971; Grote et al. 1981). These studies have been carried out using cells synchronized in G_1, or log-phase cells, i.e., a population made up mainly of cells in G_1-or S-phase.

We have determined the frequency of chromosomal aberrations in cells in their second division after irradiation in G_2 with 2.5 Gy of X-rays (Moore and Bender, in prep.). Cell death during or following the first division was prevented by exposing the cells immediately after radiation to colcemid for 16 h. During this period, cells entering mitosis remained blocked in c-metaphase for several hours, and then went back into interphase without completing division, with twice the quasi-diploid DNA content, and appeared at the next division as quasi-tetraploids. Cells in S at the time of irradiation could be distinguished from cells in other phases by pulse-labelling with tritiated thymidine (^{3}HTdR) before irradiation, and eliminated by scoring only unlabelled cells in autoradiographs.

It was anticipated that some cells in G_1 at the time of irradiation might also be blocked in their first post-irradiation division, and be in their second division at the time of fixation, but it was assumed that all aberrations in these cells would be paired, and hence be distinguishable from those in cells in G_2 at the time of irradiation, which, after a second round of replication, would be of the chromosome type, but unpaired.

The results obtained were surprising in a number of respects:

In some samples, in which cells with paired aberrations were present in significant numbers, paired dicentrics were frequently found to be accompanied by unpaired fragments.

[1] Cell Biology Unit, Peter MacCallum Cancer Institute, 481 Little Lonsdale St., Melbourne 3000, Australia

There was a significant increase in asymmetrical interchromosomal exchanges at the second division, compared with the first division, in cells irradiated in G_2.

There was a dearth of fragments in cells irradiated in G_2, and fragments were more common in cells which contained dicentrics than they were in cells without them, suggesting that a substantial number of them were the products of incomplete exchange. The sum of fragments, sister unions and exchanges was insufficient to account for the level of lethality usually encountered in cultured mammalian cells irradiated in G_2.

Several interpretations of these data are possible.

The small number of fragments at the second division, compared to breaks and gaps at the first, may be because most of the apparent breaks are simply gaps, and are healed between the first and second division. If this is so, the level of true aberrations is insufficient to explain the level of cytotoxicity usually found in cells irradiated in G_2.

The increase in exchanges between the first and second division suggested that repair might be incomplete at the first division following irradiation in G_2, and that the sort of misrepair which results in exchange formation might be accompanied by repair that eliminates fragments, that is, they may rejoin with other broken ends by the process described in classical radiobiology as restitution. If this is so, it would be expected that if cells were "nursed" through the first division with colcemid, cytotoxycity would be reduced at subsequent divisions.

On the other hand, the abundance of unpaired fragments in cells irradiated in G_1, after a second round of replication, raises the possibility that, after the first division is completed, some chromosome breaks are unrepaired, and may be digested away, with total loss of that part of the genome contained in them; that is, that cell death is in some cases due to a total lack of repair. If this is so, then it must be assumed that the DNA is protected up until the first division, since most cells do not die until they divide.

To investigate these possibilities further, clonogenic assays were carried out on cells in G_2, with and without a post-treatment with colcemid, to determine whether survival in JU56 cells, irradiated in G_2, was similar to that in other permanent cell lines, and whether further effective repair could be detected in cells prevented from dying in the immediate post-irradiation division.

As well, the dose-yield relationship for chromatid exchanges in cells irradiated in G_2 was determined. Although the yield of breaks and gaps varies throughout G_2, and hence a valid dose-yield relationship canot be determined, the yield of exchanges is constant with time throughout G_2 (Moore 1980; Moore and Bender 1987), and a dose-yield relationship can be obtained.

2 Materials and Methods

Log-phase cultures of the permanent cell line JU56 were maintained in disposable culture flasks with a surface area of 75 cm^2 (Costar), in alpha MEM (Gibco), with 20% FCS (Flow, Australia). The cell line has been described previously (Moore 1980). However, it has now undergone a spontaneous genetic change, so that its generation time has decreased from 16 to 12 h, and the new cell cycle parameters, estimated from FLM analysis, are: G_1, 3.5 h; S, 6 h; and G_2, 2.5 h.

Chromosomal Aberrations. Approximately 15 min before irradiation, ^{3}HTdR, 5 Ci/mM, 1 μCi/ml (Amersham), was added to a set of cultures. Irradiation was carried out using a Siemens Stabilipan Orthovoltage DXRT, at 180 kV, 20 mA, 0.2 mm Cu added filter, giving a half-value thickness of 0.5 mm Cu, and a dose rate of 1.24 Gy/min. All operations were carried out at 37 °C. Colcemid was added, and cultures fixed 4–8 h later, depending upon X-ray dose, to allow for radiation-induced delay in entering mitosis.

Autoradiographs were prepared from c-metaphase preparations and exposed for 6 days. Chromatid exchanges were scored in unlabelled metaphases.

The experiment was done in duplicate, with doses of 0, 1.25, 2.5, 3,75 and 5 Gy, and 200 metaphases were scored for each data point for each experiment.

Cloning Assays. A log-phase culture was incubated for 30 min with ^{3}HTdR, 85 Ci/mM, 5 μCi/ml, for 30 min, a treatment schedule which brings about reproductive cell death in S-phase cells. Further, 2×10^7 cells were loaded onto the surface of a 1–2% gradient of Bovine Albumin Fraction V (Miles Diagnostics) in a 550 ml Staput velocity sedimentation cell separator (Johns), at 4 °C. The cells were allowed to separate under unit gravity (Miller 1973) for 90 min, and 30 × 15 ml fractions were collected. The recovery of G_2- and S-phase cells was about 2×10^6, of which 60% were S-phase cells.

The cells were dispensed at suitable concentrations into disposable plastic culture vessels with a surface area of 25 cm^2. It had been found that mitosis recommenced after 2 h incubation, following the separation procedure. Therefore, the cultures were incubated for 1 h at 37 °C after they were set up, before irradiations were carried out in the warm, which were completed in less than 1 h.

After irradiation, colcemid was added to half the cultures. It was found that the minimum concentration of colcemid which allowed cells to accumulate in mitosis was 0.2 μg/ml, and this was the concentration used. It was found that concentrations of 0.01 μg/ml colcemid showed some toxicity in clonogenic assays. This was about the concentration which remained after the colcemid-containing medium was sucked off and replaced with fresh medium after 16-h incubation; it was decided to tolerate the cytotoxic effect of the low level of colcemid, rather than lose loosely attached cells by rinsing.

The plating efficiency of untreated separated cells, after correction for S-phase cells, was about 30%, which was comparable to that for log-phase cells without any pretreatment. The plating efficiency of cells treated with colcemid was 20–50% of untreated cells.

The experiment was carried out four times, using the same doses as for the measurement of chromosome aberrations, with five replicate dishes/data point.

3 Results

The yield of chromatid exchanges as a function of dose is shown in Fig. 1. The data was fitted to the expression $y = aD - bD^2$, using the method of conjugate search (Powell 1964), and assuming the errors were Poisson-distributed. The estimated values of a and b are: $a = 0.0483 \pm 0.018/\text{Gy}$, $b = 0.0279 \pm 0.004/\text{Gy}^2$.

Exchanges were also classified as interchanges (between different chromosomes), interarm exchanges (between different arms of the same chromosome), and intra-arm. The intra-arm exchanges were classified as dot deletions, sister unions, and proximal-distal interchromatid exchanges (see Fig. 2). The numbers of the different types are shown in Table 1. The dose-yield relationships for interchromosomal exchanges and intrachromosomal exchanges is shown in Fig. 3. For interchanges, $a = 0.012 \pm 0.011/\text{Gy}$, and $b = 0.0165 \pm 0.0032/\text{Gy}^2$, and for intrachanges, $a = 0.0402 \pm 0.012/\text{Gy}$ and $b = 0.0104 \pm 0.0032/\text{Gy}^2$.

Clonogenic survival of cells irradiated in G_2, with and without colcemid treatment, is shown in Fig. 4. The data were fitted to the expression $s = \text{-}\exp(aD + bD^2)$, and assuming the errors were binomially distributed, since the assumption of a Poisson distribution would lead to an underestimate of errors (see Sect. 4). The

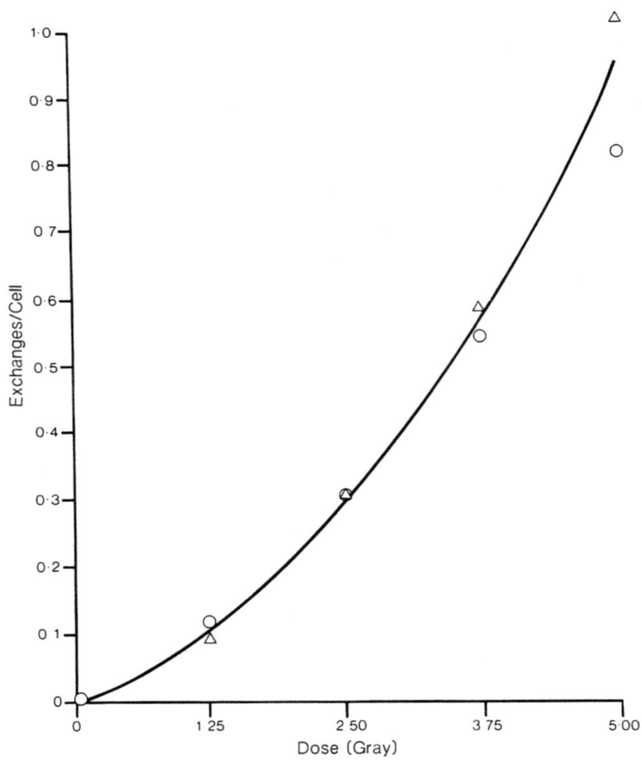

Fig. 1. Dose-yield relationship for total chromatid exchanges with X-irradiation in JU56 cells irradiated in G_2

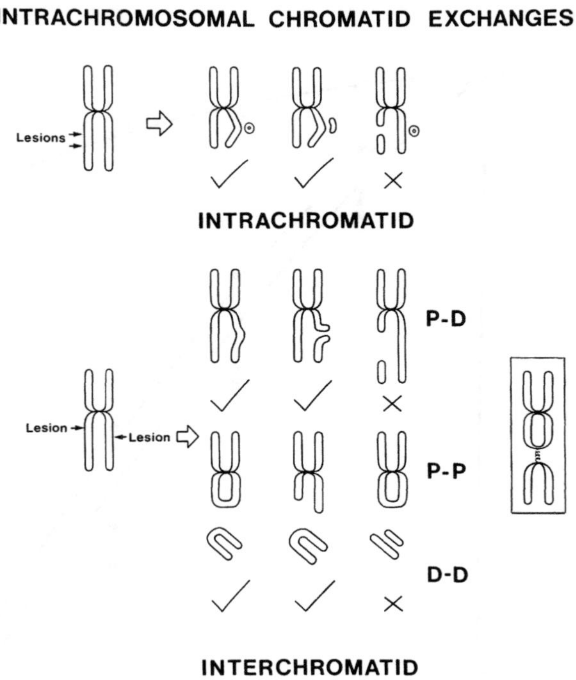

Fig. 2. Configurations of different classes of intrachanges in cells irradiated in G_2. Those marked with a *tick* were scored. Attached sister unions, which somewhat resemble true sister unions, are probably not true chromosomal rearrangements

Table 1. Numbers of different classes of chromatid exchanges in JU56 cells X-irradiated in G_2

Interchromosomal exchanges	361
Dot deletions	108
Distal sister unions	115
Proximal-distal interchromatid exchanges	103
Interarm exchanges	62
Total:	749

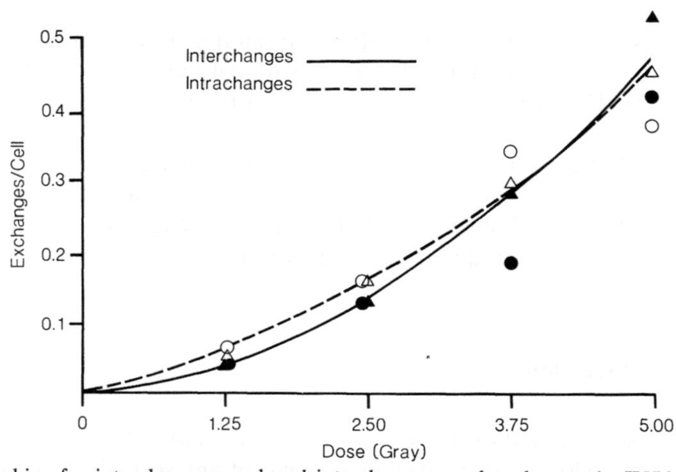

Fig. 3. Dose-yield relationships for interchromosomal and intrachromosomal exchanges in JU56 cells X-irradiated in G_2

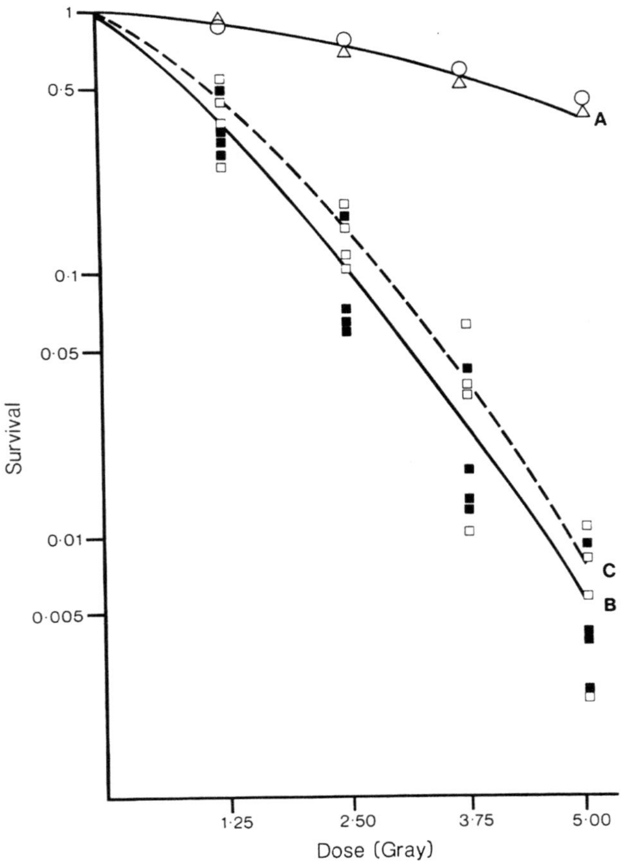

Fig. 4. Survival curves for JU56 cells irradiated in G_2. *A:* Cells without exchanges: *B (closed squares):* cells not post-treated with colcemid; *C (open squares):* cells post-treated with colcemid

values of a and b are, for cells not treated with colcemid, a = 0.906 ± 0.27, and b = 1.00 ± 0.29, and for cells treated with colcemid, a = 0.754 ± 0.53, and b = 1.01 ± 0.21. The survival curves for the two different treatments are not significantly different.

For comparison, the yield curve for exchanges has been converted to a negative exponential. The underlying assumption is that the distribution of exchanges is random, i.e., Poisson-distributed, and that cells without exchanges survive. The data points are the frequencies of cells without exchanges in the samples scored.

4 Discussion

Post-treatment of the cells with colcemid had little, if any, effect on survival. There is no evidence that substantial repair of any sort which could lead to increased cell survival takes place after the first cell division.

Quite obviously, exchanges large enough to be seen microscopically are insufficient to explain the loss of reproductive cell death.

When it comes to the contribution of chromatid breaks and the resulting fragments to lethality, no simple calculation can be made as can be done in the case of exchanges. The distribution of breaks is not Poisson; the yield varies with time after irradiation in G_2, and one would expect them to be overdispersed (Couzin and Papworth 1979). Furthermore, we found that, when examined at the second division, fragments were more frequent in cells with exchanges than they were in cells without exchanges, suggesting that some fragments arise from incomplete exchange. About half of the detectable exchanges are incomplete at the first division.

The only valid comparison is between the numbers of cells without aberrations, and the number of survivors. We found that at the second division, the frequency of cells without aberrations was 30% or higher. The frequency of survivors for cells treated with colcemid at the same dose (2.5 Gy) was 15%, and the highest individual survival was 18%. Not enough aberrations are present at the second division for their frequency to explain cell lethality.

It is still possible to postulate that some fragments have remained unrepaired, and are completely digested away. Unless a phenomenon such as this is postulated, gross chromosomal aberrations cannot be the sole cause of cell lethality in cells irradiated in G_2.

It is, however, feasible that very small changes, too small to be detected microscopically, are a cause of cell lethality. The existence of such aberrations can be inferred from an examination of the frequencies of different classes of intrachanges.

It has not infrequently been found that the ratios of inter- and intrachanges differ from those which would be expected if both exchangeable lesions and the resulting exchanges occur at random (Savage and Papworth 1982), and such is the case with exchanges observed in the experiments reported here. If exchange occurs at random between two lesions also occurring at random, then about 6% of exchanges should be intrachromosomal, interarm exchanges, and about 6% should be intra-arm exchanges. The frequency of inter-arm exchanges does not depart very markedly from that expected on the basis of randomness, but the frequency of intra-arm exchanges is much higher than would be expected on this basis. About half the exchanges are intra-arm.

The excess of intrachanges indicates a higly non-random distribution of exchangeable lesions in cells irradiated in G_2. Non-randomness of this sort was first postulated by Revell (1959, 1974), who suggested that exchanges did not arise from breakage of chromatids, but from exchange between lesions before chromatid discontinuity occurred. According to the Revell hypothesis, chromatid breakage was always the result of incomplete exchange. Although there is evidence that some true breaks occur (discussed by Savage 1989) the Revell hypothesis is supported by data such as that presented here. It is compatible with current knowledge of the molecular basis of chromosome aberrations; they may arise from double-strand DNA breaks, but DNA breaks need not necessarily lead to chromosome discontinuities, since the DNA is wrapped around nucleosomes. We have recently found evidence that the majority of DNA lesions which restitute are chemically different to the minority which undergo exchange (Moore et al. 1988), which is also evidence against a simple "breakage first" hypothesis.

Of the different classes of intra-arm exchanges, not all are completely scorable. It is not always clear whether dot deletions are complete or incomplete. Furthermore, dot deletions vary in size and some of them approach the limit of resolution, and it is likely that many small ones will go undetected.

In the case of sister unions, virtually all distal sister unions can be scored, but not all proximal ones can be seen if the aberration occurs sufficiently close to the centromere. Therefore, complete exchanges, but only half the incomplete exchanges can be scored.

In the case of proximal-distal interchromatid exchanges, complete exchanges can again be detected, and both incomplete types can sometimes be recognized. However, none of these exchanges can be identified unless they are fairly large by comparison with dot deletions; that is, unless the distance along the chromatids between the two initial lesions is fairly long. In that incomplete case where the shorter of the two chromatids resulting from exchange was incomplete, we considered them largely unscoreable, and did not attempt to record them.

If intrachanges occur at random, it would be expected that the ratio of dot deletions: sister unions: prox-dist interchromatid intrachanges would be 1:1:1. In practice, the relative ease of detectability will modify these ratios. The relative order of detectability (for the complete case and one of the incomplete cases) is sister unions > dot deletions > proximal-distal interchromatid intrachanges.

Given that this is so, it would therefore be expected that there would be more sister unions, than dot deletions, and more dot deletions than there are proximal-distal interchromosomal exchanges. In practice, there are not. It is possible therefore that proximal-distal interchromatid intrachanges occur more commonly than would be expected if exchange was not biased amongst different classes of intrachanges.

Such a bias in the distribution of exchanges would not be unexpected. Sister chromatids in G_2 cells are still interlocked, and require the action of topoisomerase to separate them (Takahashi et al. 1986). The process of repair is organized inside subcellular structures (Mullenders et al. 1983), so that the chromatids may well be held in aligned configurations. Furthermore, there is in cells in G_2 a process of repair which apparently involves both chromatids (Moore et al. 1986). It would not therefore be surprising to find that misrepair is also organized in a non-random fashion.

There is no evidence of a similar alignment involving different chromosomes in JU56 cells; symmetrical and asymmetrical interchromosomal exchanges occur with equal frequencies.

It may be, however, that we have failed to detect a large number of small intrachanges, which would result in small deletions and duplications, and that many or all of these are lethal. They may be more frequent lethal events than the failure of breaks to repair at all. These small aberrations would not, however, be detectable directly, and they would not be expected to result in micronuclei after division. In either case, cell lethality in cells irradiated in G_2 cannot be explained fully by gross chromosomal aberrations.

Acknowledgements. The authors wish to thank Mr. W. Patterson, Physical Sciences, and Mrs. J. Lobbe and the staff of D6, Radiotherapy Department. Dr. K. Koschel and Dr. I. Bertoncello were generous in their assistance. We also thank Dr. J. R. K. Savage for helpful discussions.

References

Couzin D, Papworth DG (1979) The overdispersion between cells of chromosomal aberrations. J Theor Biol 80: 249–258

Dewey WC, Stone LE, Miller HH, Giblak RE (1971) Radiosensitization with 5-bromodeoxyuridine of Chinese hamster cells X-irradiated during different phases of the cell cycle. Radiat Res 47: 672–688

Grote SJ, Joshi GP, Revell SH, Shaw SH (1981) Observations of radiation-induced chromosome fragment loss in live mammalin cells in culture, and its effect on colony-forming ability. Int J Radiat Biol 39: 395–408

Miller RG (1973) Separation of cells by velocity sedimentation. Pain R, Smith B (eds) In: New techniques in biophysics an cell biology. Wiley, London 1: 87–111

Moore RC (1980) Alteration in the frequency of different classes of chromosome aberrations throughout the first cell cycle following X-irradiation. Radiat Res 81: 267–281

Moore RC, Bender MA (1987) The synergistic effect of aphidicolin on the yield of X-ray-induced chromosome aberrations throughout the cell cycle in JU56 cells. Radiat Res 110: 385–395

Moore RC, Randell C, Bender MA (1886) Dose relationships between different effects of aphidicolin in JU 56 cells. Mutat Res 160: 103–109

Moore RC, Randell C, Bender MA (1988) An investigation using inhibition of G_2 repair of the molecular basis of lesions which result in chromosomal aberrations. Mutat Res 199: 229–233

Mullenders LH, van Zeeland AA, Natarajan AT (1983) Analysis of the distribution of DNA repair patches in the DNA-nuclear matrix complex from human cells. Biochim Biophys Acta 740: 428–435

Powell MJD (1964) An efficient method for finding the minimum of a function of several variables without calculating derivatives. Comp J 7: 162–165

Revell SH (1959) The accurate estimation of chromatid breakage and its relevance to a new interpretation of chromatid aberrations induced by ionizing radiations. Proc R Soc B 150: 563–589

Revell SH (1974) The breakage-and- reunion theory and the exchange theory for chromosomal aberrations produced by ionizing radiations: a short history. Adv Radiat Biol 4: 367–416

Savage JRK (1989) The production of chromosome structural changes by radiation: an update of Lea (1946), chapter 6. Br J Radiol 62: 507–519

Savage JRK, Papworth DG (1982) Frequency and distribution studies of asymmetrical versus symmetrical chromosome aberrations. Mutat Res 95: 7–18

Takahashi K, Kaneko I, Nishiyama C (1986) Increase in the frequency of gamma ray induced chromosomal aberrations in mammalian cells by post-treatment with novobiocin. Int J Radiat Biol 49: 657–663

Inhibitors of DNA Topoisomerases and Chromosome Aberrations

F. PALITTI[1], F. DEGRASSI[2], R. DE SALVIA[2], M. FIORE[2], and C. TANZARELLA[2]

Contents

1 Introduction

The molecular mechanisms of formation of chromosomal damages (i.e., structural chromosome aberrations and sister chromatid exchanges) are not completely known. Operationally, chromosome aberrations are believed to be the result of DNA lesions induced by physical and chemical mutagens which, whether "misrepaired" or "misreplicated", give rise to chromosomal damage (Evans 1968; Kihlman 1971; Bender et al. 1974; Natarajan et al. 1986).

Agents, such as ionizing radiations, induce DNA lesions in all stages of the cell cycle and whenever these lesions are "misrepaired" they will give rise to chromosomal aberrations. Agents, such as ultraviolet light and alkylating agents, induce DNA lesions in all stages of the cell cycle and whenever these lesions are not repaired before an intervening S-phase they will give rise to chromatid-type aberrations as result of their "misreplication". With respect to this mechanism of induction of chromosomal aberrations these agents are classified as S-dependent agents in contrast to radiomimetic agents which are classified as S-independent agents (Evans and Scott 1969). Regarding the production of chromosomal aberrations by S-dependent agents by misreplication, it is noteworthy that it has been shown that posttreatments during the G2-phase with inhibitors of DNA synthesis, such as hydroxyurea (HU), cytosine arabinoside (ARA-C), and aphidicolin (APC), and of repair, such as 3-aminobenzamide (3AB) and caffeine, enhance the frequency of chromatid aberrations induced by S-dependent agents (for review see Kihlman and Natarajan 1984). These results suggest that the lesions induced by S-dependent agents have to be altered in some

[1] Dipartimento di Agrobiologia e Agrochimica, Via San Camillo de Lellis, Università della Tuscia, I-01100 Viterbo, Italy
[2] Centro di Genetica Evoluzionistica del CNR c/o Dipartimento "C. Darwin", Università "La Sapienza", I-00185 Roma, Italy

way during the S-phase and eventually misrepaired during G2 to give rise to chromatid-type aberrations (Kihlman et al. 1981). These lesions, which have not been replicated during the S-phase, will be misreplicated in G2, giving rise to chromatid-type aberrations (Andersson 1983; Palitti et al. 1984).

On the basis of this scenario, an analysis of the induction of chromosomal damage by a quite novel class of drugs, which are used in the chemotherapy of cancer for their cytostatic and tumor inhibitory effects, will be presented.

These drugs, many of which are DNA intercalating agents, turned out to be inhibitors of DNA topoisomerases. Excellent reviews on the biochemical properties of topoisomerases have been published by Wang (1985) and Vosberg (1985) and on antitopoisomerase drugs by Zwelling (1985) and Drilca and Franco (1988).

The topoisomerases are enzymes that control the topology of DNA, regulating the superhelical structure of DNA, and play a major role in the control of a variety of biological processes: replication, transcription, and repair.

Two types of DNA topoisomerases can be distinguished: type I topoisomerase can affect the DNA superhelical structure by a single-stranded DNA passage mechanism; in contrast, type II topoisomerases alter DNA topology by catalyzing transient double-strand breaks.

The principal reactions catalyzed by type II topoisomerase are: (1) the removal of complex knots from topologically constrained DNA; (2) the ability to create or remove supercoiling in topologically constrained DNA; (3) the ability to reversibly catenate double-stranded DNA circles.

Furthermore, there is evidence that topoisomerase II is closely associated with the nuclear matrix in interphase and appears to be the major structural protein of the scaffold of mitotic chromosomes (Berrios et al. 1985; Earnshaw et al. 1985; Gasser et al. 1986).

In the mitotic chromosomes the enzyme is concentrated at the base of the radial loops in accordance with the role that this enzyme has in the architecture of the chromosomes.

The principal reaction catalyzed by type I topoisomerase is to break and then to reseal a single DNA strand, changing the linking number by steps of one. This enzyme has been found associated with transcriptionally active chromatin (Weisbrod 1982; Fleischman et al. 1984).

DNA topoisomerases, therefore, can be broadly defined as enzymes which catalyze the breaking and rejoining of DNA strands in concert.

Furthermore, it has been shown that the presence of topoisomerase II is bound to cellular proliferation activity (Sullivan et al. 1986) and that maximum activity is in the G2-phase of the cell cycle (Chow and Ross 1987). With immunofluorescence techniques it has been shown that the positive reaction to fluorescence is present only when the cells enter the S-phase, but not before. There is no topoisomerase II activity during G0 and very little during G1 (Earnshaw and Heck 1985; Heck and Earnshaw 1986).

2 Inhibitors of DNA Topoisomerases

2.1 Effects at the Molecular Level

After treatment of mammalian cells with DNA intercalating antitumor agents belonging to acridines (m-AMSA), anthracyclines (5-iminodaunorubicin), ellipticines (2-Me-9OH-E) and nonintercaling epipodophyllotoxins (VP-16 and VM-26), both DNA single- and double-strand breaks characterized by the peculiarity of being protein-linked, can be detected by DNA alkaline elution only if proteolytic enzymes are present during the lysis of the treated cells (Ross et al. 1978).

It was suggested that the protein-linked DNA breaks are the result of an enzyme-mediated process and that the DNA-linked protein is DNA topoisomerase II which represents the DNA-cleaving enzyme.

In the presence of antitopoisomerase drugs the normal chain of reactions catalyzed by topoisomerase II is subverted not as a consequence of a block of the normal enzyme function but because the antitopoisomerase drugs stabilize the "cleavable complex" normaly formed between topoisomerases and DNA, promoting consequently DNA cleavage (Zwelling 1985; Glisson and Ross 1987). This mode of action is based on the model proposed by Higgins and Cozzarelli (1982) for the effects of nalidixic acid on bacterial DNA gyrase.

2.2 Production of Chromosomal Damage

Like many other antitumor agents which are used in chemotherapy, antitopoisomerase drugs are efficient inducers of chromosomal aberrations (Ch. Abs) and sister chromatid exchanges (SCE) (Deaven et al. 1978; Crossen 1979; Pommier et al. 1985; Sakamoto-Hojo et al. 1988).

This is to be expected as chromosomal damage is one of the main causes of cell death (Cornforth and Bedford 1987). According to biochemical data, antitopoisomerase drugs produce DNA breaks which differ from those produced by ionizing radiation because they are not only associated with protein but are also protein-concealed.

In comparison with other chromosome-breaking agents these drugs do not directly induce DNA lesions but act by stabilizing the "cleavable DNA-topoisomerase complex". Therefore, the mechanism of induction of chromosomal aberrations by these drugs cannot simply be related to the classical scheme in which the DNA lesions induced by S-dependent or S-independent agents are misreplicated or misrepaired thus giving rise to chromosomal damage.

Most of the studies on the induction of chromosomal aberrations and SCE by antitopoisomerase drugs have been performed using inhibitors of DNA topoisomerase II. As it has been reported, agents capable of interfering with DNA topoisomerases produce DNA single- and double-strand breaks whose ratio varies according to the drug used. Agents known to produce double-strand breaks in DNA are capable of producing chromosomal aberrations independently of DNA replication and therefore should behave like ionizing radiations. Indeed, the results obtained by Deaven et al. (1978), who treated CHO cells with m-AMSA, suggest that the chromosomal alterations were formed with an S-phase-independent mechanism. On the other hand, it was also found that these drugs are very potent inducers of SCE, a phenomenon which occurs only in S-phase.

An investigation on the production of chromosomal alteration by two inhibitors of topoisomerases II, one DNA intercalating agent, m-AMSA, and one nonintercalating agent, VP-16, was started in collaboration with B. A. Kihlman and K. Andersson. Preliminary results concerning VP-16 have been reported previously (De Salvia et al. 1987). A detailed study on the effects of m-AMSA was published by Andersson and Kihlman (1989). As experimental material human peripheral blood lymphocytes grown in vitro were used. This material allows one to study the effects of the drugs in the different phases of the cell cycle and also in the nonproliferative G0 stage. The outcome of these studies was that 1- or 2-h treatment with m-AMSA or VP-16 in G0 or G1 induced both chromosome and chromatid-type aberrations in cells which were in their first mitosis after the treatments, and SCE in cells were fixed after two cell cycles. These results apparently suggest that these drugs behave at the same time both as S-independent and S-dependent agents. Subsequently, it was shown that the "S-dependent effects", i.e., induction of chromatid-type aberrations and SCE, after a G0 or a G1 treatment could be reduced using a washing procedure which consisted in changing the medium at various times after the first drug treatment (for details, see Andersson and Kihlman 1989). These authors compared the sensitivity of nonproliferating (G0) and proliferating cells at various stages of the cell cycle to the clastogenic effects of m-AMSA and concluded that the cell cycle sensitivity to the induction of chromosomal damage corresponds to the sensitivity of DNA cleavage induced by antitopoisomerase drugs (Estey et al. 1987) Therefore, the m-AMSA-induced chromosome alterations are compatible with the idea that the clastogenic effect of m-AMSA is mediated by DNA topoisomerase.

More recently we (Degrassi et al. 1989) investigated the chromosomal alterations induced by camptothecin (CM), an alkaloid, which has been shown to inhibit DNA topoisomerase I by stabilizing the intermediate enzyme-associated DNA single-strand breaks (Hsiang et al. 1985).

Also in this case the induction of chromosomal aberrations and SCE by CM was studied in the different stages of the cell cycle of human lymphocytes grown in vitro.

From these experiments it was concluded that CM did not induce chromosomal alterations either simply as an agent acting by an S-phase-dependent or as an S-phase-independent mechanism. Treatment of human lymphocytes with camptothecin in the G1-phase induces chromatid-type aberrations and SCE but no chromosome-type aberrations; as in the case of m-AMSA and VP-16, chromosomal aberrations are drastically reduced with additional changes of medium after the CM treatment (Table 1).

Therefore, a G1-phase treatment does not give rise to lesions in DNA which will persist until S-phase and produce SCE or chromatid-type aberrations, but, in order to induce chromosomal damage, CM has to be present during the S-phase. Furthermore, while G1 treatment does not give rise to chromosome-type aberrations, G2-phase treatment, instead, induces chromatid-type aberrations (Table 2). The effect produced in the G2-phase is unexpected as single-strand breaks should not give rise to chromatid-type aberrations when induced in G2. The production of chromatid aberrations by CM in G2 could be attributed to the peculiar type of single-strand

Table 1. Frequencies of chromosomal aberrations in human lymphocytes exposed to camptothecin for 1 h in G1 phase: effect of different washing procedures (after Degrassi et al. 1989)

Concentration of campto thecin (10-7 M)	Abnormal cells (%)	Aberrations per 100 cells					
		Gaps	Chro-matid breaks	Isochro-matid breaks	Chroma-tid exchanges	Chromo-some exchanges	Total (-gaps)
Standard washing procedure							
0	0	0	0	0	0	0	0
5	23	5	7	5	11	0	23
10	34	4	19	2	16	0	37
50		No mitoses					
100		No mitoses					
Standard washing procedure + 30 min F-10 + washing + 30 min F-10 + washing							
0	1	0	0	0	0	0	0
5	1	1	0	0	0	0	0
10	4	2	1	1	0	0	2
50	8	0	4	0	4	0	8
100	19	3	3	5	9	0	17

Table 2. Frequencies of chromosomal aberrations in human lymphocytes exposed to camptothecin for the last 2 h before harvesting

Concentration of campto-thecin (10-7 M)	Abnormal cells (%)	Aberrations per 100 cells				
		Gaps	Chromatid breaks	Isochromatid breaks	Chromatid exchanges	Total (-gaps)
0	2	1	1	0	0	1
10	8	1	4	2	2	8
50	10	2	11	1	1	13
100	19	5	17	4	1	22

breaks, i.e., protein-concealed breaks, and to the fact that the inhibitor is not removed until the cells enter mitosis.

3 Effects of G2-Phase Posttreatment with Inhibitors of DNA Replication and Repair on the Frequencies of Chromosomal Aberrations Induced by Antitopoisomerase Drugs

In order to clarify the molecular nature of the lesions produced by the antitopoisomerase drugs responsible for the production of chromosomal aberrations, investigations on the possible role of repair mechanisms in the transformation of these lesions to chromosomal aberrations have been performed.

It is well documented that posttreatments during G2-phase with inhibitors of DNA synthesis, such as HU and Ara-C, or repair, such as 3AB or caffeine, strongly potentiate the frequency of chromosomal aberrations induced by clastogenic agents.

It has been shown (Zwelling et al. 1982) that the DNA breaks resulting from the stabilization of the "cleavable complex" are rapidly resealed upon removal of the drugs without requiring DNA synthesis.

Andersson and Kihlman (1989) found that a posttreatment in the G2-phase with a combination of HU and Ara-C enhances the frequency of chromosomal aberrations induced by m-AMSA in G2-treated human lymphocytes in contrast to the biochemical data (Zwelling et al. 1982).

Recently, we have investigated the effects of combined treatments with CM and inhibitors of DNA repair or DNA synthesis on the induction of chromosomal aberrations in the G2-phase of human lymphocytes and Chinese hamster ovary cells (CHO) grown in vitro.

In both types of cells we found a potentiation of induced chromosomal aberrations when CM is combined with inhibitors of DNA synthesis (HU or Ara-C) and with 3AB in CHO cells (Figs. 1 and 2). Moreover, in both cell types the combined treatment with caffeine has no potentiating effect (Figs. 1 and 2). In order to further investigate this unexpected result Chinese hamster cells were treated for the last 1, 2, and 3 h before fixation with both CM and caffeine. As shown in Table 3 at 3-h treatment there is a potentiation of CM-induced chromosomal damage. Also in this case, caffeine reduced the mitotic delay caused by CM. Probably when CM and caffeine are given together during the last 1 or 2 h of the G2, only cells which have passed the transition point reach mitosis and are unaffected by caffeine.

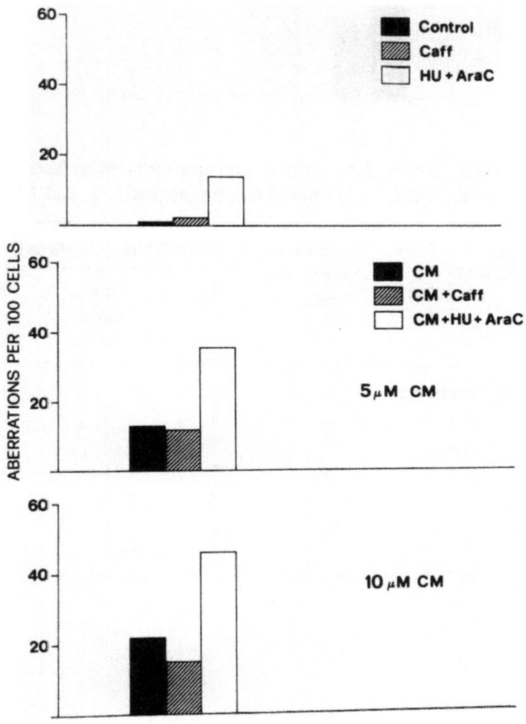

Fig. 1. Effect of camptothecin (CM) + Caff. or CM + HU + Ara-C combined treatments in the G2-phase on the frequencies of chromosomal aberrations in human lymphocytes

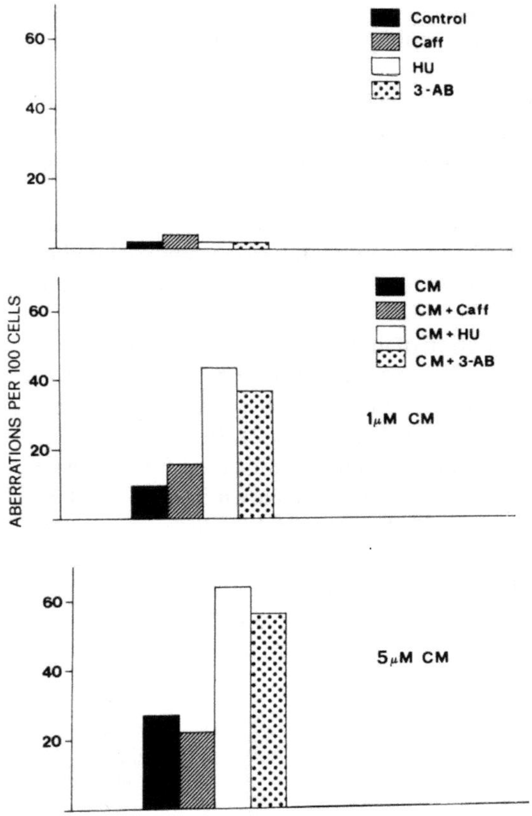

Fig. 2. Effect of combined treatments with CM + Caff., CM + HU, CM + 3AB in the G2-phase on the frequencies of chromosomal aberrations in Chinese hamster ovary (CHO) cells

Table 3. Influence of caffeine on camptothecin-induced chromosomal aberrations: effect of caffeine and camptothecin combined treatments for 1, 2, and 3 h before harvesting in CHO-cells

CM (10-6 M)	Caff.	Abnormal cells (%)	Aberrations per 100 cells					
			Gaps	Chromatid breaks	Isochromatid breaks	Chromatid exchanges	Subchromatid exchanges	Total (-gaps)
1-h Treatment								
0	−	2	1	1	0	0	0	1
0	+	6	4	2	0	0	0	2
1	−	26	17	13	5	0	0	18
1	+	18	9	8	2	0	0	10
5	−	21	12	12	3	0	0	15
5	+	22	12	11	3	0	0	14

Tabelle 3 (continued)

CM (10-6 M)	Caff.	Abnormal cells (%)	Aberrations per 100 cells					
			Gaps	Chromatid breaks	Isochromatid breaks	Chromatid exchanges	Subchromatid exchanges	Total (-gaps)
2-h Treatment								
0	+	5	2	2	1	0	0	3
1	–	15	6	6	3	1	0	10
1	+	12	8	3	3	1	0	7
5	–	29	20	21	1	0	0	22
5	+	24	9	11	6	3	3	23
3-h Treatment								
0	+	6	4	2	0	0	0	2
0.5	–	11	2	7	2	3	0	12
0.5	+	21	7	12	8	3	1	24
1	–	27	13	10	10	4	0	24
1	+	29	4	11	14	7	3	35

4 Relationship Between Topoisomerase and DNA-Repair Deficiencies

To further characterize the molecular nature of the lesions induced by these drugs several investigations have been performed using repair-deficient cell lines. Zwelling and Mattern (1982) were unable to detect any enhanced sensitivity with respect to the production of protein-associated DNA strand breaks in normal, xeroderma pigmentosum, and ataxia telangiectasia (AT) fibroblast. Evans et al. (1989) found that the different sensitivity to ionizing radiations of closely related strains of mouse lymphoma L5178Y cells correlated with the sensitivity to m-AMSA, VP-16, and ellipticine. Darroudi and Natarajan (1989), using cytogenetic end points, found that Chinese hamster ovary X-ray-sensitive mutant cells were two- to threefold more sensitive to the induction of chromosomal alterations by m-AMSA and VP-16 than the wild-type cells. Results obtained Caporossi et al. (1988) in four AT-derived lymphoblastoid lines, two non-AT (a normal and a Burkitt lymphoma), and an AT heterozygote treated in G2 with VP-16 show a dose-dependent induction of chromosomal aberrations but the AT lines were three- to fivefold more sensitive than non-AT lines and the AT heterozygote was also more sensitive but only at the lowest dose (Fig. 3). These results suggest that the DNA breaks induced by antitopoisomerase drugs, also if protein-concealed, are processed in the same way as those produced by ionizing radiation and give rise to chromosomal aberrations. In conclusion, antitopoisomerase drugs represent a new tool to further investigate the molecular mechanism of the formation of chromosomal damage, providing the opportunity to correlate the different processes which require the catâlytic action of the topoisomerase with the production of the chromosomal damage.

Acknowledegments. We would like to thank Prof. B. A. Kihlman and R. K. Andersson for valuable discussions and fruitful suggestions. The graphic assistance of Mr. V. Salviati is gratefully acknowledged. This work was partially spported by a grant from the Ministero della Pubblica Istruzione and the National Research Council (Prog. Fin. Oncologia).

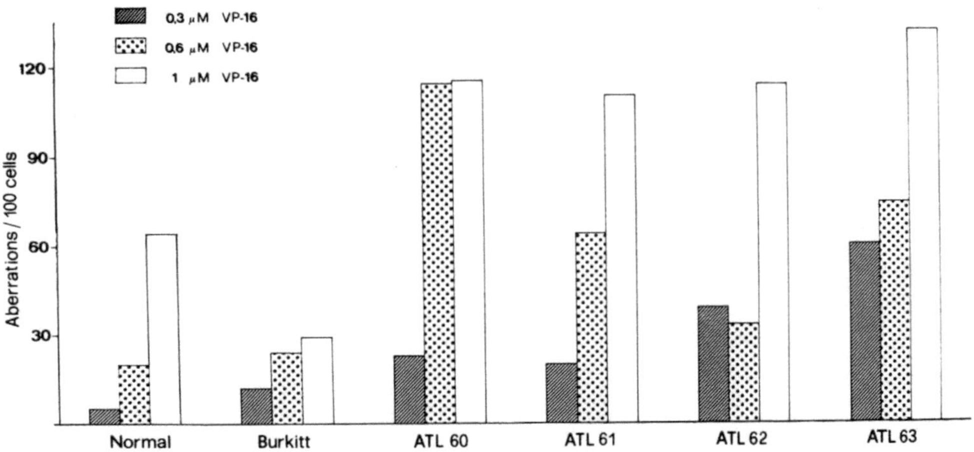

Fig. 3. Frequencies of chromosomal aberrations produced by VP-16 in the G2-phase of derived (ATL 60–63) and non-AT (*normal* and *Burkitt* lymphoma) lymphoblastoid cell lines (after Caporossi et al. 1988)

References

Andersson HC (1983) G2 repair and the formation of chromosomal aberrations. III. The effect of hydroxyurea, 5-fluorodeoxyuridine and caffeine on X-ray-induced chromosome damage in *Vicia faba* Hereditas 97: 193–209

Andersson HC, Kihlman BA (1989) The production of chromosomal alterations in human lymphocytes by drugs known to interfere with the activity of DNA topoisomerase II. I. m-AMSA. Carcinogenesis 10: 123–130

Bender MA, Griggs HG, Bedford JS (1974) Mechanisms of chromosomal aberration production. III. Chemicals and ionizing radiation. Mutat Res 23 : 197–212

Berrios M, Osheroff N, Fisher PA (1985) In situ localization of DNA topoisomerase II, a major polypeptide component of the *Drosophila* nuclear matrix fraction. PNAS (USA) 82: 4142–4146

Caporossi D, Porfirio B, Nicoletti B, Palitti F, Tanzarella C, Degrassi F, De Salvia R (1988) G2 chromosome hypersensitivity to etoposide in ataxia-telangectasia lymphoblasts. Atti AGI XXIV, Perugia, October 1988 P. 61

Chow K-C, Ross WE (1987) Topoisomerase-specific drug sensitivity in relation to cell cycle progression. Mol Cell Biol 7 p. 3119–3123

Cornforth MN, Bedford JS (1987) A quantitative comparison of potentially lehtal damage repair and the rejoining of interphase chromosome breaks in low passage normal human fibroblasts. Radiat Res 111: 385–405

Crossen PE (1979) The effect of acridine compounds on sister chromatid exchange formation in cultured human lymphocytes. Mutat Res 68: 295–299

Darroudi F, Natarajan AT (1989) Cytogenetical characterization of Chinese hamster ovary X-ray-sensitive mutant cells xrs 5 and xrs 6. III. Induction of cell killing, chromosomal aberrations and sister-chromatid exchanges by bleomycin mono- and bi-functional alkylating agents. Mutat Res 212: 123–135

Deaven LL, Oka MS, Tobey RA (1978) Cell-cycle-specific chromosome damage following treatment of cultured Chinese hamster cells with 4'- (9-acridinyl-amino-methanesulphon-m-anisidide-HCl). J Natl Cancer Inst 60: 1155–1161

Degrassi F, De Salvia R, Tanzarella C, Palitti F (1989) Induction of chromosomal aberrations and SCE by camptothecin, an inhibitor of mammalian topoisomerase I. Mutat Res 211: 125–130

De Salvia R, Degrassi F, Tanzarella C, Fiore M, Palitti F, Andersson HC, Kihlman BA (1987) Induction of chromosomal aberrations and SCE by VP-16 in human lymphocytes. In: International Congress on DNA damage and repair. Roma, July 1987, Abstracts, p 98

Drilca K, Franco RJ (1988) Inhibitors of DNA topoisomerases. Biochemistry 27: 2253

Earnshaw WC, Heck MMS (1985) Localization of topoisomerase II in mitotic chromosomes. J Cell Biol 100: 1716–1725

Earnshaw WC, Halligan B, Cooke CA, Heck MMS, Liu LF (1985) Topoisomerase II is a structural component of mitotic chromosome scaffolds. J Cell Biol 100: 1706–1715

Estey E, Adlakha RC, Hittelman WN, Zwelling LA (1987) Cell cycle stage dependent variations in drug-induced topoisomerase II mediated DNA cleavage and cytotoxicity. Biochemistry 26: 4338–4344

Evans HJ (1968) Repair and recovery at chromosome and cellular levels: similarities and differences. Brookhaven Symp Biol 20: 111–133

Evans HJ, Scott D (1969) The induction of chromosomal aberrations by nitrogen mustard and its dependence on DNA synthesis. Proc R Soc Lond B 173: 491–512

Evans HH, Ricanati M, Horng M, Menel J (1989) Relationship between topoisomerase II and radiosensitivity in mouse L5178Y lymphoma strains. Mutat Rres 217: 53–63

Fleischmann G, Pflugfelder G, Steiner EK Javaherian K, Howard GC, Wang JC, Elgin SCR (1984) *Drosophila* DNA topoisomerase I is associated with transcriptionally active regions of the genome. PNAS (USA) 41: 541–551

Gasser SM, Laroche T, Falquet J, Boy de la Tour E, Laemmli UK (1986) Metaphase chromosome structure, involvement of topoisomerase II. J Mol Biol 188: 613–629

Glisson BS, Ross WE (1987) DNA topoisomerase II: a primer on the enzyme and its unique role as a multidrug target in cancer chemotherapy. Pharm Ther 32: 89–106

Heck MMS, Earnshaw WC (1986) Topoisomerase II: a specific marker for cell proliferation. J Cell Biol 103: 2569–2581

Higgins NP, Cozzarelli PR (1982) The binding of gyrase to DNA analysis by retention by nitrocellulose filters. Nucl Acids Res 10: 6833–6847

Hsiang Y, Hertzberg R, Hecht S, Liu LF (1985) Camptothecin induces protein-linked DNA-breaks via mammalian DNA topoisomerase I. J Biol Chem 260: 14873–14878

Kihlman BA (1971) Molecular mechanism of chromosome breakage and rejoining. In: Dupraw EJ (ed) Advances in cell and molecular biology, vol 1. Academic Press, London, pp 59–107

Kihlman BA, Andersson HC, Hansson K, Hartley-Asp B, Palitti F (1981) Formation of chromatid type aberrations from chemically induced lesions: information obtained with hydroxyurea and caffeine. In: Sugimura T, Hondo S, Tahiabe H (eds) Environmental mutagens and carcinogens. Liss, New York, pp 483–493

Kihlman BA, Natarajan AT (1984) Potentiation of chromosomal alterations by inhibitors of DNA repair. In: Collins A, Johnson RT, Downes S (eds) DNA repair and its inhibition. IRL, Oxford, pp 319–339

Natarajan AT, Darroudi F, Mullanders LHF, Meijers M (1986) The nature of DNA lesions that lead to chromosome aberrations induced by ionizing radiations. Mutat Res 160: 231–236

Palitti F, Tanzarella C, Degrassi F, Salvia R De, Fiore M (1984) Potentiation of induced sister chromatid exchanges and chromatid aberrations by inhibitors of DNA synthesis and repair in G2. In: Tice RR, Hollander A (eds) Sister chromatid exchanges. Plenum, New York, pp. 313–319

Pommier Y, Zwelling LA Kao-Shan CS, Whang-Peng J, Brandley MO (1985) Correlation between intercalator-induced DNA strand breaks and sister chromatid exchanges, mutations and cytotoxicity in Chinese hamster cells. Cancer Res 45: 3143–3149

Ross WE Glaubiger DL, Kohn KW (1978) Protein-associated DNA breaks in cells treated with adriamycin or ellipticine. Biochim Biophys Acta 519: 23–30

Sakamoto-Hojo ET, Takahashi CS, Ferrari I, Motidome M (1988) Clastogenic effect of a plant alkaloid ellipticine on bone marrow cells of Wistar rats and on human peripheral blood lymphocytes. Mutat Res 199: 11–19

Sullivan DM, Glisson BS, Hodges PK, Smallwood-Kentro S, Ross WE (1986) Proliferation dependence of topoisomerase II mediated drug action. Biochemistry 25: 2248–2256

Vosberg HP (1985) DNA topoisomerases: enzymes that control DNA conformation. Curr Microbiol Immunol 114: 19–85

Wang JC (1985) DNA topoisomerases, Annu Rev Biochem 54: 665–697
Weisbrod ST (1982) Properties of active nucleosomes as revealed by HMG 14 and 17 chromatography. Nucl Acid Res 10: 2017–2042
Zwelling LA (1985) DNA topoisomerase II as a target of antineoplastic drug therapy. Cancer Metastasis Rev 4: 263–276
Zwelling LA, Kerrigan D, Pommier Y, Michaels S, Steren A, Kohn KV (1982) Formation and resealing of intercalator-induced DNA strand-breaks in permeabilized L1210 cells without the stimulated synthesis of poly (ADP-ribose). J Biol Chem 257: 8957–8963
Zwelling LA, Mattern MR (1982) DNA repair deficiencies do not effext intercalator-induced cytotoxicity or DNA scission in human cells. Mutat Res 104: 295–304

Restriction Endonuclease- and Radiation-Induced DNA Double-Strand Breaks and Chromosomal Aberrations: Similarities and Differences

P. E. Bryant[1]

Contents

1 Introduction

It is a characteristic of ionizing radiations that the energy per absorption event is greatly in excess of the energy required to dissociate covalent bonds. In an irradiated cell, therefore, damage can occur at any point in the structure of DNA. As a result of this high energy per ionizing event, radiation is able to induce a variety of lesions in the DNA of cells which can be grouped into those involving damage to the base moeity without breakage of the sugar-phosphate chain (base damage), cross-links (DNA-DNA and DNA-protein) and those involving breakage of one or both strands of the double helix (single- and double-strand breaks respectively). Accumulating evidence points to the double-strand break (dsb) as the major lesion in causing cellular, chromosomal, mutagenic and oncogenic effects of ionizing radiation. It is for example, quite characteristic of ionizing radiation and certain chemical agents (e.g. bleomycin) that they induce chromosome aberrations at first mitosis following exposure of a cell in the G1 phase of the cell cycle, whereas agents such as UV that induce almost entirely single-stranded lesions (principally thymine dimers, which are converted to single-strand breaks during repair) only chromatid aberrations are seen, resulting from misreplication of DNA during S-phase. Furthermore, derived mutants of mammalian cells which are deficient in the repair of dsb show a high chromosomal sensitivity (Kemp and Jeggo 1986; Darroudi and Natarajan 1987; Bryant et al. 1987).

Evidence from experiments in which restriction endonucleases (RE) were used to generate dsb in the DNA of cells (Bryant 1984; Natarajan and Obe 1984; see Bryant 1988 for a review) similarly argue strongly in favour of the dsb as the lesion leading to chromosomal aberrations in irradiated cells.

However, the question arises as to how far treatment with RE is a representative model of the action of ionizing radiation. In this chapter I shall attempt to answer

---

[1] Department of Biology and Preclinical Medicine, University of St. Andrews, St. Andrews, Fife KY16 9TS, UK

this question; drawing comparisons between the clastogenic effects of restriction endonucleases and those of ionizing radiation. In order to address this question we need to examine the relationship between dsb and chromosomal aberrations. However, we must first define what we mean by a dsb, then take into account how frequently these are induced. Finally, we must examine how dsb may be repaired or misrepaired, leading to chromosomal aberrations.

2 What Is a Double-Strand Break?

As stated above, a dsb comprises a break in both covalently bonded sugar-phosphate chains of the double helix of DNA. However, the points on each strand at which the DNA chain is broken appears to be an important parameter in determining the biological consequence of a dsb. It was for example shown (Bryant 1984) that when cells, permeabilized with Sendai virus, were treated with the restriction endonuclease (RE) Pvu II, which generates blunt-ended dsb at the six-base specific recognition sequence CAG'CTG, high frequencies of chromosomal aberrations were observed. On the other hand, Bam H1, which induces a cohesive-ended dsb with a four-base overlap at the sequence G'GATCC was found to cause a very low frequency of aberrations. A similar effect, i.e. a difference in the effectiveness of blunt- and cohesive-ended dsb in causing aberrations has been demonstrated for other RE, e.g. Eco RV (blunt) and Eco R1 (cohesive) (Bryant et al. 1987; Bryant and Christie 1989) and also for cell killing (Bryant 1985).

Dsb induced by RE have termini with 3'hydroxyl and 5'phosphoryl groups. These can be termed "clean" dsb since they require no modification prior to ligation, although the mechanism by which this occurs in a cellular situation is not yet known. The end structures of radiation-induced dsb are not known but by analogy with the structure of single-strand breaks (ssb) the termini of dsb are likely to be highly variable and comprise some with "clean" ends and some with "dirty" ends, i.e. where the rupture of covalent bonds is not between the 3'hydroxyl and the 5'phosphoryl groups and where exonuclease cleaning prior to ligation would be required. Based on our knowledge of ssb termini (Landbeck and Hagen 1973; Coquerelle et al. 1973) the majority of radiation induced dsb will have "dirty" termini (e.g. with a 3'phosphoryl group) and some may have one clean and one dirty terminus in respective strands. Similarly, although nothing is yet known about the relative frequencies of blunt- and cohesive-ended dsb induced by radiation, it is very likely that the dsb induced will comprise a mixture of blunt- and cohesive-ended dsb (Bryant and Christie 1989).

With low LET radiation such as X-rays, the probability of inducing a dsb will be very low since only the ends of the electron tracks will have a significant probability of producing the spacing of ionizations sufficient to induce dsb. A plot (Fig. 1) of intrinsic efficiency (probability per unit track of an electron or particle traversing the nucleus of a cell producing a chromosome aberration) as a function of the mean free path (MFP: mean distance between ionizations in the path of an ionizing particle) of ionizing particles (Watt 1989) shows an inflection point at approximately 2 nm, the intrinsic efficiency levelling off at values of MFP of about 2 nm or less. This inflection point was interpreted as the approximate distance between the two strands of DNA.

Fig. 1. The three-dimensional structure of DNA showing the distances measured between carbon atoms in the 3' position in opposite strands. The distance (1.4548 nm) between ends of a hypothetical blunt-ended double-strand break, and the distance (1.8061 nm) between ends of a hypothetical cohesive-ended double-strand break are shown. Distances and diagram were derived using the Chem-X Molecular Modelling Package (Chemical Design Ltd. 1987)

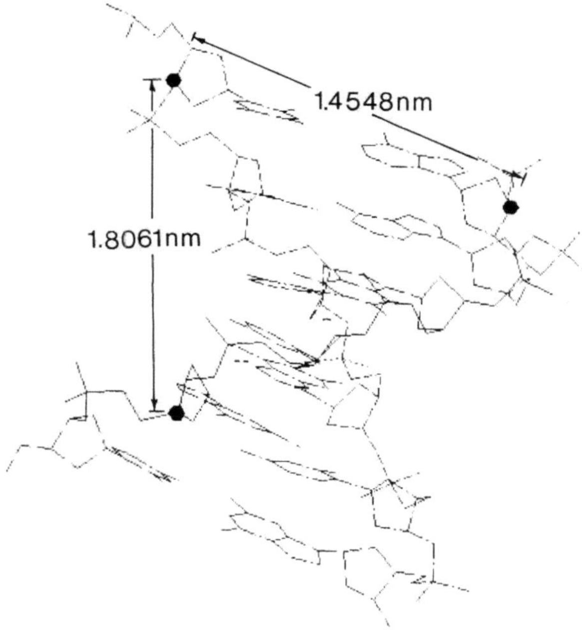

X-rays of 250 kV, which produce ionizing particles (electrons) at equilibrium with an MFP of approximately 10 nm, would therefore be very inefficient compared with high LET radiations in causing double-strand breaks. Thus, for low LET radiations such as X-rays the efficiency of inducing dsb is very low indeed, some one to two orders of magnitude lower than for alpha particles.

Measurements of the interstrand distances between the breakage points of a blunt- and a cohesive-ended dsb were calculated, using a Chem-X Molecular Modelling Package (Chemical Design Ltd., 1987), to be 1.4548 nm and 1.8061 nm respectively (Fig. 2). In the absence of energy transfer within the DNA molecule, it seems unlikely that the difference between these two values (0.3513 nm) could significantly influence the probability of induction of either a blunt- or a cohesive-ended dsb. However, the probability of inducing cohesive-ended dsb with variable degrees of base overlap will be several times higher than for the specific case of a blunt-ended dsb; therefore, it is likely that the predominant dsb types in X-irradiated cells will be those with cohesive termini.

Thus, we can be reasonably confident in predicting that cohesive-ended dsb with a 4-base or more overlap are much more frequent than the blunt-ended type in the DNA of X-irradiated cells. The precise "cut-off" point, where the degree of base overlap of a cohesive-ended dsb is less than 4 bases and where these dsb act in the same way as blunt-ended dsb has not yet been determined, although preliminary work (Bryant 1986) suggests that there is a steady progression in efficiency of causing aberrations as the degree of overlap decreases from 4 to 1. Work on this is currently being performed in our laboratories to determine the effectiveness of a variety dsb types. The implications of the above are that X-rays are relatively poor inducers of

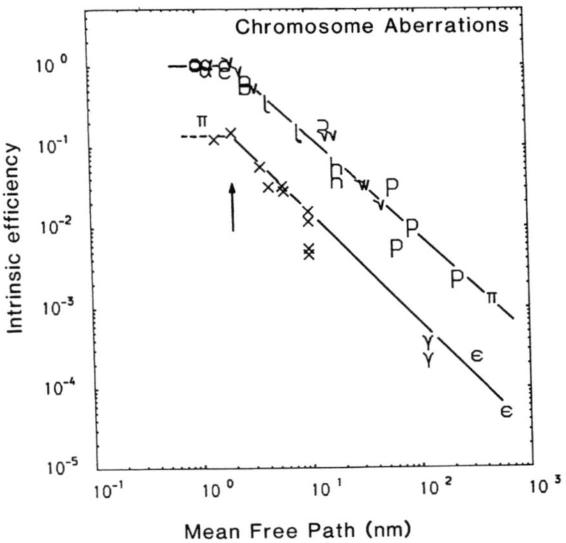

Fig. 2. Intrinsic efficiency (probability of an ionizing track entering a cell nucleus, causing a chromosome aberration) as a function of the mean free path (average distance between ionizing events in the tracks of ionizing particles). Graph redrawn from Watt (1989). The *arrow* indicates the inflection point, at about 2 nm, the approximate distance between the two strands of the DNA double helix. The *lower line* represents data for X-rays and fast electrons, indicating the lower efficiency of these radiations in inducing aberrations (due to the relative shortness of the densely ionizing electron-track ends; the *upper line* represents data for heavy particles

chromosomal aberrations with only a small fraction of the dsb induced (those with blunt or near blunt ends) able to give rise to chromosomal aberrations.

This could be an explanation for the discrepancy between the frequency of dsb induced in mammalian cells by X-rays; 40 dsb/Gy/diploid genome (Bryant and Blöcher 1980; Blöcher 1982) and the frequencies of chromatid aberrations (approximately seven per cell per Gy) observed (e.g. Mozdarani and Bryant 1989). The frequency of PCC fragments: 6.3 per Gy per cell induced by X-rays in human fibroblasts (Cornforth and Bedford 1983), is similarly much lower than the frequency of dsb measured by biochemical techniques.

In short, although we are faced with uncertainty about the end structure of radiation-induced dsb, we know that they are likely to be a mixture of blunt- and cohesive-ended types with a preponderance of the latter, and consequently relatively inefficient inducers of chromosomal aberrations, based on our knowledge of the effectiveness of RE-induced dsb. An important difference between radiation- and RE-induced dsb is that the former are always accompanied by a variety of other lesions which might conceivably affect the local stability of sites of double-strand breakage, rendering dsb less easily repaired; this could affect the efficiency with which these dsb are converted into chromosomal aberrations. Another important factor which will determine the observed frequencies of chromosomal aberrations is the repair of dsb. Clearly, this will affect numbers of dsb remaining to cause observable breaks and will be an important determinant of interaction probabilities. This will be considered later.

3 Induction of Double-Strand Breaks by Restriction Endonucleases and Radiation

Low LET radiation induces dsb in mammalian cells as a linear function of dose and at a frequency of approximately 40 dsb per Gy per diploid genome (Blöcher 1982).

These dsb are induced at the time of irradiation. With RE there are potentially a very large number of recognition sites where an enzyme could cut the DNA. The theoretical number, assuming a random distribution of the four bases, would be one cutting site every 4096 base pairs (for a six-base recognition site) and one site every 256 base pairs (for a four-base recognition site). However, these frequencies may be modified by a non-random distribution of bases, some combinations occurring more frequently than others (Bishop et al. 1983). The methylation pattern may also influence cutting frequency. For example it has been calculated that the cutting frequencies of Eco RV (blunt) and Eco R1 (cohesive) are 1 in 5000 base pairs and 1 in 3000 base pairs respectively, and for Pvu II (blunt) and Bam H1 (cohesive) 1 in 3200 and 1 in 5500 respectively (Bishop et al. 1983; Bryant 1988). There are only two previous studies in which attempts have been made to measure the frequencies of dsb in RE-treated cells (Bryant 1984; Natarajan et al. 1985). The former study employed the unwinding technique and cells were lysed after 30-min incubation with enzymes and Sendai virus. Low frequencies of breaks were observed and the frequency appeared to be the same in both Pvu II- and Bam H1-treated cells. Five hundred units of enzyme were found to produce the amount of DNA unwinding found after about 2 Gy. In a similar study (Natarajan et al. 1985), but using the nucleoid sedimentation technique (Cook and Brazell 1975), breaks induced in cells by Pvu II were measured. A similar result obtained to that of Bryant (1984), i.e. that 60–120 units of Pvu II induced breaks equivalent to about 0.25 Gy.

Recently in our laboratories we have used electroporation to treat cells with Pvu II and studied the frequencies of dsb as a function of time after exposure, using neutral filter elution (Costa and Bryant 1990). The data is shown in Fig. 3. It shows that the frequency of dsb increases with time of incubation over a period of hours and is still increasing gradually, even after 24 h following exposure. Work is currently in progress to measure the effect of other RE, especially those inducing cohesive-ended dsb. We interpreted the data in Fig. 3 as suggesting that the frequency of dsb induced by RE is determined by competing processes: (1) incision by the RE and (2) cellular repair (ligation). The frequency observed at any one time will therefore depend on the relative rates of these two competing processes. The enzyme Pvu II appears to persist for very long periods inside the cell.

These results indicate that the induction of dsb by Pvu II is not like that after an acute dose of X-rays. It might, however, be likened to an exposure of cells to X- or γ-rays delivered at a low dose rate where breaks would be induced over a long period of incubation and are concomitantly repaired. The persistant induction of dsb in RE-treated cells over many hours means that cells harvested at 20–24 h after treatment will probably receive damage in more than one, and perhaps all phases of the cell cycle. Several studies have shown that, like radiation, RE induce a mixture of chromosomal aberration types; chromosome aberrations are observed in cells treated in G1 phase and chromatid aberrations in cells treated in S of G2 phase (Obe and Winkel 1985). Other studies have observed mixtures of aberration types (chromosome and chromatid) in the same cells (e.g. Natarajan and Obe 1984). This may be explained by the persistence of RE inside the cell for long periods after treatment. Clearly, further work needs to be carried out before the kinetics of induction of dsb and aberrations by RE are fully understood.

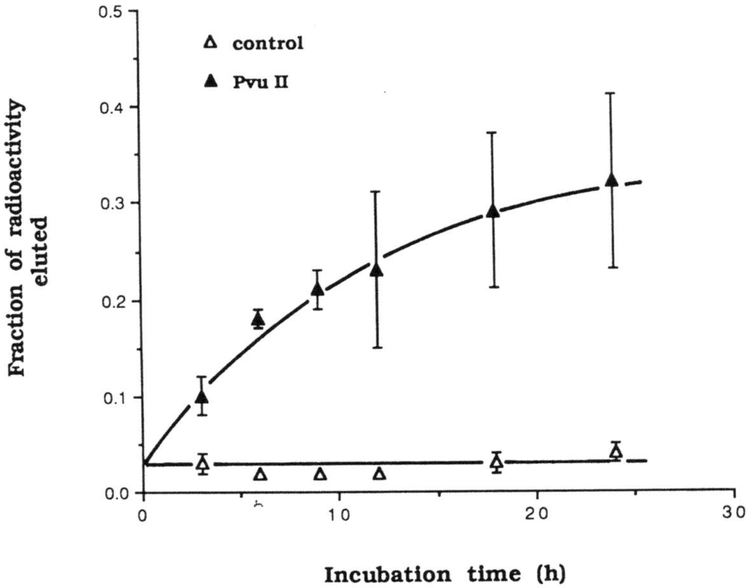

Fig. 3. Fraction of radioactivity eluted (proportional to the frequency of DNA double-strand breaks) in the neutral filter elution assay, as a function of incubation time after electroporation in the presence or absence of Pvu II (200 U/ml). Errors are standard errors of means of two experiments. Graph redrawn from Costą and Bryant (1990)

4 Repair of Double-Strand Breaks

The repair of radiation-induced dsb is a controversial topic since the results of measurements depend critically on the type of assay used. Present space does not allow a full discussion of this controversy, but briefly, measurements made with the neutral velocity sedimentation technique indicate that dsb are repaired with a T1/2 of some 1.5–4 h, depending on conditions (Bryant and Blöcher 1980; Blöcher and Pohlit 1982). In contrast, measurements made with the neutral filter elution technique show a very rapid decrease in dsb with time and repair appears to be complete in about 60 min following radiation exposure (e.g. Weibezahn and Coquerelle 1981). It seems at present more likely that measurements with the former method, which measures absolute numbers of dsb, would be intrinsically more reliable. Also, the kinetics of rejoining of chromatid breaks (Mozdarani and Bryant 1987; 1989) and PCC fragments (Cornforth and Bedford 1983) appear to correlate with the repair of dsb as measured by neutral filter elution. As described above, the repair of RE-induced breaks does not appear to be directly measurable because of the continued induction of dsb over many hours following treatment.

5 Induction of Chromosomal Aberrations in Radiosensitive Mutants

Radiosensitive mutant Chinese hamster ovary lines (*xrs*) show enhanced frequencies of aberrations in response to X-rays (Kemp and Jeggo 1986; Bryant et al. 1987;

Darroudi and Natarajan 1987) over those in the CHO K1 parent line. These include both chromosome and chromatid aberrations (Bryant et al. 1987). A similar picture emerges from experiments with RE. It was shown (Bryant et al. 1987) that RE which induce dsb with cohesive termini (e.g. Bam H1 and Eco R1) showed much reduced levels of aberrations when compared with those causing blunt termini (Pvu II and Eco RV), however, the levels of aberrations caused by cohesive-ended dsb were enhanced in *xrs* 5 cells as compared to the parent CHO K1 line (Fig. 4). This suggests that a proportion of cohesive-ended dsb may be recognized by cells in the same way as blunt-ended dsb and may be repaired by the *xrs* pathway.

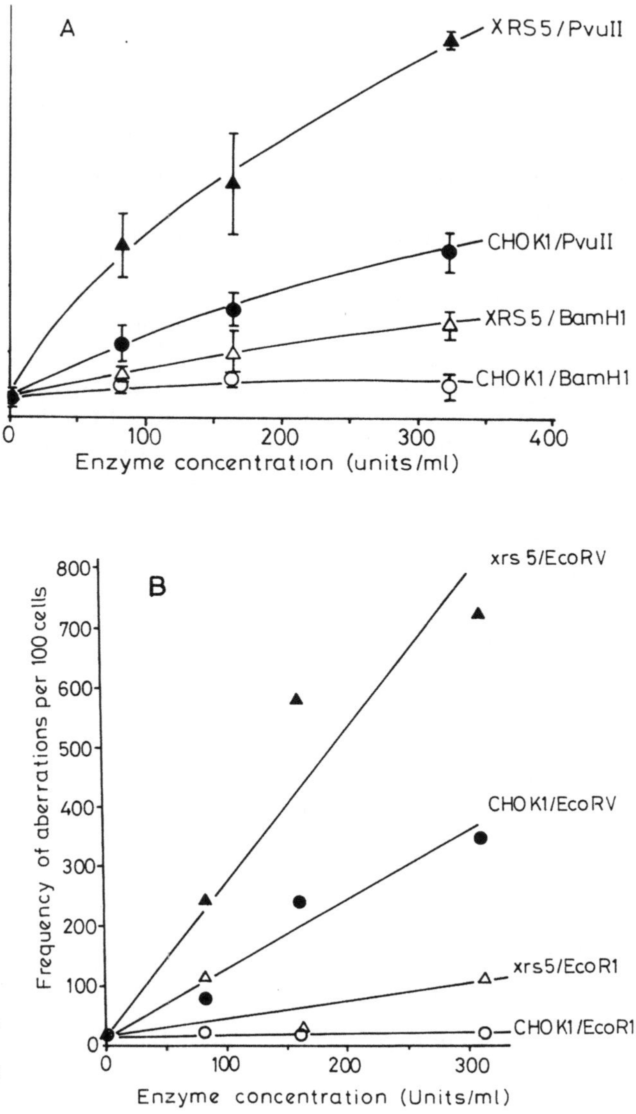

Fig. 4 A, B. Frequencies of chromosomal aberrations per 100 cells as a function of RE concentration in *xrs* 5 and parent CHO K1 cell lines. **A** Pvu II and Bam H1; **B** Eco R1 and R1 (Bryant et al. 1987)

One paradoxical feature of the induction of aberrations of the exchange type, noted previously (Bryant 1988; Mozdarani and Bryant 1989), was that under conditions where the repair of dsb is defective or inhibited (in the *xrs* 5 line or in the presence of the inhibitor ara A), the frequency of exchanges usually increases We suggested that this might indicate a second dsb joining (misjoining) mechanism possibly involving simple ligation of ends of non-homologous DNA. While this still appears a possible mechanism for exchange formation, recent data indicates that in X-irradiated G2 human peripheral blood lymphocytes an increase in exchanges occurs in the presence of ara A at a level of drug where the rejoining of chromosomal breaks was not significantly effected (MacLeod and Bryant, submitted).

6 Conclusions

From the data presently available on the effects of RE we can say that the chromosomal aberrations induced by RE are essentially of the same types as those induced by X-rays; the pattern of induction at various stages of the cell cycle also appears similar to X-rays, yielding chromosome aberrations in G1 and chromatid aberrations in S and G2. Although RE mimic the effect of ionizing radiation, the kinetics of induction of dsb by RE appear to be more like that of a chronic irradiation and this should be taken into account when making comparisons between RE and X-ray data. There are clearly strong differences in the effectiveness of dsb generated by blunt- and cohesive-ended dsb.

X-rays are thought to induce a mixture of these two types with a preponderance of cohesive-ended dsb. RE-induced dsb differ from those induced by X-rays in that RE dsb have "clean" ends, whereas most X-ray-induced dsb will have "dirty" end structures and will therefore require enzymatic cleaning before joining can take place. In addition, in the case of radiation-induced dsb, these are induced in the presence of various other DNA lesions which might affect the fate of the dsb, e.g. with respect to the formation of chromosomal aberrations.

In spite of these differences in type and kinetics of dsb induced by X-rays and RE, the work with RE should yield further valuable information about the underlying mechanisms involved in the induction of chromosomal aberrations.

Acknowledgements. The author's work is supported by the Cancer Research Campaign. The author wishes to acknowledge the contribution made by Mr. Alan Tennant in the operation of the Chem-X computer programme and calculation of inter-atomic distances in DNA.

References

Bishop DT, Williamson JA, Skolnick MH (1983) A model for restriction fragment length distributions. Am J Hum Genet 35: 795–815

Blöcher D (1982) DNA double-strand breaks in Ehrlich ascites tumour cells at low doses of X-rays. I. Determination of induced breaks by centrifugation at low speed. Int J Radiat Biol 42: 317–328

Blöcher D, Pohlit W (1982) DNA double-strand breaks in Ehrlich ascites tumour cells at low doses of X-rays. II. Can cell death be attributed to double-strand breaks? Int J Radiat Biol 42: 329–338

Bryant PE (1984) Enzymatic restriction of mammalian cell DNA using Pvu II and Bam H1: evidence for the double-strand break origin of chromosomal aberrations. Int J Radiat Biol 46: 57–65
Bryant PE (1985) Enzymatic restriction of mammalian cell DNA: evidene for double-strand breaks as potentially lethal lesions. Int J Radiat Biol 48: 55–60
Bryant PE (1986) The DNA double-strand break origin of chromosomal aberrations (abstract). Br J Cancer 54: 374
Bryant PE (1988) Use of restriction endonucleases to study relationships between DNA double-strand breaks, chromosomal aberrations and other end-points in mammalian cells. Int J Radiat Biol 54: 869–890
Bryant PE, Blöcher D (1980) Measurement of the kinetics of DNA double-strand break repair in Ehrlich ascites tumour cells using the unwinding method. In. J Radiat Biol 38: 335–347
Bryant PE, Christie AF (1989) Induction of chromosomal aberrations in CHO cells by restriction endonucleases: effects of blunt and cohesive-ended double-strand breaks in cells treated by "pellet" methods. Mutat Res 213: 233–241
Bryant PE, Birch D, Jeggo PA (1987) High chromosomal sensitivity of Chinese hamster *xrs* 5 cells to restriction endonuclease induced double-strand breaks. Int J Radiat Biol 52: 537–554
Cook PR, Brazell I (1975) Determination and repair of single-strand breaks in nuclear DNA. Natue 263: 679–682
Coquerelle T, Bopp A, Kesseler B, Hagen U (1973) Strand breaks and 5' end groups of irradiated thymocytes. Int J Radiat Biol 24: 397
Cornforth MN, Bedford JS (1983) X-ray induced breakage and rejoining of human interphase chromosomes. Science 222: 1141–1143
Costa ND, Bryant PE (1990) The induction of DNA double-strand breaks in CHO cells by Pvu II: Kinetics using neutral filter elution (pH 9.6) Int. J. Radiat. Biol., 57, 933–938.
Darroudi F, Natarajan AT (1987) Cytological characterisation of Chinese hamster ovary X-ray sensitive mutants *xrs* 5 and *xrs* 6. I: Induction of chromosomal aberrations by X-irradiation and its modulation with 3-amino benzamide and caffeine. Mutat Res 177: 133–148
Kemp LM, Jeggo PA (1986) Radiation induced chromosome damage in X-ray sensitive mutants (*xrs*) of the Chinese hamster ovary cell line. Mutat Res: DNA Repair Rep 166: 255–263
Landbeck A, Hagen U (1973) Action of DNA polymerase I on gamma irradiated DNA. Biochem Biophys Acta 331: 318–327
Mozdarani H, Bryant PE (1987) The effect of 9-β-arabinofuranosyladenine on the formation of X-ray induced chromatid aberrations in X-irradiated human cells. Mutagenesis 2: 371–374
Mozdarani H, Bryant PE (1989) Kinetics of chromatid aberrations in G2 ataxia-telangiectasia cells exposed to X-rays and ara A. Int J Radiat Biol 55: 71–84
Natarajan AT, Obe G (1984) Molecular mechanisms involved in the production of chromosomal aberrations. Chromosoma (Berl) 90: 120–127
Natarajan AT, Mullenders LHF, Meijers M, Mukherjee M (1985) Induction of sister chromatid exchanges by restriction endonucleases. Mutat Res 144: 33–39
Obe G, Winkel E (1985) The chromosome breaking activity of the restriction endonuclease Alu I in CHO cells is independent of the S-phase of the cell cycle. Mutat Res 152: 25–29
Watt DE (1989) On absolute biological effectiveness and unified dosimetry. J Radiol Prot 9: 33–49
Weibezahn KF, Coquerelle T (1981) Radiation induced double-strand breaks are joined by ligation and recombination processes. Nucl Acids Res 9: 3139–3150

The Use of Restriction Endonucleases to Study the Mechanisms of Chromosome Damage

W. F. Morgan and R. A. Winegar[1]

Contents

1 Introduction

Bacterial restriction endonucleases recognize specific, short (4–8 base pair) nucleotide sequences in DNA. They bind at or near these sequences, then cleave DNA to generate either blunt-end or cohesive-end double-strand breaks. The role of DNA double-strand breaks in chromosome damage has long been a matter of some controversy (see Bryant 1988). By introducing restriction endonucleases into cells and examining their cytogenetic effects, investigators have now been able to define a more precise role for DNA double-strand breaks in chromosome damage and repair. In this report we review current methods for introducing restriction enzymes into cells, and the effects of restriction enzymes on the induction of DNA double-strand breaks, chromosome aberrations, and sister chromatid exchange (SCE).

[1] Laboratory of Radiobiology and Environmental Health, University of California, San Francisco, CA 94143-0750, USA

2 Methods for the Introduction of Restriction Endonucleases into Cells

2.1 Inactivated Sendai Virus

The inactivated Sendai virus technique was first used by Tanaka et al. (1975) to introduce T4 endonuclease into mammalian cells, and later by Natarajan et al. (1980) to introduce *Neurospora* endonuclease into cells. This method involves treatment of attached cells with inactivated Sendai virus on ice for 10 min followed by incubation at 37 °C for 30 min. Bryant (1984) and Natarajan and Obe (1984) used this procedure to assess the ability of restriction enzymes to induce chromosome aberrations. Although this technique demonstrates that restriction enzymes are capable of inducing chromosome aberrations, problems with the technique include the low percentage of cells taking up enzyme, high variability, and the fact that a relatively high number of units of enzyme is required to elicit a cytogenetic response.

2.2 Pellet Technique

The pellet technique, developed by Obe and co-workers (Obe and Winkel 1985; Obe et al. 1985, 1986), involves either scraping or trypsinizing cells, washing with complete medium, then washing with serum or buffered saline. After centrifuging, as much of the supernatant as possible is removed. A restriction enzyme is added directly to the pellet in a total volume of 8 μl, and the cells are incubated at 37 °C for 30 min. This method has been used to confirm that restriction enzymes induce chromosome aberrations. Using this technique, Vasudev and Obe (1988) have speculated that AluI uptake in Chinese hamster ovary (CHO) cells involves receptor-mediated endocytosis, i.e., that AluI binds to membrane surface receptors and then becomes internalized endocytically. Although it is difficult to conceive of the inherent function of such receptors in eukaryotic cells, one must assume they function for a wide variety of enzymes, not just AluI. Bryant and Christie (1989), however, have provided evidence that it is the storage buffer (consisting of 50% glycerol) that is responsible for cells' taking up enzyme in this technique. Like the inactivated Sendai virus technique, the pellet method also suffers from high variability, a low percentage of aberrant cells, and the fact that relatively high concentrations of enzyme are required.

2.3 Osmolytic Shock

The osmolytic shock technique, developed by Okada and Rechsteiner (1982), was used by Winegar and Preston (1988) to introduce several different restriction enzymes into CHO cells. Osmolytic shock involves treating attached cells with a hypertonic buffer containing sucrose and polyethylene glycol and restriction enzymes. This procedure induces the production of pinocytic vesicles containing restriction enzymes. The cells are then briefly exposed to a slightly hypotonic buffer, causing the pinocytic vesicles to burst and release restriction enzymes into the cells. The method proved quite effective at introducing restriction enzymes into cells, and a dose response was seen with all four enzymes tested. However, like other techniques, a low percentage of cells are aberrant and there is some variability between experiments.

2.4 Expression Vectors

A completely different approach was taken by Morgan et al. (1988), who constructed a plasmid (pMENs) containing the EcoRI gene under the control of the heavy metal-inducible metallothionine promoter. A CHO clone transfected with pMENs displayed a high percentage of chromosome aberrations after induction with cadmium. This technique has the advantage that >90% of cells contain aberrations. There are, however, some disadvantages. Long periods of cadmium treatment (22–28 hours) are required before chromosome aberrations are observed, and there is little control over the amount of enzyme produced. Furthermore, much work is needed to construct expression vectors and to verify and characterize their function in eukaryotic cells. Therefore, only a limited number of enzymes can be examined.

2.5 Electroporation

Electroporation was used by Winegar et al. (1989) to introduce restriction enzymes into CHO cells. This technique involves exposing a suspension of cells to a brief electrical pulse. Although the mechanisms of electroporation are not completely understood, it is believed that thinning and then localized breakdown of the membrane (pore formation) occurs when an electrical field induces a potential difference across the cell membrane. After the electrical pulse, the pores close, and the cells continue to grow and proliferate. Electroporation has several advantages: up to 100% of cells show aberrations, a linear dose-response curve is obtained, and very small quantities of enzyme can be used. A fraction of cells appear to burst immediately upon electroporation; of those cells still intact, there is a high relative plating efficiency. There is preliminary evidence that by modifying the electroporation buffer, it is possible to obtain very high cell survival with an even higher efficiency of enzyme incorporation (G. Hook and G. Horesovsky, pers. comm.). Although there is some small variation between experiments (Morgan et al. 1989a), the main drawback of electroporation appears to be the expense involved in obtaining an electroporation unit. Nevertheless, electroporation is a convenient, extremely rapid, and reliable method for the introduction of restriction enzymes into cells.

3 DNA Breakage Induced by Restriction Endonucleases

To date, few attempts have been made to measure DNA strand breakage in mammalian cells after introduction of restriction enzymes. Bryant (1984) measured breaks induced by PvuII and BamHI by using the alkali unwinding method of Ahnström and Erixon (1973). Both enzymes induced breaks at a similar frequency despite dramatic differences in chromosome aberration yields. Bryant calculated that 500 units of enzyme yielded an unwinding value equivalent to that induced by 2 Gy of X-rays; extrapolating from this figure, he concluded that the number of breaks induced by one unit of enzyme was approximately equivalent to that induced by 0.11 Gy of X-rays. Natarajan et al. (1985) used nucleoid sedimentation (Cook and Brazell 1976) to determine the frequency of breaks induced by either 60 or 120 units of PvuII. No dose-response relationship was observed, and the effect of PvuII was equivalent to that of 0.25 Gy of X-rays, a value reasonably close to that observed by Bryant (1984). As Bryant (1988) points out, however, these values must be treated with caution. The cytogenetic data indicated a high degree of overdispersion of damage in treated populations (see Sect. 4); thus, strand breaks were measured in a heterogeneous mixture of heavily damaged and undamaged cells and may not provide an accurate estimate of break frequency.

Recently, we have used pulsed-field gel electrophoresis to analyze DNA strand breakage after electroporation of restriction enzymes into cells (D. Ager, J. W. Phillips, E. Abella Columna, R. A. Winegar, and W. F. Morgan, submitted). By examining break yields at different times after electroporation, it was possible to determine when enzymes cleaved DNA and for how long they were active. AluI was active almost immediately, with most breakage observed 30 min after electroporation and a rapid decline in break yield 45–60 min after its introduction into cells. Using enzymes that generate different types of DNA double-strand breaks, i.e., blunt-end or cohesive-end breaks with either 3′ oder 5′ overhangs, we have studied break frequency in relation to chromosome aberration formation (Ager, Phillips, Abella Columna, Winegar, and Morgan, submitted). As Bryant (1984) observed with BamHI, we found that high levels of DNA strand breakage do not necessarily indicate high yields of aberrations.

4 Chromosome Aberrations

Irrespective of the method used to introduce restriction endonucleases into cells, there is a consensus that, once in a cell, an enzyme will induce chromosome aberrations. There is a lingering controversy, however, concerning the relative efficiency of blunt-end versus cohesive-end double-strand breaks as inducers of aberrations. The initial experiments done by Bryant (1984) and Natarajan and Obe (1984) compared PvuII (blunt-end) and BamHI (cohesive-end) for their ability to induce aberrations. In each case, PvuII was found to be more effective than BamHI as an aberration inducer. It is of interest that Bryant and co-workers (Bryant 1984, 1988; Bryant et al. 1987) have consistently found that BamHI does not induce aberrations in wild-type CHO cells, whereas other investigators find that BamHI can induce chromosome aberrations (Gustavino et al. 1986; Zhang and Dong 1987). The reasons for this discrepancy aside, there is now abundant evidence that both blunt- and cohesive-end breaks can lead to aberration formation, but that enzymes inducing blunt-end breaks are more effective (Obe et al. 1985; Winegar and Preston 1988; Winegar et al. 1989).

An important and often overlooked factor in comparing the cytogenetic effects of restriction endonucleases is the potential number of recognition sequences available for cleavage (Bishop et al. 1983). Winegar and Preston (1988) examined two enzymes that induce blunt-end breaks (RsaI and AluI) and two enzymes that induce cohesive-end breaks (Sau3AI and MspI). All four enzymes were effective inducers of aberrations. The total number of aberrations appeared to increase with the calculated frequency of recognition sites for each enzyme. Winegar et al. (1989) examined the effect of equal concentrations of AluI and Sau3AI on aberration yields. Although both enzymes have a calculated recognition site frequency of approximately 1/200 base pairs, AluI induced two – three times as many aberrations as Sau3AI.

In addition to the potential number of sites available for cleavage, a number of other factors may affect the observed aberration frequency. These include the size and activity of the enzyme used, the methylation status of the DNA, the repair capacity of the cell, and the stage of the cell cycle at which cells are treated.

5 Cell Cycle Effects

In contrast to the controversy over cut-end structure and the efficiency of aberration induction, there is excellent agreement that restriction enzymes act as radiomimetic agents. Like ionizing radiation, restriction enzymes induce chromosome-type aberrations during the G_1 phase and chromatid-type aberrations during S and G_2 (Natarajan and Obe 1984; Obe and Winkel 1985; Obe et al. 1985; Winegar and Preston 1988).

6 X-Ray-Sensitive Mutants

One area of special interest is the characterization of the nature of the defects in radiosensitive cell lines. Bryant et al. (1987) examined the sensitivity of the radiosensitive cell line xrs-5 and its normal parental cell line CHO-K1 to blunt- and cohesive-end break-producing enzymes. Exposure to enzymes producing blunt-end breaks caused three times the level of aberrations in xrs-5 as in CHO-K1. Both cell lines yielded fewer aberrations with enzymes producing cohesive-end breaks. Although these results are of interest, the degree of variability in the study was extremely high. In fact, with EcoRI, CHO-K1 was found to be 6.5-fold more sensitive than xrs-5.

Darroudi and Natarajan (1989) also examined the sensitivity of xrs-5 and xrs-6 to restriction enzymes. They too found the xrs lines to be more sensitive than CHO-K1. Both groups of investigators noted the presence of a high level of exchange-type aberrations in the cell lines despite their defective repair of double-strand breaks. This suggests that these cells do rejoin double-strand breaks but that some misrejoining of cleaved DNA is involved.

7 Sister Chromatid Exchange

The effect of restriction enzymes on SCE frequency is controversial. Natarajan and co-workers (Natarajan et al. 1985; Stoilov et al. 1986) reported that DNA double-strand breaks induced by restriction enzymes during the S-phase led to an increase in the frequency of SCEs, despite the fact that agents that cause DNA strand breakage have traditionally been thought to be poor inducers of SCEs (Wolff 1978; Morgan et al. 1985). Recently, we found that expression of EcoRI within CHO cells induced chromosome aberrations but not SCEs (Morgan et al. 1988). When we carried out experiments similar to those of Natarajan et al. (1985), using the highly efficient method of electroporation (Winegar et al. 1989) to introduce restriction enzymes into cells, we found no evidence for the induction of SCEs by restriction enzymes (Morgan et al. 1989b). The reason for these conflicting results is not known. Incorporation of the bromodeoxyuridine necessary for visualization of SCEs does not affect enzyme function (Stoilov et al. 1986; Morgan et al. 1989b). Furthermore, in all of the studies mentioned, restriction enzymes did induce aberrations. It is our opinion that the discrepancy is due to the differing techniques for introducing restriction enzymes into cells.

Whether using an expression vector or electroporation, we consistently find no effect of restriction enzymes on SCE frequency. However, Darroudi and Natarajan (1989) have continued and expanded on earlier work from their laboratory showing that both wild-type CHO and X-ray-sensitive mutant CHO cell lines xrs-5 and xrs-6 showed increased SCE frequencies after enzymes were introduced by the pellet method. The role of double-strand breaks in SCE formation remains unresolved.

8 Modulation of Restriction Endonuclease Activity Within Cells

Restriction enzymes are supplied in storage buffers containing 50% glycerol, but they usually differ widely in their combinations of salts and detergents. These storage buffers can also affect aberration yields, e.g., Zhang and Dong (1987) found that 6.5 to 9.5% of their storage buffer-treated cells contained aberrations, and Vasudev and Obe (1987) found that 11% of their storage buffer-treated cells were aberrant. Subsequent work from Obe and co-workers demonstrated that various salts can modulate enzyme-induced aberration yields, e.g., ammonium sulfate (Obe and Kamra 1986) and magnesium chloride, calcium chloride, and sodium chloride (Tuschy and Obe 1988). These investigators speculated that salts may lead to partial dehistonization of chromatin and that the resulting conformational changes in chromatin make more recognition sites available for enzyme binding and cleavage. In agreement with this interpretation, Vasudev and Obe (1987) showed that heat (46 °C, 3–7 min) pretreatment led to a reduction in AluI-induced chromosome-type aberrations. Heat induced an accumulation of accessory proteins in the chromatin (Roti Roti 1982), and this excess protein apparently protected DNA from enzyme cleavage.

Chemical compounds that interfere with cellular responses to DNA damage also modulate restriction enzyme-induced chromosome damage. Cytosine arabinoside, an inhibitor of polymerase α, increased enzyme-induced aberration yields (Natarajan and Obe 1984; Obe and Natarajan 1985), as did another polymerase α inhibitor, aphidicolin, and the poly(ADP-ribose) polymerase inhibitor 3-aminobenzamide (H. W. Chung, J. W. Phillips, R. A. Winegar, and W. F. Morgan, in prep.). Caffeine, on the other hand, did not affect the yield of enzyme-induced aberrations (Natarajan and Obe 1984; Chung, Phillips, Winegar, and Morgan, in prep.).

9 Conclusions

Restriction enzymes can be introduced into mammalian cells by a number of techniques, with varying effectiveness. Once in cells, enzymes cleave DNA. Whether or not restriction enzymes are faithful in their sequence recognition and generate only DNA double-strand breaks is not known. The consensus is that, despite the complexity of cellular DNA, enzymes do bind to their defined nucleotide sequence and give rise to DNA double-strand breaks as their only perturbation in DNA. DpnI, which recognizes GATC sequences in DNA but cuts only if the adenine position is methylated, does not lead to aberration formation in CHO cells (R. A. Winegar, unpubl. obs.). This is as expected for this enzyme, because methylated adenine, to the best of our knowledge, has not been found in eukaryotic cells. Consequently, there should be no recognition or cleavage after treatment with DpnI. Its isoschizomer Sau3AI, however, which is not methylation-dependent, induces high levels of aberrations (Winegar and Preston 1988; Winegar et al. 1989). Furthermore, we have found that the isoschizomers MspI and HpaII vary significantly in their ability to produce chromosome damage. Both enzymes recognize CCGG sequences in DNA, but HpaII will cleave only if the internal cytosine of this sequence is unmethylated, whereas MspI cleaves irrespective of the methylation status of the internal cytosine. This cytosine, however, is highly methylated in mammalian cells (Cooper 1983) and, as predicted, MspI is much more efficient in inducing chromosome damage than HpaII at all stages of the cell cycle (R. A. Winegar, J. W. Phillips, L. H. Lutze, and W. F. Morgan, Somat. Cell Mol. Genet. 16: 251–256, 1990).

There is unanimous agreement that treatment with restriction enzymes leads to the formation of aberrant chromosomes independent of cell type or the stage of the cell cycle at which the cells are treated. There is good agreement that enzymes producing blunt-end breaks are more effective in generating aberrant chromosomes than enzymes producing cohesive-end breaks. The role of restriction enzymes in SCE formation remains controversial.

Future studies should shed light on the mechanisms of chromosome aberration formation and the reasons that some restriction enzyme-induced breaks are rejoined without giving rise to aberrations, others are misrejoined to give rise to exchange-type aberrations, and still others are never rejoined and lead to deletions in metaphase chromosomes.

Acknowledgements. This work was supported by the Office of Health and Environmental Research, U.S. Department of Energy, contract DE-AC03-76-SF01012, and by the National Institutes of Health National Research Service Award 5-T32-ES07106 from the National Institute of Environmental Health Sciences.

We thank Drs. S. Wolff, H. W. Chung, D. Ager, L. H. Lutze, and J. W. Phillips for helpful discussions, P. Winegar for typing the manuscript, and M. McKenney for editing it.

References

Ahnström G, Erixon K (1973) Radiation induced strand breakage in DNA from mammalian cells: strand separation in alkaline solution. Int J Radiat Biol 23: 285–289

Bishop DT, Williamson JA, Skolnick MH (1983) A model for restriction fragment length distributions. Am J Hum Genet 35: 795–815

Bryant PE (1984) Enzymatic restriction of mammalian cell DNA using Pvu II and Bam HI: evidence for the double-strand break origin of chromosomal aberrations. Int J Radiat Biol 46: 57–65

Bryant PE (1988) Use of restriction endonucleases to study relationships between DNA double-strand breaks, chromosomal aberrations and other endpoints in mammalian cells. Int J Radiat Biol 54: 869–890

Bryant PE, Christie AF (1988) Induction of chromosomal aberrations in CHO cells by restriction endonucleases: effects of blunt- and cohesive-ended double-strand breaks in cells treated by "pellet" methods. Mutat Res 213: 233–241

Bryant PE, Birch DA, Jeggo PA (1987) High chromosomal sensitivity of Chinese hamster xrs 5 cells to restriction endonuclease induced DNA double-strand breaks. Int J Radiat Biol 52: 537–554

Cook PR, Brazell IA (1976) Detection and repair of single-strand breaks in nuclear DNA. Nature 263: 679–682

Cooper DN (1983) Eukaryotic DNA methylation. Hum Genet 64: 315–333

Darroudi F, Natarajan AT (1989) Cytogenetical characterization of Chinese hamster ovary X-ray-sensitive mutant cells, xrs 5 and xrs 6. IV. Study of chromosomal aberrations and sister-chromatid exchanges by restriction endonucleases and inhibitors of DNA topoisomerase II. Mutat Res 212: 137–148

Gustavino B, Johannes C, Obe G (1986) Restriction endonuclease Bam H I induces chromosomal aberrations in Chinese hamster ovary (CHO) cells. Mutat Res 175: 91–95

Morgan WF, Djordjevic MC, Jostes RF, Pantelias GE (1985) Delayed repair of DNA single-strand breaks does not increase cytogenetic damage. Int J Radiat Biol 48: 711–721

Morgan WF, Fero ML, Land MC, Winegar RA (1988) Inducible expression and cytogenetic effects of the EcoRI restriction endonuclease in Chinese hamster ovary cells. Mol Cell Biol 8: 4204–4211

Morgan WF, Ager D, Chung HW, Ortiz T, Phillips JW, Winegar RA (1989a) The cytogenetic effects of restriction endonucleases following their introduction into cells by electroporation. Metabolism, Testing Methods and Chromosomes (M. L. Mendelsohn, ed.), Wiley-Liss, New York (in press)

Morgan WF, Chung HW, Phillips JW, Winegar RA (1989b) Restriction endonucleases do not induce sister-chromatid exchanges in Chinese hamster ovary cells. Mutat Res 226: 203–209

Natarajan AT, Obe G (1984) Molecular mechanisms involved in the production of chromosomal aberrations. III. Restriction endonucleases. Chromosoma (Berl) 90:120–127

Natarajan AT, Obe G, Zeeland AA van, Palitti F, Meijers M, Verdegaal-Immerzeel EAM (1980) Molecular mechanisms involved in the production of chromosomal aberrations. II. Utilization of *Neurospora* endonuclease for the study of aberration production by X-rays in G_1 and G_2 stages of the cell cycle. Mutat Res 69: 293–305

Natarajan AT, Mullenders LHF, Meijers M, Mukherjee M (1985) Induction of sister-chromatid exchanges by restriction endonucleases. Mutat Res 144: 33–39

Obe G, Kamra OP (1986) Elevation of AluI-induced frequencies of chromosomal aberrations in Chinese hamster ovary cells by *Neurospora crassa* endonuclease and by ammonium sulfate. Mutat Res 174: 35–46

Obe G, Natarajan AT (1985) Chromosomal aberrations induced by the restriction endonuclease AluI in Chinese hamster ovary cells: influence of duration of treatment and potentiation by cytosine arabinoside. Mutat Res 152: 205–210

Obe G, Winkel E-U (1985) The chromosome-breaking activity of the restriction endonuclease AluI in CHO cells is independent of the S-phase of the cell cycle. Mutat Res 152: 25–29

Obe G, Palitti F, Tanzarella C, Degrassi F, Salvia R De (1985) Chromosomal aberrations induced by restriction endonucleases. Mutat Res 150: 359–368

Obe G, Hude W Von der, Scheutwinkel-Reich M, Basler A (1986) The restriction endonuclease Alu I induces chromosomal aberrations and mutations in the hypoxanthine phosphoribosyltransferase locus, but not in the Na^+/K^+ ATPase locus in V79 hamster cells. Mutat Res 174: 71–74

Okada CY, Rechsteiner M (1982) Introduction of macromolecules into cultured mammalian cells by osmolytic lysis of pinocytic vesicles. Cell 29: 33–41

Roti Roti JL (1982) Heat-induced cell death and radiosensitization: molecular mechanisms. In: Dethlefsen LA (ed) Cancer therapy by hyperthermia, drugs, and radiation. Natl Cancer Inst Monogr 61: 3–10

Stoilov L, Mullenders LHF, Natarajan AT (1986) Influence of bromodeoxyuridine substitution of thymidine on sister-chromatid exchanges and chromosomal aberrations induced by restriction endonucleases. Mutat Res 174: 295–301

Tanaka K, Sekiguchi M, Okada Y (1975) Restoration of ultraviolet-induced unscheduled DNA synthesis of xeroderma pigmentosum cells by the concomitant treatment with bacteriophage T4 endonuclease V and HVJ (Sendai virus). Proc Natl Acad Sci USA 72: 4071–4075

Tuschy S, Obe G (1988) Potentiation of Alu I-induced chromosome aberrations by high salt concentrations in Chinese hamster ovary cells. Mutat Res 207: 83–87

Vasudev V, Obe G (1987) Effect of heat treatment on chromosomal aberrations induced by the alkylating agent trenimon or the restriction endonuclease Alu I in Chinese hamster ovary (CHO) cells. Mutat Res 178: 81–90

Vasudev V, Obe G (1988) Evidence for a receptor-mediated endocytosis of Alu I in Chinese hamster ovary cells. Mutat Res 197: 109–116

Winegar RA, Preston RJ (1988) The induction of chromosome aberrations by restriction endonucleases that produce blunt-end or cohesive-end double-strand breaks. Mutat Res 197: 141–149

Winegar RA, Phillips JW, Youngblom JH, Morgan WF (1989) Cell electroporation is a highly efficient method for introducing restriction endonucleases into cells. Mutat Res 225: 49–53

Wolff S (1978) Chromosomal effects of mutagenic carcinogens and the nature of the lesions leading to sister chromatid exchange. In: Evans HJ, Lloyd DC (eds) Mutagen-induced chromosome damage in man. Edinburgh University Press, Edinburgh, pp 208–215

Zhang S, Dong W (1987) Chromosomal aberrations induced by the restriction endonucleases EcoRI, Pst I, Sal I, and Bam HI in CHO cells. Mutat Res 180: 109–114

Induction of Chromosomal Aberrations by the Restriction Endonuclease AluI in Chinese Hamster Ovary (CHO) Cells: Influence of Glycerol on Aberration Frequencies

C. JOHANNES and G. OBE[1]

Contents

1 Introduction

Restriction endonucleases (REs) induce chromosomal aberrations in living cells indicating that the ultimate lesion for the production of chromosomal aberrations is a DNA double-strand break (Obe et al. 1987; Bryant, this Vol.; Morgan et al., this Vol.). It is still not clear how living cells can uptake REs. Different methods have been used to treat cells with REs (for discussion see Bryant 1988; Morgan et al., this Vol.) and all of them are more or less successful when measured as frequencies of chromosomal aberrations induced by REs. From our own results concerning the induction of chromosomal aberrations with AluI following treatment of pellets of 4×10^6 cells (about 10 μl volume) with 8 μl of a suspension of REs and newborn calf serum, we proposed that REs are internalized via an energy-dependent receptor-mediated mechanism (Johannes et al. 1989; Obe et al. 1989; Vasudev and Obe 1987). Analyses of Bryant and Christie (1989) have shown that the uptake of REs using the "pellet technique" is dependent on the shipping buffer in which the REs are suspended. The authors assume that "this may result from a combination of hypertonic shock and the detergent present in the shipping buffer" (Bryant and Christie 1989). In order to avoid a possible influence on the uptake of AluI by the relatively high concentrations of buffer components we analyzed the chromosome breaking effects of AluI by treatment of monolayers in such a way that the concentrations of the buffer components are considerably lower when compared to our earlier method. By variation of the amount of buffer or buffer components in the treatment solutions we found that glycerol has a considerable influence on the uptake of AluI by living Chinese hamster ovary (CHO) cells. In all experiments the intercellular distributions of polycentric chromosomes induced by AluI were overdispersed ($s^2/\overline{x} > 1$).

[1] Department of Genetics, University-GH Essen, P.O. Box 103 764, 4300 Essen 1, FRG

2 Materials and Methods

2.1 Cell Cultures

Chinese hamster ovary (CHO) cells were grown in 10 ml McCoy's 5A medium containing 10% newborn calf serum and antibiotics in plastic Petri dishes in a humidified atmosphere containing 5% CO_2 at 37 °C.

2.2 Treatment of Cells with the Restriction Endonuclease AluI

Two different sets of experiments were performed: (1) treatment of monolayers and (2) treatment of pellets of trypsinized cells.

1. Monolayer Treatment. Cells were trypsinized, washed once with medium and 1×10^6 cells were seeded in 35 mm Petri dishes containing 2 ml of complete medium. The medium was discarded 8 h later and the monolayers were washed once with prewarmed medium. The cells were treated with 100 μl of a mixture of medium with AluI (5 units/μl) in shipping buffer (10 mM Tris-HCl, pH 7.5; 50 mM KCl; 0.1 mM EDTA; 1 mM mercaptoethanol; 0.5 mg/ml bovine serum albumin, BSA; 50% v/v glycerol), additional shipping buffer or components of shipping buffer. Shipping buffer and components of the shipping buffer were prepared with deionized water as stock solutions in concentrations according to the manufacturer.

The mixture was spread evenly over the monolayers and the cells were exposed to the treatment mixture for 30 min at 37 °C, washed twice with prewarmed medium and recovered for 16 h in complete medium supplemented with 10% fetal calf serum and 2×10^{-5} M bromodeoxyuridine. The cultures received 0.08 μg/ml colcemid 2 h prior to fixation. Mitotic cells were collected by shake off and hypotonically treated with 1% sodium citrate for 8 min. Fixation in methanol, acetic acid (3:1) was done twice and the cells were dropped on wet, cold, grease-free slides.

2. Treatment of Pellets of Trypsinized Cells. Monolayers of cells were trypsinized and 4×10^6 or 1×10^6 cells were pelleted in 12 ml sterile plastic tubes. The cells were washed once with prewarmed medium, pelleted again, and incubated for 30 min at 37 °C with 100 μl treatment mixtures as described in (1). The cells were washed once with prewarmed medium and recovered in two 60-mm Petri dishes with 5 ml medium (2×10^6 cells each from pellets of 4×10^6 cells) or in one 35-mm Petri dish with 2 ml medium (pellets with 1×10^6 cells) for 18 h. Preparations were made as described in (1).

The preparations were stained differentially and first posttreatment metaphases (uniformly stained) were scored for chromosomal aberrations. All experiments were done at least twice independently.

The following aberrations were scored: polycentric chromosomes, ring chromosomes (rings), double minutes (min), isochromatid/chromosome breaks (B''), chromatid translocations (RB'), metaphases with complex RB' are indicated separately (RB' complex), triradials (RB'B''), chromatid intrachanges (CI), chromatid breaks (B'). Achromatic lesions or gaps were scored but not given in the tables. The polycentric chromosomes were calculated as dicentric chromosomes (DIC) by substracting 1 from the number of centromeres in a polycentric and taking the number of centromers left as number of DIC (e.g., a tetracentric chromosome results in 3 DIC). The relative variance $s^2/\overline{x}$ of the intercellular distribution of DIC was calculated, where s^2 is the standard deviation and $\overline{x}$ the mean number of DIC per metaphase. The distribution is overdispersed in cases where $s^2/\overline{x} > 1$.

3 Results

Treatment of monolayers (Tables 1–3) and pellets (Table 4) with 100 μl reaction mixtures containing AluI and recovery times of 16 (monolayers) or 18 h (pellets) led to chromosomal aberrations which were mainly of the chromosome type, i.e., polycentric chromosomes, ring chromosomes, minutes, and isochromatid/chromosome breaks. In addition, some chromatid-type aberrations were found such as chromatid interchanges, chromatid intrachanges and chromatid breaks. Since polycentric chromosomes are typical aberrations found in all experiments, and since they can be easily distinguished from all the other aberration types, the polycentric chromosomes are given together with the percentages of aberrant metaphases in Tables 1, 3, and 4. The polycentric chromosomes are given as dicentric chromosomes (DIC) and for the intercellular distribution of DIC we calculated the relative variances $s^2/\overline{x}$.

The aberration frequencies are positively correlated with the amount of shipping buffer in the reaction mixture (Table 1). Apart from glycerol, no other component of the shipping buffer shows stimulating effects on the frequencies of chromosomal aberrations induced by AluI (Table 2). The stimulating effect of glycerol is positively correlated with the amount of glycerol in the reaction mixture (Tables 3 and 4). Neither shipping buffer nor glycerol itself induced chromosomal aberrations (Tables 1 and 3).

With only two exceptions (5 units AluI with and without additional glycerol) in which $s^2/\overline{x}$ is 1.66 and 1.95, respectively, the intercellular distributions of DIC were clearly overdispersed ($s^2/\overline{x} > 2$) when the cells were treated with AluI.

Table 1. Percent aberrant metaphases, dicentric chromosomes (DIC, calculated from all polycentric chromosomes) per 100 metaphases and relative variances ($s^2/\overline{x}$) of the intercellular distributions of DIC in CHO cells. Treatment of monolayers for 30 min with mixtures (100 μl) of different composition (recovery time 16 h)

Dose of AluI in units (amount in μl)	Amount of AluI shipping buffer added in addition to AluI in μl[a]	Final concentration of glycerol in the reaction mixture (mM)	Percent aberrant metaphases (number of metaphases analyzed/number of independent experiments)	DIC per 100 metaphases	Relative variance $s^2/\overline{x}$
20 (4)	4	547	20.0 (200/2)	19.0	2.14
20 (4)	12	1095	59.0 (300/3)	73.3	2.86
40 (8)	8	1095	48.0 (200/2)	88.5	2.86
–	16	1095	7.5 (200/2)	4.0	1.72

[a] In addition to AluI and shipping buffer the reaction mixture contained medium.

Table 2. Chromosomal aberrations induced in monolayers of CHO cells following exposure to a mixture of 100 μl containing 20 units AluI, medium, and components of the shipping buffer (recovery time 16 h). Pooled data from two independent experiments are given (100 metaphases were analyzed per entry per experiment)

AluI[a] (units)	Buffer components or whole buffer added to the reaction mixture[b]	Aberrant metaphases (%)	Aberrations per 100 metaphases								
			Polycentrics (No. of centromeres)	Rings	Min	B''	RB'	RB' complex	RB'	B''CI	B'
20	No[c]	13.5	6.5 (2) 2.0 (3)	7.5	11.5	36.0	0.5	–	0.5	0.5	0.5
20	Tris (6 mM)	7.0	3.0 (2)	2.0	6.5	11.5	1.0	–	–	–	2.0
20	KCl (6 mM)	4.5	1.5 (2)	2.0	0.5	6.0	–	–	–	–	0.5
20	Na-EDTA (0.012 mM)	7.0	5.5 (2) 0.5 (3)	4.5	9.0	20.5	–	0.5	0.5	–	–
20	Mercaptoethanol (0.12 mM)	7.5	7.5 (2) 0.5 (3)	5.5	5.5	23.0	0.5	0.5	–	–	–
20	BSA[d] (0.06 mg/ml)	14.0	11.5 (2) 0.5 (3) 0.5 (4)	6.0	4.0	21.5	0.5	0.5	0.5	–	0.5
20	Glycerol (6 %, v/v)	70.5	106.5 (2) 16.5 (3) 2.5 (4) 0.5 (5)	73.5	98.0	466.5	5.5	11.5	5.5	1.0	–
20	Whole buffer[e]	56.5	53.5 (2) 7.0 (3) 4.0 (4) 1.0 (5)	46.5	73.5	280.5	10.0	9.0	13.5	2.0	2.5

[a] 4 μl of AluI in shipping buffer.
[b] The reaction mixture contained 84 μl medium and 12 μl solutions of buffer components. The final concentrations of the buffer components in the reaction mixtures are given in parentheses.
[c] The reaction mixture contained 96 μl medium.
[d] Bovine serum albumin.
[e] 12 μl of a mixture of all components of the shipping buffer was added. The concentrations of the single compounds correspond to the ones tested separately in the other entries.

Table 3. Percent aberrant metaphases, dicentric chromosomes (DIC, calculated from all polycentric chromosomes) per 100 metaphases and relative variances ($s^2/\overline{x}$) of the intercellular distributions of DIC in CHO cells. Treatment of monolayers for 30 min with mixtures (100 μl) of different composition (recovery time 16 h)

Dose of AluI in units (amount of glycerol in the reaction mixture resulting from the shipping buffer)[a]	Amount of 50 % glycerol added to the reaction mixture in μl (amount of glycerol in the reaction mixture)[b]	Final concentration of glycerol in the reaction mixture (mM)	Percent aberrant metaphases (number of metaphases analyzed/ number of independent experiments	DIC per 100 metaphases	Relative variance $s^2/\overline{x}$
5 (0.5 %, v/v)	–	68	4 (200/2)	2	1.66
5 (0.5 %, v/v)	15 (7.5 %, v/v)	1095	25 (300/3)	21	1.95
5 (0.5 %, v/v)	31 (15.5 %, v/v)	2189	47 (200/2)	64	2.87
10 (1.0 %, v/v)	–	137	8 (200/2)	5	3.57
10 (1.0 %, v/v)	14 (7.0 %, v/v)	1095	38 (300/3)	57	3.57
10 (1.0 %, v/v)	22 (11.0 %, v/v)	1642	55 (200/2)	100	2.47
10 (1.0 %, v/v)	30 (15.0 %, v/v)	2189	49 (200/2)	82	2.25
15 (1.5 %, v/v)	13 (6.5 %, v/v)	1095	34 (200/2)	58	2.36
20 (2.0 %, v/v)[c]	–	274	7 (700/7)	4	3.15
20 (2.0 %, v/v)[c]	12 (6.0 %, v/v)	1095	62 (400/4)	121	2.58
40 (4.0 %, v/v)	–	547	12 (300/3)	14	2.83
60 (6.0 %, v/v)	–	821	20 (200/2)	22	3.80
80 (8.0 %, v/v)	–	1095	54 (400/4)	114	2.96
200 (2.0 %, v/v)[d]	–	274	13 (200/2)	9	3.17
–	32 (16.0 %, v/v)	2189	9 (200/2)	2	1.49

[a] 5 units = 1 μl.
[b] In addition to AluI and glycerol, the reaction mixture contained medium.
[c] Data from two experiments in Table 2 are included.
[d] High concentrated AluI (50 units/μl) was used in this entry.

4 Discussion

Our results show that glycerol and not the other components of the shipping buffer has a considerable positive influence on the chromosome breaking effectivity of AluI in CHO cells. Two possibilities could explain these results: (1) the high glycerol concentrations lead to relaxed specificity of AluI or (2) glycerol facilitates the cellular uptake of AluI.

Relaxed specificity is not known for AluI and probably can not explain our findings.

Fraley et al. (1981) have shown that liposome-mediated uptake of SV40 DNA by African green monkey cells is enhanced (up to 200-fold) when the cells were postincubated with buffer containing 25% (v/v) glycerol. Similar results were obtained with mouse L-cells and HeLa cells. When cells were pretreated with sodium azide and 2-deoxyglucose, the glycerol-mediated uptake of liposomes was suppressed, indicating that the mechanism is energy-dependent. Similar results were obtained when cells were exposed to liposomes containing carboxyfluorescein. In the absence of glycerol nearly no fluorescence was observed inside the cells, subsequent treatment

Table 4. Percent aberrant metaphases, dicentric chromosomes (DIC, calculated from all polycentric chromosomes) per 100 metaphases and relative variances ($s^2/\bar{x}$) of the intercellular distributions of DIC in CHO cells. Pellets of 4×10^6 or 1×10^6 cells were treated for 30 min with mixtures (100 μl) of different composition (recovery time 18 h). Pooled data from two independent experiments are given (100 metaphases were analyzed per entry per experiment)

Dose of AluI in units (amount of glycerol in the reaction mixture resulting from the shipping buffer)	Amount of 50 % glycerol added to the reaction mixture in μl (amount of glycerol in the reaction mixture)	Final concentration of glycerol in the reaction mixture (mM)	Aberrant metaphases (%)	DIC per 100 metaphases	Relative variance $s^2/\bar{x}$
Pellet size: 4×10^6 cells					
10 (1 %, v/v)	–	137	5.0	4.0	2.19
10 (1 %, v/v)	14 (7 %, v/v)	1095	11.5	20.0	3.35
10 (1 %, v/v)	30 (15 %, v/v)	2189	34.0	69.0	3.53
– (1 %, v/v)[a]	–	137	2.5	0	–
– (1 %, v/v)[a]	14 (7 %, v/v)	1095	4.0	0.5	1.00
– (1 %, v/v)[a]	30 (15 %, v/v)	2189	7.0	2.0	1.49
Pellet size: 1×10^6 cells					
10 (1 %, v/v)	–	137	8.5	6.5	2.33
10 (1 %, v/v)	14 (7 %, v/v)	1095	17.5	32.0	4.66
10 (1 %, v/v)	30 (15 %, v/v)	2189	43.0	77.0	2.74
– (1 %, v/v)[a]	–	137	9.0	2.5	0.98
– (1 %, v/v)[a]	14 (7 %, v/v)	1095	10.5	1.5	0.99
– (1 %, v/v)[a]	30 (15 %, v/v)	2189	10.0	5.0	1.16

[a] 2 μl of AluI shipping buffer was added according to the volume of 10 units AluI.

with glycerol led to strong intracellular fluorescence. The latter effect was not observed when the cells were pretreated with sodium azide and 2-deoxyglucose. Uptake of free (i.e., not encapsulated in liposomes) carboxyfluorescein was not influenced by glycerol, indicating that glycerol does not influence the membrane permeability for fluorescein. These data of Fraley et al. (1981) indicate that glycerol stimulates the uptake of liposomes via an energy-dependent endocytotic mechanism. This assumption is supported by the effect of chloroquine and NH_4Cl on the glycerol-mediated uptake of liposomes (Fraley et al. 1981). Both compounds lead to an elevation of the pH in lysosomes (see also Obe et al. 1989). Treatment of cells with these compounds induced a considerable (up to six fold) enhancement of the infectivity of liposome-enclosed SV40 DNA. This finding is interpreted by the authors to mean that an elevation of the pH in lysosomes prevents degradation of endocytotically internalized SV40 DNA. Electron microscope analyses show that the cell membranes of glycerol-treated cells are intensively ruffled, indicating an increase in membrane activity. In the cytoplasm multilamellar vesicles can be seen within membrane-bound vesicles (Fraley et al. 1981).

With respect to the data of Fraley et al. (1981) our own results can be interpreted to mean that AluI is internalized by the cells via a glycerol-mediated endocytotic mechanism.

Since glycerol passes the cell membrane freely, an osmotic imbalance cannot be expected to occur, which may explain why treatment of CHO cells for 4 h with up to 1 M glycerol does not induce chromosomal aberrations (Galloway et al. 1987). In our experiments 30-min treatment of CHO cells with more than 2 M glycerol did not lead to chromosome aberrations.

Irrespective of the very effective uptake of AluI in the presence of glycerol, the intercellular distribution of DIC is overdispersed, a typical finding in experiments with REs (see Johannes et al. 1989). This phenomenon could be explained by the assumption that the uptake of AluI shows intercellular variability (Johannes et al. 1989).

Treatment of CHO cells with AluI using the method of electroporation leads to higher frequencies of aberrant metaphases when compared to our "glycerol method" (Winegar et al. 1989). This can be explained by assuming that the mechanism of uptake of AluI is different with the two methods.

Acknowledgments. This work was financially supported by the European Community (Contract CI1-0435-D).

References

Bryant PE (1988) Use of restriction endonucleases to study relationships between DNA double-strand breaks, chromosomal aberrations and other end-points in mammalian cells. Int J Radiat Biol 54: 869–890

Bryant PE, Christie AF (1989) Induction of chromosomal aberrations in CHO cells by restriction endonucleases: effects of blunt- und cohesive-ended double-strand breaks in cells treated by "pellet" method. Mutat Res 213: 233–241

Fraley R, Straubinger RM, Rule G, Springer EL Papahadjopoulos D (1981) Liposome-mediated delivery of deoxyribonucleic acid to cells: enhanced efficiency of delivery related to lipid composition and incubation conditions. Biochemistry 20: 6978–6987

Galloway SM, Deasy DA, Bean CL, Kraynak AR, Armstrong MJ, Bradley MO (1987) Effects of high osmotic strength on chromosome aberrations, sister-chromatid exchanges and DNA strand breaks, and the relation to toxicity. Mutat Res 189: 15–25

Johannes C, Tuschy S, Lamprecht I, Obe G (1989) The intercellular distributions of chromosomal aberrations induced by the restriction endonucleases AluI and Bam HI in Chinese hamster ovary cells are overdispersed. Biol Zentralbl 108: 445–451

Obe G, Vasudev V, Johannes C (1987) Chromosome aberrations induced by restriction endonucleases. In: Obe G, Basler A (eds) Cytogenetics, basic and applied aspects. Springer, Berlin Heidelberg New York, pp 300–314

Obe G, Besoke R, Schmidt S (1989) Receptor-mediated uptake of AluI by Chinese hamster ovary cells: evidence from experiments with various treatment times and treatment temperatures, and with chloroquine, methylamine and monensin. Biol Zentralbl 108: 439–444

Vasudev V, Obe G (1987) Evidence for a receptor-mediated endocytosis of AluI in Chinese hamster ovary cells. Mutat Res 197: 109–116

Winegar RA, Philips JW, Youngblom JH, Morgan W (1989) Cell electroporation is a highly efficient method for introducing restriction endonucleases into cells. Mutat Res 225: 49–53

Patterns of Chromosome Variation in Neoplasia

F. MITELMAN[1]

Contents

1 Introduction

One hundred years ago, von Hansemann (1890) first drew attention to the mitotic irregularities that characterize malignant cells, and suggested that this phenomenon could be used as a criterion for diagnosing the malignant state. A quarter of a century later, Boveri (1914) proposed that malignancy might in fact result from this disturbance of the normal chromosomal balance, essential for normal cell function. This very attractive idea is still today the basic question in cancer cytogenetics. It is now established beyond doubt that most neoplasms, benign and malignant, in both experimental animals and man, have karyotypic changes detectable with existing cytogenetic techniques. The unresolved question is: Are these abnormalities the cause or the consequence of the malignant state?

Immense amounts of data about neoplasia-associated chromosome aberrations have been collected during the last two decades, and karyotypic abnormalities have now been described in more than 12000 human neoplasms analyzed by means of chromosome banding (Mitelman 1988). The most important fact to emerge from recent cytogenetic studies is the realization that the aberrations are strictly nonrandom, i.e., different chromosomes and chromosome regions are preferentially involved in different neoplasms. Even more important, consistent and specific karyotypic changes have been described in an increasing number of tumor types, thus providing convincing evidence for the fundamental role of chromosomal rearrangements in the carcinogenic process. The considerable aberration pattern specificity strongly argues that the underlying DNA changes are important steps in tumorigenesis. The precise characterization of the various breakpoints of acquired chromosome anomalies is therefore a crucial step in the identification of those genomic sites that harbor genes essential in the pathogenesis of neoplasia. So far, two functionally different classes of cancer-relevant genes have been detected: the dominant onco-

[1] Department of Clinical Genetics, University Hospital, S-221 85 Lund, Sweden

genes and their recessive, tumor-suppressing counterparts, the antioncogenes. Both gene classes are located at neoplasia-associated chromosomal breakpoints (Mitelman and Heim 1988) and can be activated or inactivated by acquired chromosomal aberrations.

In spite of the increased knowledge obtained during the last few years, it is important to bear in mind that the information presently available is still in many respects incomplete. First, the data is heavily biased in favor of hematologic disorders. Solid tumors comprise only 20% of the total data base, which is totally disproportionate to the relative contribution of these disorders to human cancer morbidity and mortality. Second, our knowledge about early tumor stages is very limited. For example, the great majority of the solid tumors that have been studied cytogenetically have been metastatic lesions or effusions, i.e., highly advanced tumors with usually a large number of complex structural and numerical chromosome aberrations acquired as part of the clonal evolution during tumor progression. Obviously, many more such neoplasms will have to be studied before the primary, pathogenetically essential changes can be distinguished from the confusing variety of secondary abnormalities that may dominate the karyotype.

This review attempts to summarize the most important, recurrent, chromosomal rearrangements presently known in human neoplasia. More extensive discussion of the significance of chromosome aberrations in neoplasia may be found in several recent reviews (e.g., Bloomfield and de la Chapelle 1987; Croce et al. 1987; Heim and Mitelman 1987, 1989; Rowley 1988a, b; Sandberg et al. 1988; Atkin 1989, Greaves 1989; Nowell 1989; Teyssier 1989; Mitelman and Heim 1990a, b) and detailed references to each particular abnormality are to be found in Mitelman (1988) and Trent et al. (1989).

2 Myeloproliferative Disorders

The bulk of tumor cytogenetic data available at present has been gathered from the myeloproliferative disorders. Karyotypic abnormalities have now been reported in about 6300 cases, which constitutes 53% of the entire tumor cytogenetics data base.

Acute Nonlymphocytic Leukemia (ANLL)

Clonal chromosome abnormalities are found in two-thirds of all cases at diagnosis, and more than 3000 patients with chromosome changes have been described in the literature. More than 50 karyotypic abnormalities, the majority structural, have now been identified as primary changes in ANLL. Several of these abnormalities are associated with particular morphological ANLL subtypes as defined by the French-American-British (FAB) classification (Bennett et al. 1985).

The best characterized primary structural abnormalities are presented in Table 1. No less than 80% of cytogenetically abnormal ANLL patients have one of these rearrangements. Almost 20% of all ANLL patients with identifiable abnormalities have a simple numerical aberration as the only cytogenetic abnormality. The distribution of these changes is clearly nonrandom, although the high degree of morpho-

Table 1. Primary structural chromosome aberrations in ANLL

Type of rearrangement	Hematologic characteristics [a]
t(1;3)(p36;q21)	M1 and M4 with dysmegakaryocytopoiesis
t(1;7)(p11;q11)	M4
t(1;11)(q21;q23)	M4 and M5
t(2;3)(p13-22;q26-29)	M2
t(2;11)(p21;q23)	M4 and M5
ins(3;3)(q26;q21q26)	Abnormal megakaryocytes and thrombocytosis
inv(3)(q21q26)	
t(3;3)(q21;q26)	
t(3;5)(q21;q31)	M2, M4 and M6
t(4;11)(q21;q23)	M4 or mixed phenotype
t(6;9)(p23;q34)	M2 and M4 with basophilia
t(6;11)(q27;q23)	M5, mostly M5a
t(7;11)(p15;p15)	M2 and M4
del(7q)	M1, M2 and M4
t(8;16)(p11;p13)	M5 with phagocytosis
t(8;21)(q22;q22)	M2 with Auer rods and eosinophilia
t(9;11)(p21-22;q23)	M5, mostly M5a
t(9;22)(q34;q11)	M1 and M2
t(10;11)(p11-15;q23)	M4 and M5
t(11;17)(q23;q25)	M4 and M5
del/t(11)(q13-23)	M4 and M5
t(11;19)(q23;p13)	M4 and M5
del/t(12p)	M2 and M4
t(15;17)(q22;q11-12)	M3
inv(16)(p13q22)	M4 with eosinophilia
t(16;16)(p13;q22)	
del(16)(q22)	
i(17q)	M1-M2
del(20q)	M6

[a] M1–M6 refer to the FAB classification of ANLL: M1, undifferentiated acute myeloid leukemia; M2, well-differentiated acute myeloid leukemia; M3, acute promyelocytic leukemia; M4, acute myelomonocytic leukemia; M5, acute monocytic leukemia; M6, acute erythroleukemia.

logic-cytogenetic specificity seen between some structural abnormalities and hematological subtypes is never achieved. Thus, none of the numerical rearrangements is restricted to any particular FAB subgroup. The most common primary numerical rearrangements are +4, −5, −7, +8, +21, and −Y.

Myelodysplastic Syndromes (MDS)

This group comprises several hematopoetic dysfunction states that carry an increased risk for the ultimate development of acute nonlymphocytic leukemia. The FAB study group (Bennett et al. 1982) recognizes five different MDS subtypes: refractory anemia, refractory anemia with ringed sideroblasts, chronic myelomonocytic leukemia, refractory anemia with excess of blasts, and refractory anemia with excess of blasts in transformation.

Clonal chromosome abnormalities have been reported in about 1200 MDS patients. The overall frequency of aberrations in unselected series has in most large studies been one-third to one-half of all successfully karyotyped cases. The fact that karyotypic aberrations may be found in MDS strongly supports the presently held view that these syndromes do in fact represent truly neoplastic disorders. Overall, the karyotypic anomalies in MDS resemble ANLL changes, thus underscoring the basic biological similarity between the two modes of disease presentation. On the other hand, some of the most consistent ANLL rearrangements – t(8;21), t(15;17), and inv(16) – are not seen in MDS, indicating that patients with these abnormalities pass through a very brief, if any, preleukemic phase. The reverse situation, changes frequently occurring in MDS, but not in ANLL, has not been reported. About 80% of cytogenetically abnormal MDS patients have one of the following changes: t(1;3)(p36;q21), t(1;7)(p11;q11), inv(3)(q21q26), del(5q), t(6;9)(p23;q34), −7, +8, t/del(11q), t/del(12p), del(13q), i(17q), del(20q), and −Y.

Chronic Myeloid Leukemia (CML)

Bone marrow cells from approximately 90% of patients with clinically typical CML contain the Philadelphia (Ph^1) chromosome, which originates through the reciprocal translocation t(9;22)(q34;q11). This rearrangement is usually the only karyotypic abnormality during chronic phase CML. At the time of blastic crisis, 75–80% of all patients have acquired additional chromosome aberrations. These secondary aberrations are also nonrandom: at least one of four changes – +8, i(17q), +19, and an extra Ph^1 marker – occurs in 80% of all cases.

In a subgroup of Ph^1 positive patients, the Ph^1 does not result from the usual 9;22 translocation. These variant translocations appear to be of two kinds: simple translocations in which the deleted segment of chromosome 22 has been translocated onto a chromosome other than No. 9, and complex translocations involving one or more chromosomes besides 9 and 22. Investigations utilizing the recombinant DNA technique have now demonstrated that all variant translocations are in fact complex and involve interchanges between chromosomes 9 and 22 (Dreazen et al. 1988; Kurzrock et al. 1988).

About 5% of CML patients have Ph^1-negative disease. These patients may be cytogenetically normal or may have other changes, mostly trisomy 8 and i(17q). Some Ph^1-negative cases have now been shown to contain submicroscopic rearrangements that involve 9q34 and 22q11 and are pathogenetically equivalent to the visible t(9;22).

Polycythemia Vera (PV)

The frequency of clonal abnormalities in PV can be estimated at about 15% at diagnosis and somewhat higher, perhaps 40%, among treated patients in later disease phases, thus giving an overall estimate of 20%. Chromosome changes identified by banding techniques have been reported in about 200 PV patients. The five most common aberrations are, in decreasing order of frequency: del(20q), +8, +9, del(13q), and structural abnormalities of chromosome 1 resulting in partial trisomy

1q. Two-thirds of all cytogenetically abnormal cases have at least one of these aberrations.

Idiopathic Myelofibrosis (IMF)

The low number of banding studies reported in IMF reflects the technical difficulties in obtaining good cytogenetic preparations from fibrotic bone marrow. The overall frequency of abnormal clones is difficult to estimate and varies considerably in different series, ranging from 20 to 70%. Chromosome aberrations have been characterized by banding techniques in about 150 patients. The most characteristic abnormalities have been structural rearrangements of the long arm of chromosome 1, −7, +8, +9, del(13q), and del(20q). About 60% of cytogenetically abnormal patients have one or more of these aberrations.

3 Lymphoproliferative Disorders

Karyotypic abnormalities have been reported in about 3300 cases, which constitutes 27% of all neoplasms studied by chromosome banding. The most prominent aberrations in this tumor group are presented in Table 2.

Acute Lymphocytic Leukemia (ALL)

About two-thirds of patients with ALL have clonal chromosome abnormalities in their bone marrow cells at diagnosis, and karyotypic abnormalities have now been reported in roughly 1200 cases. As in ANLL, the karyotypic changes are distributed in a markedly nonrandom manner. More than 40 consistently occurring rearrangements are presently known, all of which are prime candidates for a crucial role in leukemogenesis. These aberrations are present in two-thirds of all cytogenetically abnormal ALL cases.

The most characteristic abnormalities (Table 2) associated with particular morphological and immunological ALL subtypes are: t(1;19)(q23;p13), t(2;8)(p12;q24), t(4;11)(q24;q23), del(6q), t(8;14)(q24;q32), t(8;22)(q24;q11), t/del(9p), t(9;22)(q34;q11), t/del(12p), and abnormalities of 14q11.

Chronic Lymphocytic Leukemia (CLL)

Chromosome abnormalities have been reported in about 300 patients. The most common aberration in B-CLL is an extra chromosome 12, found in one-third of all cases. The second most common aberration is a 14q+ marker chromosome, present in one-fourth of cytogenetically abnormal patients. The exact mode of origin of the 14q+ marker has been established in only half of the cases; the most commonly identified translocation, giving rise to a 14q+, is t(11;14)(q13;q32). The most consistent chromosome change in T-CLL is an inversion of the long arm of chromosome

Table 2. Primary chromosome aberrations in lymphoproliferative disorders

Type of rearrangement	Disease type [a]
t(1;11)(p23;q23)	Pre-B-ALL
t(1;14)(q21-25;q32)	B-NHL
t(1;19)(q23;p13)	pre-B-ALL
t(2;5)(p23;q35)	T-NHL
t(2;8)(p12;q24)	BL, B-ALL
+3	ATL, T-NHL
t(4;11)(q21;q23)	Early B-precursor ALL
t/del(6)(p21-23)	T-NHL
del(6q)	ALL, ATL, HCL, NHL, PLL
i(8q)	T-CLL, T-PLL
t(8;14)(q24;q32)	BL, B-ALL
t(8;14)(q24;q11)	T-ALL
t(8;22)(q24;q11)	BL, B-ALL
del(9)(p21)	T-ALL
t(9;12)(p11-12;p12)	ALL
t(9;22)(q34;q11)	pre-B or early B-precursor ALL
t(10;14)(q24;q11)	T-ALL
t(11;14)(p13;q11)	T-ALL
t(11;14)(q13;q32)	B-CLL, B-NHL, B-PLL, PCL
t(11;19)(q23;p13)	NHL
+12	B-CLL
inv(14)(q11q32)	ATL, T-CLL, T-PLL
del(14)(q22-24)	B-CLL
t(14;17)(q32;q23)	B-CLL
t(14;18)(q32;q21)	B-NHL
t(14;19)(q32;q13)	B-CLL
t(14;22)(q32;q11)	B-ALL
14q+	ALL, ATL, HCL, NHL, PCL, PLL

[a] ALL, acute lymphocytic leukemia; ATL, adult T-cell leukemia/lymphoma; BL, Burkitt's lymphoma; CLL, chronic lymphocytic leukemia; HCL, hairy cell leukemia; NHL, non-Hodgkin's lymphoma; PCL, plasma cell leukemia; PLL, prolymphocytic leukemia.

14, inv(14)(q11q32), which is found in one-third of all cytogenetically abnormal patients.

Adult T-Cell Leukemia/Lymphoma (ATL)

The karyotypic changes are often complex and variable. Structural changes involving 14q11 and 14q32 appear to be the most characteristic anomalies, and the single most common change is actually a recombination between 14q11 and 14q32.

Prolymphocytic Leukemia (PLL)

The most common abnormality in B-cell PLL is a 14q+ marker with a breakpoint in band 14q32. In the majority of cases the extra material has not been identified. The most frequent abnormality in T-cell PLL, observed in 50% of the cases, is an inv(14)(q11q32) or other rearrangements involving 14q11.

Multiple Myeloma (MM)

The incidence of abnormal karyotypes varies greatly among published series, from 20 to 60%. The most consistent abnormality is a 14q+ marker, present in one-third of the roughly 100 cytogenetically abnormal patients so far studied. Most commonly the 14q+ originates through a t(11;14)(q13;q32), i.e., apparently the same translocation that is frequently found in B-CLL.

Hodgkin's Disease (HD)

Only about 40 cases of HD with clonal karyotypic abnormalities have been reported. No consistent aberration pattern has been detected.

Burkitt's Lymphoma (BL)

The characteristic aberration in BL is t(8;14)(q24;q32), present in 75–85% of all patients. In the remaining tumors one of two variant translocations, t(2;8)(p12;q24) or t(8;22)(q24;q11), is present. Only a few cases without these typical translocations have been reported.

Non-Hodgkin's, Non-Burkitt's Lymphomas (NHL)

More than 1200 cytogenetically abnormal cases of NHL have been reported. Although none of the aberrations characteristic of NHL are associated with histopathologic subtypes with the same consistency as the three translocations in BL, several nonrandom associations (Table 2) have been detected. The following aberrations represent the best candidates for primary rearrangements: t(2;5)(p23;q35), +3, t/del(6p), del(6q), t(11;14)(q13;q32), and t(14;18)(q32;q21).

4 Solid Tumors

The acquisition of reliable knowledge about the essential genomic rearrangements in solid tumors has been hampered by several problems: First, few tumors have been investigated. Chromosome aberrations have been identified in less than 2500 solid tumors or only 20% of all neoplasms so far studied with banding techniques. Secondly, most investigations have been made on highly advanced tumors, often metastatic lesions or effusions. In these analyses many additional abnormalities acquired during tumor progression have contributed to the frequently bizarre karyotypic complexity encountered. However, often the chromosomal anomalies are highly complex even when the analysis is performed on biopsies taken at diagnosis. Distinguishing between primary, pathogenetically essential changes and secondary aberrations is therefore as a rule even more difficult in solid tumors than in leukemias. Finally, metaphases obtained from solid tumors, in particular from epithelial neoplasms, are sometimes of such low technical quality that the true nature of the chromosomal markers cannot be determined with certainty.

All these difficulties notwithstanding a growing number of consistent cytogenetic abnormalities are now being revealed in both benign and malignant tumors. Some of the associations are as specific as the various cytogenetic-morphologic correlations known from hematologic malignancies; the best characterized primary abnormalities are presented below and summarized in Table 3.

Meningioma

Meningiomas are the most extensively investigated benign tumors in man. Data on cytogenetic abnormalities in more than 200 tumors are now available, and the vast majority of cases are characterized by the loss of one chromosome 22. Occasionally, partial monosomy in the form of del(22q) is found.

Table 3. Primary chromosome aberrations in solid tumors

Type of rearrangement	Disease type
del(1)(p32-36)	Neuroblastoma
t(1;6)(q11-12;q15-21)	Malignant melanoma
t(2;13)(q35-37;q14)	Rhabdomyosarcoma
del(3)(p14-23)	Lung carcinoma
	Kidney carcinoma
t(3;8)(p21;q12)	Pleomorphic adenoma
t(3;12)(q27-28;q13-14)	Lipoma
i(5p)	Bladder carcinoma
del(6)(q22-25)	Salivary gland carcinoma
del(7)(q21q31)	Leiomyoma
del(9)(p13-24)	Astrocytoma
t(9;12)(p13-22;q13-15)	Pleomorphic adenoma
del(11)(p13)	Wilms' tumor
t(11;22)(q24;q12)	Ewing's sarcoma
	Askin tumor
	Neuroepithelioma
+12	Leiomyoma
i(12p)	Germ cell tumors
t(12)(q13-14)	Lipoma
t(12)(q13-15)	Pleomorphic adenoma
t(12;16)(q13;p11)	Liposarcoma
t(12;14)(q14-15;q23-24)	Leiomyoma
del(13)(q12-13)	Lipoma
del(13)(q14)	Retinoblastoma
t(19)(p13)	Ovarian carcinoma
	Malignant fibrous histiocytoma
−22/del(22)(q12-13)	Meningioma
t(X;18)(p11;q11)	Synovial sarcoma

Pleomorphic Adenoma

The only benign epithelial neoplasm studied in a sufficient number to permit reasonably well-founded conclusions is the mixed tumor of the salivary gland (pleomorphic adenoma). Information is available on more than 100 tumors. The chromosomal involvement is strikingly nonrandom: three cytogenetic subgroups may be distinguished, namely, tumors with aberrations involving 3p21, 8q12, and 12q13–15. The two first breakpoints are often recombined in a t(3;8)(p21;q12).

Lipoma

Chromosome abnormalities have been reported in about 80 lipomas. Three major cytogenetic subgroups can be identified: (1) reciprocal, apparently balanced translocations between 12q13–14 and various other chromosomes, most frequently as t(3;12)(q27–28;q13–14); (2) one or more supernumerary ring chromosomes in addition to an otherwise apparently normal karyotype; and (3) other aberrations than ring chromosomes or rearrangements of 12q13–14.

Uterine Leiomyoma

Clonal chromosome abnormalities may be detected in approximately 20% of benign uterine leiomyomas. Abnormal karyotypes have been identified in about 80 tumors, and the reciprocal translocation t(12;14)(q14–15;q23–24) has been found as the sole aberration in a substantial proportion of these tumors, and hence obviously represents a primary cytogenetic change. Other cytogenetic subtypes include del(7q) and trisomy 12.

Carcinoma of the Breast

Chromosome abnormalities have been identified in about 100 tumors, the majority studied as cell lines or effusions. Because of the extreme numerical and structural complexity encountered, the number of cases yet studied is clearly insufficient for any reliable evaluation of the pathogenetic significance of the various changes. Some information about nonrandom structural rearrangements can nevertheless be gained from the confusing picture. Chromosome 1 is preferentially involved; in fact, almost four–fifths of all cytogenetically abnormal cases have recognizable structural changes of this chromosome. The majority are various translocations affecting 1q, most commonly the region 1q12–23. About 25% are rearrangements of 1p, usually deletions or translocations of 1p11–21. Other chromosomes nonrandomly involved in structural aberrations are 3, 13, 16, and 17.

Carcinoma of the Uterus

A majority of the approximately 100 cytogenetically characterized uterine carcinomas have been squamous cell carcinomas of the cervix; only 20% have been endometrial adenocarcinomas. The chromosomal picture in both tumor types appears to be extremely variable and complex, allowing complete karyotyping in only a minority of cases.

No single cytogenetic abnormality has been consistently associated with either tumor type. Chromosome 1 is involved in a nonrandom manner, and has been affected in 80% of all cytogenetically abnormal cases. The aberrations of chromosome 1 comprise both numerical and all kinds of structural rearrangements, without obvious breakpoint clustering.

Carcinoma of the Ovary

Chromosome 1 abnormalities appear to be the most common changes and have been noted in over three-fourths of the about 130 successfully studied cases. The aberrations include deletions, duplications, translocations, and inversions of both the short and the long arms, as well as formation of 1q isochromosomes. The breakpoints have mostly been mapped to 1p36 and the distal half of the long arm. Other chromosomes commonly involved in structural abnormalities are Nos. 3, 6, 11, 14, and 19. An apparently identical 19p+ marker chromosome has recently been reported in a subset of ovarian seropapillary cystadenocarcinomas.

Carcinoma of the Salivary Glands

At least one-third of the approximately 30 malignant salivary gland tumors that have been characterized cytogenetically show clonal rearrangements involving the long arm of chromosome 6, mostly terminal deletions with breakpoints in 6q22-25. The 6q abnormalities have been seen in several histologic subtypes, e.g., adenoid cystic carcinomas, acinic cell tumors, mucoepidermoid tumors, and epidermoid carcinomas.

Carcinoma of the Lung

Chromosome abnormalities have been identified in about 100 cases. Of the four major histologic subtypes, only small cell lung carcinomas have been cytogenetically investigated in any numbers. More than 90% of these tumors have a deletion of the short arm of chromosome 3, with loss of 3p14-23 as the least common denominator. The 3p- markers are not confined to the small cell subtype, however, as they have also been found in other bronchogenic carcinomas, including adenocarcinoma. More data are obviously needed before the prevalence of 3p- in lung cancers of various histologices can be established.

Carcinoma of the Large Bowel

Adenocarcinomas of the colon and rectum are the most extensively studied malignant epithelial neoplasms in man. Chromosome abnormalities have been identified in more than 150 tumors. In spite of the large number of cases studied, the cytogenetic results have been extremely hard to interpret. This is due mainly to the often low technical quality of the preparations, but also to the complexity of the changes. Half of all tumors have stemline karyotypes in the hypotriploid-hypotetraploid region, and most have a number of unidentified marker chromosomes. At least four cytogenetic subgroups may be distinguished: trisomy 7, and structural changes of chromosomes 1, 17, and 18.

Carcinoma of the Kidney

Structural rearrangements of the short arm of chromosome 3, with breakpoints mapping to 3p11–21, are the changes most consistently seen among the 100 cases with aberrations so far reported. The one feature common to all these 3p aberrations appears to be the net loss of chromosomal material: mostly the changes have been interstitial or terminal deletions, on other occasions reciprocal translocations between 3p and other chromosomes but with concomitant loss of variable 3p segments.

Wilms' Tumor

Wilms' tumor (nephroblastoma) is caused by the deletion or inactivation of a tumor suppressor gene located in 11p13 (Knudson 1989). Cytogenetic studies of tumor tissue have been reported in more than 50 patients with nonhereditary Wilms' tumor. The most common visible changes, found in half of all cases, are various structural rearrangements of chromosome 1, in particular of the long arm. Acquired deletions of 11p13 or translocations affecting the same band are found in slightly more than 25%. Evidence is now mounting that even those tumors in which no del(11)(p13) is cytogenetically detectable may nevertheless have equivalent submicroscopic changes.

Carcinoma of the Bladder

Less than 100 tumors have been studied. The most common anomalies are structural changes of chromosomes 1 and 11, each present in about one–third of all cases. The rearrangements of chromosome 1 include deletions and duplications of both the long and short arms, as well as a variety of translocations with breakpoints in most cases localized to the central segments of the long arm. The chromosome 11 abnormalities include isochromosome formation for both 11p and 11q, translocations with a diversity of breakpoints on both arms, and deletions, usually del(11p). The abnormalities affecting both chromosomes are thus highly variable, and possibly represent secondary, nonspecific changes. In contrast, three frequently recurring abnormalities – an isochromosome for the short arm of chromosome 5, trisomy 7, and monosomy 9 –

have all been repeatedly described as sole anomalies, and are therefore credible candidates for a primary role in transitional cell carcinogenesis.

Carcinoma of the Skin

Although very few skin cancers have been cytogenetically investigated, recent studies of both preinvasive and infiltrating squamous cell carcinomas have produced interesting and unexpected results. In most tumors with abnormalities, a number of clones have been found, and often the clones have been cytogenetically unrelated. Admittedly the number of cases so far studied are few, but the remarkably diverse karyotypic pictures hint that the tumors may be poly- rather than monoclonal (Heim et al. 1988). Additional cytogenetic evidence indicates similar polyclonality in cancers of the pharynx and larynx. Hence, the combined data seem to point at a fundamental pathogenetic difference between these epithelial tumors and most hematologic and mesenchymal neoplasms, where the karyotypic findings strongly support a monoclonal origin.

Malignant Fibrous Histiocytoma

Although malignant fibrous histiocytoma is the most common malignant soft tissue tumor, cytogenetic abnormalities in have been reported less than 30 cases. The complexity of most karyotypes makes it a difficult task to identify primary and secondary abnormalities. The breakpoint distribution of the various rearrangements is, however, clearly nonrandom, with preferential involvement of chromosomes 1, 2, 3, 7, 11, and 19, in particular bands 1q11, 1p36, 3p12, 11p11, and 19p13. Of particular clinical interest may be the observation that tumors with 19p13 abnormalities seem to have a pronounced tendency to recur locally (Mandahl et al. 1989).

Liposarcoma

Chromosome aberrations have been described in approximately 20 cases. A very characteristic and apparently highly specific chromosomal rearrangement, t(12;16)(q13;p11), characterizes myxoid liposarcomas. In the few malignant lipogenic tumors of other histologies reported, no systematic involvement of chromosomes 12 and 16 has been identified.

Synovial Sarcoma

The translocation t(X;18)(p11;q11) has been found in all cases of synovial sarcoma so far studied. Since less than 20 tumors have been investigated, it is of course too early to know both how frequent and how specific this rearrangement will prove to be. It is noteworthy that t(X;18) is the first cancer-associated rearrangement to involve one of the sex chromosomes.

Rhabdomyosarcoma

The reciprocal translocation t(2;13)(q35–37;q14) has been described in more than 40 rhabdomyosarcomas. Although the majority of these tumors were of the alveolar type, the translocation has also been found in cases classified as undifferentiated and embryonal. Since the t(2;13) has so far not been reported in any other tumor type, it may well turn out to be pathognomonic for rhabdomyosarcomas.

Ewing's Sarcoma

A characteristic chromosomal abnormality, t(11;22)(q24;q12), is a consistent finding in Ewing's sarcoma. An apparently identical reciprocal translocation has also been described in neuroepithelioma and in the Askin tumor, both of which may consequently be closely pathogenetically related to Ewing's sarcoma.

Malignant Glioma

No specific karyotypic abnormalities have been detected in the 150 malignant gliomas so far investigated. The most frequent findings, present in half of all tumors, have been double minutes (dmin). In addition, deletions and translocations of 1p, 7q, 9q and 19q, trisomy 7, and loss of chromosomes 10 and 22 are often seen.

Neuroblastoma

More than 100 neuroblastomas have been investigated cytogenetically. The predominant abnormality are deletions and other rearrangements of the terminal portion of 1p. Although the breakpoints have varied in different reports, the segment from 1p31 to 1pter seems to have been deleted in the great majority of cases. Using a panel of DNA probes derived from a library of microdissected distal 1p fragments, Weith et al. (1989) very recently located a consensus deletion to 1p36. In addition to the 1p abnormalities, a significant fraction, perhaps two-thirds of all investigated tumors, display homogeneously staining regions (HSR) and/or dmin.

Retinoblastoma

The essential genomic change in retinoblastoma (RB) is the inactivation of the *RB1* locus in 13q14 (Knudson 1989). The most consistent acquired cytogenetic abnormality, found in one-third of the 80 cases with identified aberrations, is an isochromosome for the short arm of chromosome 6. Less consistent but even more frequent (50%) are various structural rearrangements of chromosome 1, mostly leading to a net increase of 1q material. It should be noted that monosomy 13 and 13q deletions are discernible in less than 20% of all tumors. Apparently, visible chromosome deletion is a relatively minor mechanism for the production of homozygous inactivation of the RB locus.

Germ Cell Tumors

Chromosome abnormalities have been reported in more than 60 germ cell tumors, the great majority of the testis. An isochromosome for the short arm of chromosome 12, i(12p), has consistently been found in various histologic types, including seminomas, teratomas, embryonal cell carcinomas, as well as tumors of mixed histology.

5 Conclusions

Clonal chromosome abnormalities are a characteristic feature of both benign and malignant neoplasms. Apart from being clinically useful as a diagnostic and prognostic tool, tumor cytogenetics also plays an important role in identifying those genomic sites which harbor genes essential in the pathogenesis of neoplasia. It is important to bear in mind, however, that not all chromosomal anomalies found in neoplastic cells are of pathogenetic importance. Numerous examples exist of neoplasms, in particular solid tumors, that display seemingly unique, usually complex structural and/or numerical karyotypic changes. This complexity of the cytogenetic abnormalities has in the past unquestionably added to the confusion regarding their significance in tumorigenesis. Greater clarity may be obtained if it is kept in mind that any chromosome aberration in a tumor cell population may in principle be referred to one of three categories: (1) Primary abnormalities, which may occur as solitary cytogenetic changes already in the earliest disease phase. These are essential in establishing the neoplasm, and are as a rule strongly correlated with tumor type. (2) Secondary abnormalities, which develop only after the tumor has been established, and reflect the clonal evolution during tumor progression. These aberrations may, or may not, be important in tumor progression; and (3) cytogenetic noise, which is the background level of nonconsequential aberrations, detectable as nonclonal aberrations. These changes are, in contrast to the primary and secondary changes, randomly distributed throughout the genome. When the chromosomal instability in a tumor cell population is pronounced, such noise abnormalities may completely dominate the karyotype and obscure the pathogenetically important changes. Obviously, a most important task in cancer cytogenetics is to determine which of the confusing variety of karyotypic changes present in neoplastic cells are of decisive importance in tumor development, and which are secondary abnormalities important in tumor progression. Only then will the true significance of chromosomal changes in neoplasia be fully appreciated.

Acknowledgements. Original work described in this chapter was supported by grants from the Swedish Cancer Society and the Children's Cancer Fund of Sweden.

References

Atkin NB (1989) Solid tumor cytogenetics. Progress since 1979. Cancer Genet Cytogenet 40: 3–12

Bennett JM, Catovsky D, Daniel MT, Flandrin G, Galton DAG, Gralnick HR, Sultan C (1982) Proposals for the classification of the myelodysplastic syndromes. Br J Haematol 51: 189–199

Bennett JM, Catovsky D, Daniel MT, Flandrin G, Galton DAG, Gralnick HR, Sultan C (1985) Proposed revised criteria for the classification of acute myeloid leukemia. Ann Intern Med 103: 626–629

Bloomfield CD, Chapelle A de la (1987) Chromosome abnormalities in acute nonlymphocytic leukemia: clinical and biological significance. Semin Oncol 14: 372–383
Boveri T (1914) Zur Frage der Entstehung maligner Tumoren. Fischer, Jena
Croce CM, Erikson J, Tsujimoto Y, Nowell PC (1987) Molecular basis of human B- and T-cell neoplasia. Adv Viral Oncol 7: 35–51
Dreazen O, Canaani E, Gale RP (1988) Molecular biology of chronic myelogenous leukemia. Semin Hematol 25: 35–49
Greaves MF (1989) The contribution of molecular genetics to the study of leukemia. Cancer Genet Cytogenet 40: 217–230
Hansemann D von (1890) Ueber asymmetrische Zelltheilung in Epithelkrebsen und deren biologische Bedeutung. Virchows Arch A Pathol Anat Histol 19: 299–326
Heim S, Mitelman F (1987) Cancer cytogenetics. Liss, New York
Heim S, Mitelman F (1989) Primary chromosome abnormalities in human neoplasia. In: Weinhouse S, Klein G (eds) Advances in cancer research, vol 52. Academic Press, London, pp 1–43
Heim S, Mandahl N, Mitelman F (1988) Genetic convergence and divergence in tumor progression. Cancer Res 48: 5911–5916
Knudson AG (1989) Hereditary cancers disclose a class of cancer genes. Cancer 63: 1888–1891
Kurzrock T, Gutterman JU, Talpaz M (1988) The molecular genetics of Philadelphia chromosome-positive leukemias. N Engl J Med 319: 990–998
Mandahl N, Heim S, Willén H et al. (1989) Characteristic karyotypic anomalies identify subtypes of malignant fibrous histiocytoma. Genes Chrom Cancer 1: 9–14
Mitelman F (1988) Catalog of chromosome aberrations in cancer, 3rd edn. Liss, NewYork
Mitelman F, Heim S (1988) Consistent involvement of only 71 of the 329 chromosomal bands of the human genome in primary neoplasia-associated rearrangements. Cancer Res 48: 7115–7119
Mitelman F, Heim S (1990a) Chromosome abnormalities in cancer. Cancer Detect Prev (in press)
Mitelman F, Heim S (1990b) Nonrandom chromosome changes in human cancer and leukemia. Chromosomes Today (in press)
Nowell PC (1989) Chromosomal and molecular clues to tumor progression. Semin Oncol 16: 116–127
Rowley JD (1988a) Chromosome abnormalities in leukemia. J Clin Oncol 6: 194–202
Rowley JD (1988b) Chromosome studies in the non-Hodgkin's lymphomas: the role of the 14;18 translocation. J Clin Oncol 6: 919–925
Sandberg AA, Turc-Carel C, Gemmill RM (1988) Chromosomes in solid tumors and beyond. Cancer Res 48: 1049–1059
Teyssier JR (1989) The chromosomal analysis of human solid tumors. A triple challenge. Cancer Genet Cytogenet 37: 103–125
Trent JM, Kaneko Y, Mitelman F (1989) Report of the committee on structural chromosome changes in neoplasia. Human Gene Mapping 10 (1989) Cytogenet Cell Genet 51: 533–562
Weith A, Martinsson T, Cziepluch C, Brüderlein S, Amler LC, Berthold F, Schwab M (1989) Neuroblastoma consensus deletion maps to 1p36.1–2. Genes Chrom Cancer 1: 159–166

Tumorigenesis and Tumor Response: View from the (Prematurely Condensed) Chromosome

W. N. HITTELMAN, N. CHEONG, H. Y. SOHN, J. S. LEE, J.-D. TIGAUD, and S. VADHAN-RAJ[1]

Contents

1 Introduction

Chromosome changes have long been known to be associated with the process of tumor development (Boveri 1914). For example, individuals with chromosome breakage syndromes such as ataxia telangiectasia and xeroderma pigmentosum or constitutional chromosomal abnormalities are found to be at high risk for the development of malignancy (Lejeune et al. 1959; Ray and German 1981). Similarly, individuals who have increased exposure to clastogenic agents (e.g., cigarette smoke, radiation emitters, chemicals) are also at higher risk for developing tumors, especially in the tissue fields of high exposure (e.g., aerodigestive tract) (Auerbach et al. 1961; Vaughan et al. 1980). In some cases, it has also been found that individuals can exhibit an inherent propensity for developing a particular type of tumor, perhaps resulting from a combination of inherent sensitivity, specific fragile sites, and carcinogen exposure. Since much of this damage to the genome is translated into alterations in chromosome structure and number, most tumors examined so far exhibit chromosome changes (Heim and Mitelman 1987).

The specific changes found in tumors often have been found to have biological as well as clinical implications. For example, the number of types of chromosome changes in human leukemia are usually limited, and the specific changes are often related to the type of disease and have prognostic significance (LeBeau and Rowley 1986). The molecular basis for the relationship between specific chromosome changes and biology of disease is not well understood, however, the sites of changes in the

[1] Departments of Medical Oncology and Clinical Immunology and Biological Therapy, The University of Texas M.D. Anderson Cancer Center, 1515 Holcombe Blvd., Houston, Tx 77030, USA

chromosomes should lead investigators to focus on the genes residing in these specific chromosome regions. Some of these considerations are discussed in Chapter 10.

The pattern of chromosome changes found in tumors might also reflect some characteristic of tumor development. For example, many solid tumors exhibit multiple and complex chromosome changes (Heim and Mitelman 1987). It is now becoming apparent that in some tissues, tumor development is a multistep process (Knudson 1985). For example, some specific chromosome changes are observed in premalignant conditions prior to any evidence for tumor development, while other chromosome changes are preferably found in the established tumor. Thus, in colon cancer, premalignant lesions such as papillomas often exhibit a chromosome 5q abnormality while colonic tumors contain additional alterations in chromosome 17 and 18 (Vogelstein et al. 1988). Thus, it is proposed that these sequential chromosome changes can serve as chromosomal hallmarks of tumor development. However, if these tumors arise in regions of clastogenic exposure, one would suspect that many chromosome changes are occurring in the tissue at risk, and tumors might only develop in those cells accumulating specific chromosome changes.

Once a malignant condition has become established, it is of interest to analyze the response of the tumor cells to therapeutic intervention. For example, one of the predominant phenotypes of malignant cells is the inability to differentiate into a mature functioning cell. In human leukemia, the abnormal clone apparently is unable to differentiate into mature peripheral blood cells and remains as an immature blast cell. While most types of therapy try to eliminate the abnormal cells through cytotoxic activity (i.e., inducing chromosome aberrations), alternative approaches are being attempted that would involve manipulating the leukemic cell to act normally (i.e., differentiation therapy). Thus, one would like to develop a way to examine tumor cells during treatment in order to better understand the biology of response. Such an understanding could perhaps lead to the development of new therapeutic strategies.

2 Problems Associated with Cytogenetic Evaluations of Tissues

One of the problems facing the cytogenetic approach for studying the biology of tumor formation and response to therapeutic intervention is that conventional techniques require that the cells be proliferating so that the chromosomes can be visualized at mitosis. However, many of the tissues that one would like to analyze are either, growing very slowly or not growing at all (Teyssier 1989). For example, if one would like tocytogenetically map the multistep process of tumorigenesis, one is limited in that the genetic changes are occurring in tissues that are not proliferating to a sufficient extent for direct mitotic analysis (e.g., lung). Similarly, if one were trying to determine whether a particular type of therapy were causing differentiation of leukemic cells, one would be limited in using a conventional cytogenetic approach because proliferation decreases as maturation increases.

To overcome the problems associated with conventional cytogenetic approaches to such questions, many researchers have resorted to placing tumor cells derived from tissues into short- or long-term culture in the hope of stimulating the cells of interest to proliferate. Several problems confound this approach. First, the cells derived from the tumor, premalignant condition, or mature population of cells might not grow in vitro. Second, if the cells do grow, subclones might preferentially grow yielding an altered cytogenetic view of the original tissue population. Third, tumor specimens represent heterogeneous mixtures of cells, and the normal cells might outgrow the tumor cells. Thus, the finding of diploidy in a population might be difficutl to interpret (i.e., diploid tumor cells or diploid normal cells?). Finally, new karyotypic alterations might occur in culture due to the stress involved with tissue dissociation and growth in artificial ex vivo conditions. Thus, karyotypic patterns might be observed that did not exist in the original tissue.

What is needed is a technique that can be used to identify chromosome alterations without requiring the cells to proliferate. When the specific chromosome changes to be examined are already known, then various molecular approaches can be utilized (e.g., DNA analysis of polymorphic regions (Vogelstein et al. 1989), or in situ hybridization using chromosome-specific probes (Pinkel et al. 1986)). However, if one is trying to determine the identity of the particular cytogenetic change without prior knowledge of the lesion involved, one needs an alternative method.

The present chapter describes an alternative approach to the conventional cytogenetic evaluation of the tumorigenesis process as well as the response of tumors to therapeutic intervention. Instead of examining the chromosomes of cells at mitosis, these cells are examined while still in interphase utilizing the technique of premature chromosome condensation.

3 The Technique of Premature Chromosome Condensation

Fifteen years ago, Johnson and Rao (1970) reported that when mitotic cells are fused with interphase cells, the nucleus of the interphase cell undergoes an immediate prophasing-like reaction such that the nuclear envelope is disassembled and the chromatin condenses into chromosomes (i.e., prematurely condensed chromosomes or PCC; Fig. 1A). While the mitotic factors involved in this process are still not well characterized (Dunphy and Newport 1988), their activity is found to cross species barriers (Johnson et al. 1970). Thus, cells that grow rapidly in tissue culture and are easily synchronized to mitosis can be utilized to induce PCC in a variety of other cell types. The morphology of the PCC reflects the cell cycle phase of the interphase cell at the time of fusion (Johnson and Rao 1970). For example, cells from G1 phase exhibit a single chromatid per chromosome, and cells from G2 phase exhibit two chromatids per chromosome. Cells from S-phase exhibit pulverized-appearing PCC, with a mixture of single and double chromatid fragments.

Since cells from G1 and G2 yield discrete chromosomes, these preparations can be subjected to conventional methodologies for karyotypic evaluation, including the determination of chromosome number per cell as well as evaluation of structural rearrangements using G- and C-banding (Hittelman et al. 1988b; Fig. 1B). Moreover, using a combination of immunocytochemical and cytofunctional methodologies along with the PCC technique, one can visualize the immunocytochemical or functional marker and the PCC on the same fused cell (Petkovic et al. 1989). This offers the capability to make a one-to-one correlation between a cell characteristic and karyotype.

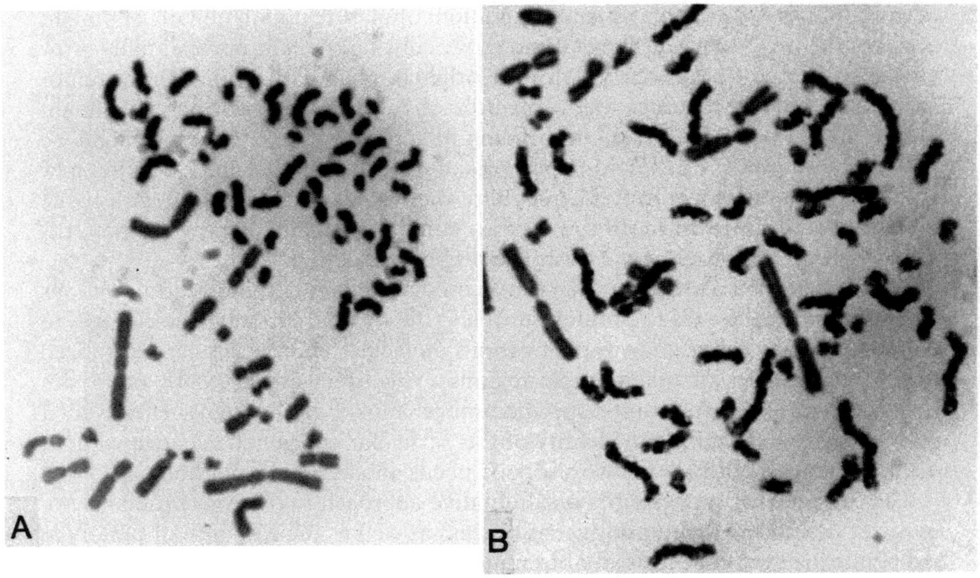

Fig. 1. A G1 PCC arising from fusion between a human interphase cell and a CHO mitotic cell. A differential staining technique makes the PCC dark and the mitotic chromosomes light. **B** G-banded G1 PCC

The technique of premature chromosome condensation therefore offers several advantages over conventional techniques for analyzing the chromosome constitution of cell populations. First, the chromosomes of the target cell population can be visualized immediately without requiring a period of cell culture. This overcomes the problem of artifacts associated with cell culture. In addition, it permits the cytogenetic analysis of cell populations which are very slowly proliferating (e.g., lung tissue) or are not proliferating at all (e.g., terminally differentiated blood cells). Third, the fusion products can be analyzed for the cell type involved in the fusion using immunocytochemical or cytofunctional methodologies. This latter property is important for the analysis of heterogeneous cell populations where any one of several cell types could be involved in cell fusion.

4 Cytogenetic Evaluations of Aerodigestive Tract Tumors and Normal Lung Tissue

4.1 "Field Cancerization" and Multistage Carcinogenesis

Two types of concepts have been utilized to describe the process of tumorigenesis in the aerodigestive tract. First, the process is thought to reflect a process of "field cancerization" whereby a whole field of tissue is exposed to carcinogens (e.g., tobacco smoke or tobacco juice) and is at increased risk for developing tumors (Slaughter et al. 1953). At the clinical level, this notion was first proposed since multiple, independent primary cancers are often found in the exposed field, occurring either at the same time or separated in the time. The notion of "field cancerization" is supported by the frequent finding of premalignant lesions in the tissue field at risk for developing a tumor. For example, leukoplakia is found to be a premalignant condition of oral cancer in both humans and animal model systems (Shklar 1986), and bronchial metaplasia/dysplasia is a premalignant condition for lung cancer.

A second concept to describe tumorigenesis in general is that of the multistage process, whereby a set of genotypic and phenotypic changes must occur prior to the formation of a malignant lesion (Farber 1984). The driving force behind this multistep process is thought to be genetic damage caused by continuing carcinogenic insult and mutagenic accumulation in the target tissue. The genetic alterations then give rise to phenotypic alterations in the tissue such as dysregulation of proliferation and differentiation. Evidence for the presence of a driving genetic force is provided by the findings of increased micronuclei frequency in the damaged and premalignant tissue (Stich 1987). Evidence for the resultant phenotypic alterations is provided by immunocytochemical and biochemical analyses of premalignant lesions where increased proliferation (e.g., increased frequencies of cells expressing proliferation markers) and altered differentiation (e.g., inappropriate expression of keratin subtypes) is observed (Gerdes et al. 1986; Moll et al. 1982; Robbins et al. 1987).

To better define which genetic alterations are important in the formation of aerodigestive tract tumors, it has been of interest to examine the chromosome changes associated with various steps of the process. We have therefore applied the technique of premature chromosome condensation to aerodigestive tract tissues in order to examine the concepts of "field cancerization" and the "multistage process" of tumor development.

4.2 Carcinogenesis in the Syrian Hamster Cheek Pouch

When the cheek pouch of the Syrian hamster is directly exposed to carcinogenic agents (e.g., DMBA) over a prolonged period, the tissue undergoes a number of alterations prior to tumor development, including hyperplasia, metaplasia/dysplasia, and carcinoma in situ (Shklar et al. 1979). Examination of representative tissues obtained at various times during this process using the PCC technique demonstrated an accumulation of cytogenetic damage with time (Cheong et al. 1989). As shown in Fig. 2, the chromosome numbers in cells derived from exposed cheek pouch tissue were found to increase with time. Interestingly, at 8 weeks of treatment, chromosome changes were observed in the contralateral cheek pouch as well, possibly a result of the movement of food by the animal from the exposed cheek pouch to the other cheek pouch. This result supports the notion of "field cancerization" and the role of genetic damage driving the process.

Karyotypic analysis of PCC obtained during the various stages of carcinogenesis demonstrated that both changes in chromosome number as well as the development of chromosome structural rearrangements occurred during the period of carcinogenic exposure. This would suggest that chromosome changes accompany and accumulate during the multistep tumorigenesis process. It will now be of interest to determine

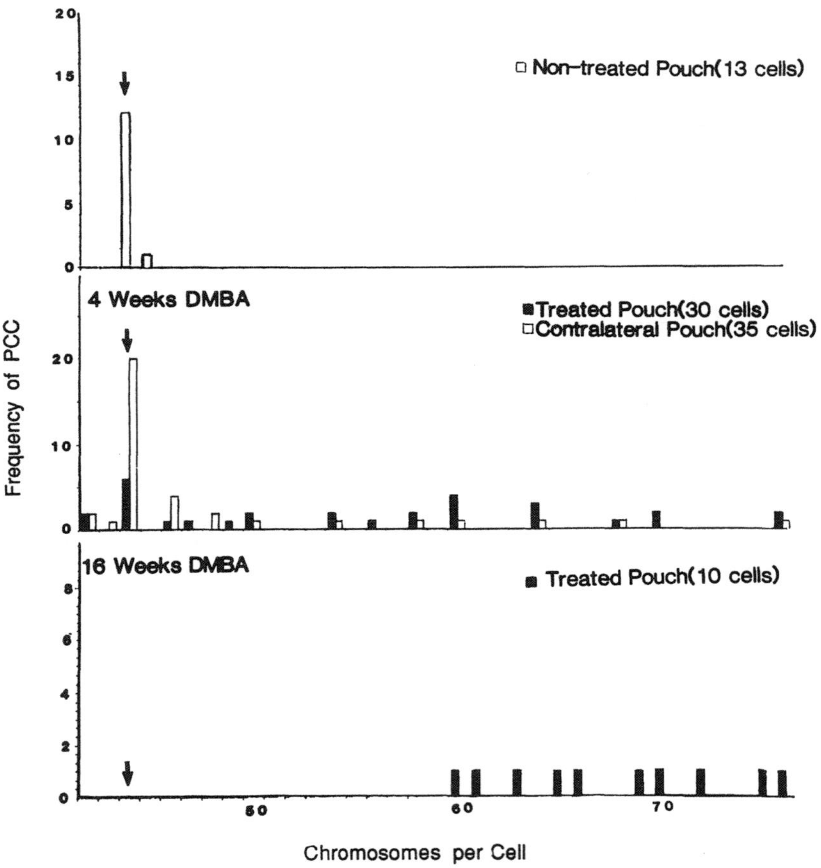

Fig. 2. Chromosome number per cell in Syrian hamster cheek pouch cells after varying times of DMBA exposure in vivo. Note the increase in chromosome number and variability with increased duration of exposure

which specific chromosome changes are involved in this process and how specific changes are associated with specific phenotypic changes in the target tissue.

4.3 Evidence for "Field Cancerization" in the Human Lung

Another way to investigate the carcinogenesis question at the chromosome level is to examine both the tumor as well as the morphologically normal tissue at risk for developing a second primary tumor. It would be of interest to determine whether normal tissue exhibits any of the cytogenetic manifestations of the tumor in that same tissue. We have recently examined, using the PCC technique, both the primary lung tumor as well as normal lung tissue obtained from seven patients with lung tumors (Sohn et al. 1989). In these cases, material was obtained at lobectomy or pneumonectomy, and thus normal lung tissue could be obtained at a distant site from the tumor. For the lung tumors, two cases showed predominantly hypodiploid populations, one case showed both a hypodiploid and a hypotetraploid population, three cases showed hyperdiploidy, and one case showed a near diploid peak. Interestingly, in all but one of the normal lung samples, the distribution of chromosome numbers per cell resembled that observed in the corresponding tumor specimen (Fig. 3). These results support the notion of "field cancerization" because chromosome changes were observed in normal lung cells in the field of carcinogenic exposure. The fact that similar chromosome number distributions were observed in the normal and tumor populations might suggest that prolonged carcinogenic exposure results in the accumulation of chromosome damage throughout the tissue at risk.

While banding analyses were limited in these studies, intriguing results were obtained where such studies were successsful. In one case involving a hypodiploid tumor, for example, both the tumor and the normal lung cells exhibited monosomy 18, 19, 20, and 21. While the tumor cell also contained chromosome changes not observed in the normal cell (as might be expected from genetic evolution), the normal cell contained chromosome changes not observed in the tumor. These results support the multistep tumorigenesis notion where multiple genetic changes must occur prior to the formation of the tumor. However, these results also suggest that not all the chromosome changes, previously called secondary changes, occurred after the tumor was established. In fact, the normal tissue might have accumulated as many changes as did the tumor tissue; however, the specific chromosome changes in the tumor cells might have involved those loci important for proliferative and differentiative dysregulation.

Such PCC studies might be useful in the future to shed light on several aspects of tumorigenesis. First, comparisons of changes between tumor and normal tissue might help to identify which specific chromosome changes are important for tumor

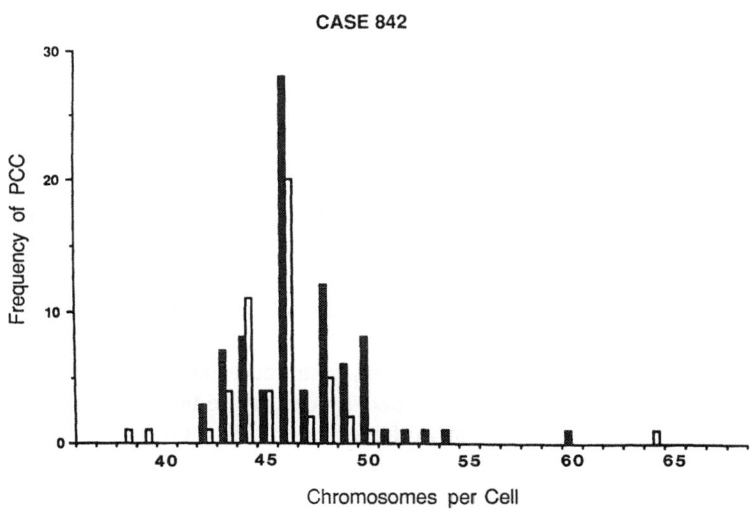

Fig. 3. Chromosome number per cell in human lung tumor and normal lung tissue obtained from a distance at the time of tumor excision. Note the similarity in chromosome number distributions between tumor and normal tissue

formation. Second, by combining the PCC technique with immunocytochemical characterization of cell proliferation or differentiation markers, one should be able to determine which specific chromosome changes are associated with specific phenotypic alterations (e.g., high expression of a growth factor receptor). Third, since genetic alterations are proposed to drive the multistep process, one might expect that the greater the degree of genetic damage accumulated in a tissue, the higher the likelihood of tumor formation. Thus, the PCC determination of the amount of accumulated genetic damage in a target tissue (e.g., lung) might shed insight into the risk of development of a second primary tumor in the remaining tissue. Such information might be useful in the interpretation of chemopreventive trials in such individuals at high risk for developing second primaries.

5 Cytogenetic Evaluation of Differentiation and Response in Hematopoietic Disorders

The development of leukemia is thought to involve alterations in the ability of hematopoietic elements to regulate their proliferation and differentiation. These regulatory difficulties, however, are not considered absolute, and evidence for maturation of leukemic elements has been accumulating over the years (Sachs 1987). For example, during the treatment of leukemia with cytotoxic agents, one occasionally observes a fall in the absolute number of immature elements accompanied by a rise in the absolute number of mature peripheral blood elements without evidence of aplasia in the bone marrow. Analysis of the mature peripheral blood granulocytes of several patients utilizing the PCC technique has shown that these cells were derived from the abnormal clone (Hittelman et al. 1988a).

Myelodysplastic syndromes are a set of preleukemic hematopoietic disorders characterized by refractory cytopenia resulting from dysregulated maturation and functional abnormalities of mature cells (Pierre 1974). As in the case of acute leukemia, more than 50% of patients exhibit the presence of an abnormal cytogenetic clone associated with the disease (Heim and Mitelman 1986). Since this disease represents a bone marrow failure syndrome, several patients with myelodysplastic syndrome were entered on a clinical trial of recombinantly derived human granulocyte-macrophage colony stimulating factor (GM-CSF) in an attempt to restore bone marrow function. Interestingly, a dramatic dose-dependent increase in peripheral blood white cell count was observed in nearly all patients, consisting predominantly of mature granulocytes (Vadhan-Raj et al. 1987). It was therefore of interest to determine whether GM-CSF stimulated the proliferation and maturation of the normal clone or the abnormal clone.

To examine this question, peripheral blood was obtained after GM-CSF treatment, and the mature granulocytes were enriched by density sedimentation techniques. The chromosomes of these terminally differentiated cells were visualized by the PCC technique and analyzed for karyotype (Fig. 4). The results of this study indicated that the mature granulocytes increasing after GM-CSF treatment were derived from both normal and abnormal elements, and the ratio of normal to abnormal cells varied from patient to patient (Hittelman et al. 1988c; Table 1). However, when the pretreatment normal clone : abnormal clone ratio was compared to that

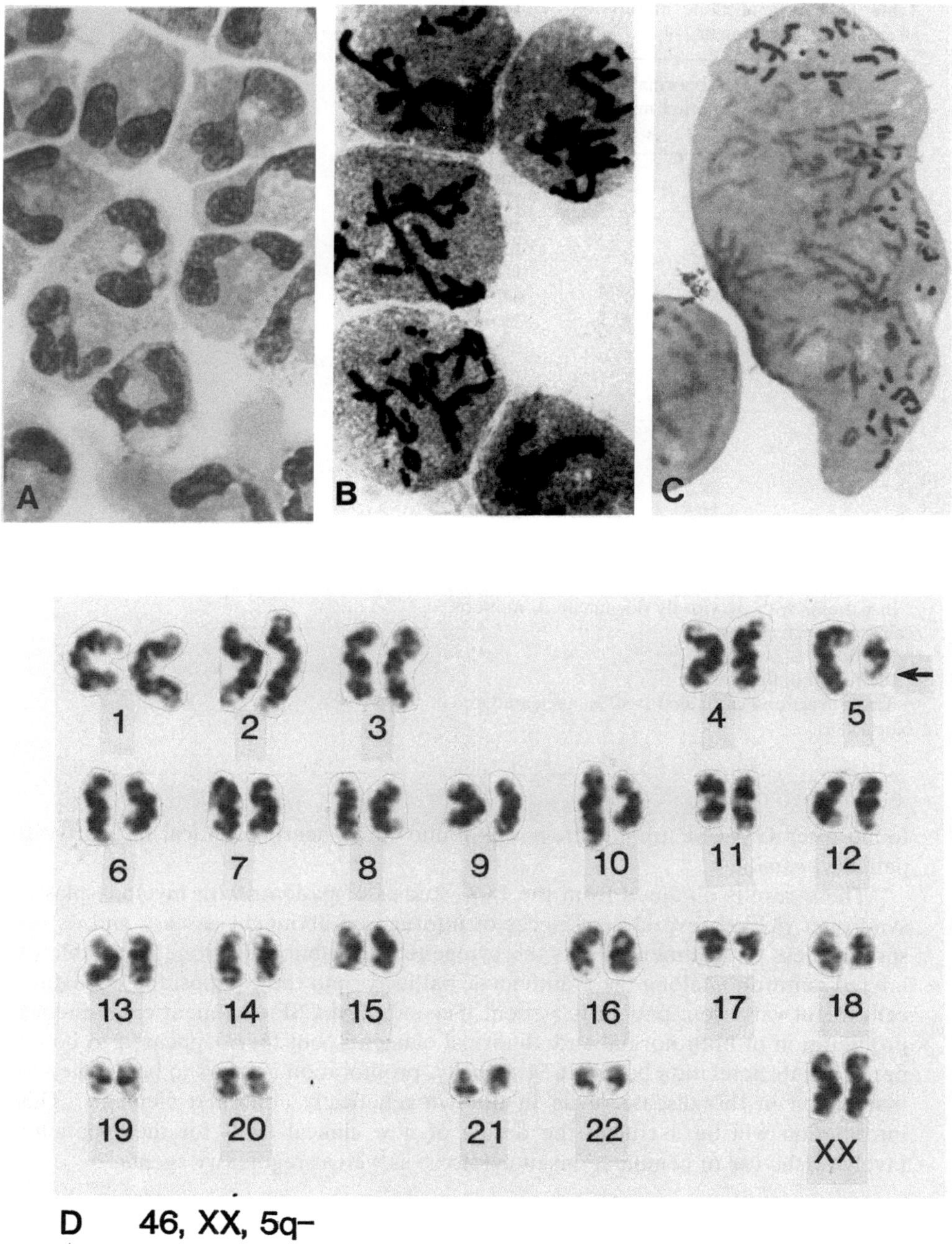

Fig. 4 A–D. PCC analysis of mature peripheral blood granulocytes in a patient with myelodysplastic syndrome. **A** Polymorphonuclear granulocytes prior to fusion; **B** mitotic CHO cells; **C** fusion product of human granulocyte and CHO mitotic cell resulting in G1 PCC; **D** G-banded G1 PCC karyogram of patient granulocyte exhibiting a 5q chromosome anomaly

Table 1. Abnormal clone maturation before and after GM-CSF treatment

Patient No.	Abnormal cells in mature myeloid fraction[a] (%)	
	Pre	Post
1	ND[b]	67.3
2	ND	68.8
3	ND	73.3
4	ND	11.1
5	ND	67.5
7	TFC[c]	39.6
9	78.5	85.7
10	2.0	7.1
11	89.4	84.4
14	95.8	92.0
15	57.5	55.0
16	47.7	ND
17	68.2[d]	39.7[d]
18	36.5	ND
19	91.0	ND
21	18.8	ND

[a] In patients with previously documented, numerically abnormal clones.
[b] Not done.
[c] Too few granulocytes.
[d] These fractions contained >90% nucleated red blood cells.

found after GM-CSF treatment, it was found to be nearly identical in nearly all patients examined.

These results obtained from the PCC studies of patients with myelodysplastic syndrome yielded several new pieces of information about the disease and its responsiveness to treatment. First, the cytogenetically abnormal clone is capable of limited maturation along the granulocytic pathway, and the composition of mature cells might vary from patient to patient. Second, GM-CSF treatment could induce proliferation of both normal and abnormal elements, but there appeared to be no preferential stimulation by GM-CSF. Finally, proliferation appears to be coupled to maturation in this disease, even in the cytogenetically abnormal elements. This information will be useful in the design of new clinical trials for these patients involving the use of combinations of cytotoxic as well as regulatory agents.

6 Conclusions

The purpose of this chapter was to illustrate the usefulness of the PCC technique for understanding the tumorigenesis process as well tumor response to therapeutic intervention. By being able to visualize the chromosomes of slowly dividing and nonproliferating cell populations, one cannot only directly monitor the accumulation in target tissues, but one can also begin to tabulate which specific chromosome changes are important for tumor development. This will allow the future generation of specific molecular probes for these genetic changes which can then be applied to premalignant tissue to map the multistep tumorigenesis process. Moreover, the PCC technique is also useful for monitoring response to therapeutic management of malignant tumors, whether this means evaluating the degree of chromosome damage inflicted by cytotoxic agents or evaluating the nature of response in therapies designed to reregulate proliferation and differentiation. Research along these lines is now only in its infancy and should yield significant new information in the future about the biology of tumor development und response.

References

Auerbach O, Stout AP, Hammond EC et al. (1961) Changes in bronchial epithelium in relation to cigarette smoking and in relation to lung cancer. New Eng J Med 265: 253–267

Boveri T (1914) Zur Frage der Entstehung Maligner Tumoren. Fischer, Jena

Cheong N, Sohn HY, Shin D, Hong WK, Hittelman WN (1989) Cytogenetic changes associated with Syrian hamster cheek pouch tumorigenesis induced by DMBA. Cancer Genet Cytogenet 41: 251

Dunphy WG, Newport JW (1988) Unraveling of mitotic control mechanisms. Cell 55: 925–928

Farber E (1984) The multistep nature of cancer development. Cancer Res 44: 4217–4223

Gerdes J, Lelle RJ, Pickhartz H et al. (1986) Growth fractions in breast cancers determined in situ with monoclonal antibody Ki67. J Clin Pathol 39: 977–980

Heim S, Mitelman F (1987) Cancer genetics. Liss, New York

Heim S, Mitelman F (1986) Chromosome abnormalities in the myelodysplastic syndromes. Clin Haematol 15: 1003–1021

Hittelman WN, Agbor P, Petkovic I et al. (1988a) Detection of leukemic clone maturation in vivo by premature chromosome condensation. Blood 72: 1950–1960

Hittelman WN, Petkovic I, Agbor P (1988b) Improvements in the premature chromosome condensation technique for cytogenetic analysis. Cancer Genet Cytogen 30: 301–312

Hittelman WN, Tigaud J-D, Estey E, Gutterman JU, Vadhan-Raj S (1988c) Relative maturation of aneuploid and diploid elements in patients with myelodysplasia (MDS) treated with granulocyte-macrophage colony stimulating factor (GM-CSF). Blood 72: 121a

Johnson RT, Rao PN (1970) Mammalian cell fusion: induction of premature chromosome condensation in interphase nuclei. Nature 226: 717–722

Johnson RT, Rao PN, Hughes SD (1970) Mammalian cell fusion. II. A HeLa cell inducer of premature chromsome condensation active in cells from a variety of animal species. J Cell Physiol 77: 151–158

Knudson AG (1985) Hereditary cancer, oncogenes, and antioncogenes. Cancer Res 45: 1437–1443

Le Beau MM, Rowley JD (1986) Chromosome abnormalities in leukemia and lymphoma: clinical and biological significance. Adv Hum Genet 15: 1–54

Lejeune J, Gautier M, Turpin R (1959) Etude des chromosomes de neuf enfants mongoliens. C R Hebd Seances. Acad Sci (Paris) 248: 1721–1729

Moll R, Franke WW, Schiller DL, Geiger B, Kreplea R (1982) The catalog of human keratins: patterns of expression in normal epithelia, tumors and cultured cells. Cell 31: 11–24

Petkovic I, Agbor P, Hittelman WN (1989) Simultaneous cell type identification and premature chromsome condensation analysis in a case of multiple myeloma. Cancer Genet Cytogenet 39: 203–217

Pierre RV (1974) Preleukemic states. Semin Hematol 11: 73–92

Pinkel D, Straume T, Gray JW (1986) Cytogenetic analysis using quantitative, high sensitivity, fluorescence hybridization. Proc Natl Acad Sci USA 83: 2934–2938

Ray JH, German J (1981) The chromosome changes in Bloom's syndrome, ataxia-telangiectasia and Fanconi's anemia. In: Arrighi FE, Rao PN, Stubblefield E (eds) Genes, chromosomes and neoplasia. Raven, New York, pp 351–378

Robbins BA, Vega D de la, Ogata K, Tan EM, Nakamura RM (1987) Immunohistochemical detection of proliferating cell nuclear antigen in solid human malignancies. Arch Pathol Lab Med 111: 841–845

Sachs L (1987) Cell differentiation and bypassing of genetic defects in the suppression of malignancy. Cancer Res 47: 1981–1986

Shklar G (1986) Oral leukoplakia. New Eng J Med 315: 1544–1546

Shklar G, Eisenberg E, Flynn E (1979) Immunoenhancing agents and experimental leukoplakia and carcinoma of the hamster buccal pouch. Prog Exp Tumor Res 24: 269–282

Slaughter DP, Southwick HW, Smejkal W (1953) "Field Cancerization" in oral stratified squamous epithelium: clinical implications of multicentric origin. Cancer 6: 963–968

Sohn HY, Cheong N, Wang Z-W, Hong WK, Hittelman WN (1989) Detection of aneuploidy in normal lung tissue adjacent to lung tumor by premature chromosome condensation. Cancer Genet Cytogenet 41: 250

Stich HF (1987) Micronucleated exfoliated cells as indicators for genotoxic damage and as markers in chemoprevention trials. J Nutrit Growth Cancer 4: 9–18

Teyssier JR (1989) The chromosomal analysis of human solid tumors: a triple challenge. Cancer Genet Cytogenet 37: 103–2125

Vadhan-Raj S, Keating M, LeMaistre A et al. (1987) Effects of recombinant human granulocyte-macrophage colony-stimulating factor in patients with myelodysplastic syndromes. New Eng J Med 317: 1545–1552

Vaughan CW, Homburger F, Shapshay S et al. (1980) Carcinogenesis in the upper aerodigestive tract. Otolaryngol Clin N Am 13: 403–412

Vogelstein B, Fearon ER, Hamilton SR (1988) Genetic alterations during colorectal-tumor development. New Eng J Med 319: 525–535

Vogelstein B, Fearon ER, Kern SE, Hamilton SR, Presinger AC, Nakamura Y, White R (1989) Allelotype of colorectal carcinomas. Science 244: 207–211

Detection of Cancer-Prone Individuals Using Cytogenetic Response to X-Rays

K. K. SANFORD[1] and R. PARSHAD[2]

Contents

1 Introduction

A number of genetic disorders manifesting widely different clinical symptoms predispose the affected individual to a high risk of cancer. These include, among others, ataxia telangiectasia, Bloom's syndrome, familial polyposis, Fanconi's anemia, Gardner's syndrome, and xeroderma pigmentosum (Setlow 1978). We have shown that skin fibroblasts from affected individuals or individuals with a family history of cancer, compared to cells from clinically normal controls, with few exceptions, show a significantly higher frequency of chromatid breaks and gaps following X-irradiation during the G_2 cell cycle phase (Parshad et al. 1983, 1985a; Sanford et al. 1989). Furthermore, it appears that this enhanced chromatid damage results from deficient DNA repair during G_2 phase (Parshad et al. 1983, 1985b; Gantt et al. 1986, 1989), that it has a genetic basis (Sanford et al. 1986), and that it is associated with genetic susceptibility to cancer (Gantt et al. 1989; Sanford et al. 1989).

The enhanced chromatid damage observed during the period after G_2 phase X-irradiation has potential use in detecting individuals genetically predisposed to cancer. In this report, we describe the use of this assay to identify individuals carrying cancer-prone genes in families with ataxia telangiectasia (A-T), hereditary cutaneous malignant melanoma (HCMM) and its precursor, the dysplastic nevus syndrome (DNS), or xeroderma pigmentosum (XP). Whereas the clinically normal carriers of the A-T gene(s), like A-T patients, are at high cancer risk, the cancer proneness of carriers of XP gene(s) is uncertain (Swift and Chase 1979; Pippard et al. 1988). Hereditary DNS is a dominant trait with high penetrance and high cancer risk (Greene et al. 1985).

[1] Laboratory of Cellular and Molecular Biology, National Cancer Institute, Bethesda, MD 20892, USA
[2] Pathology Department, Howard University College of Medicine, Washington, DC 20059, USA

2 Detection of Cancer-Prone Individuals in A-T Families

In collaboration with Y. Shiloh, Department of Human Genetics, Tel Aviv University, Tel Aviv, Israel, we have quantified the extent of chromatid damage in skin fibroblasts from 30 members of three Israeli A-T families. Coded samples of cells were sent to Bethesda for testing and decoded only after the study was completed. The cells, grown on coverslips in Leighton tubes, were X-irradiated (68 R in air in culture vessel), culture medium renewed after 0.5 h, and metaphase cells collected with colcemid from 0.5 to 1.5 h after irradiation before processing for chromosomes (Sanford et al. 1989). Chromatid aberrations scored as breaks showed discontinuity and displacement of the broken segment; aberrations scored as gaps showed an apparent discontinuity longer than the chromatid width with no displacement.

Cells from all six obligate heterozygotes (parents of A-T patients) maintained the high frequency of chromatid breaks and gaps typical of cancer-prone individuals. Of the 24 cell lines from unaffected family members at various risks of being A-T carriers, 7 responded as normal, while the remaining 17 responded like A-T heterozygotes (Fig. 1). On the assumption that the G_2 repair deficiency cosegregates with the A-T gene, and behaves as an incomplete dominant trait, the proportion of cell lines showing the A-T heterozygous phenotype is in accord with the expected value based on Mendelian chance calculations. Subsequent to this analysis, two couples of the 17 individuals responding like A-T heterozygotes each had a child clinically diagnosed as having A-T. An additional 11 obligate heterozygotes were correctly

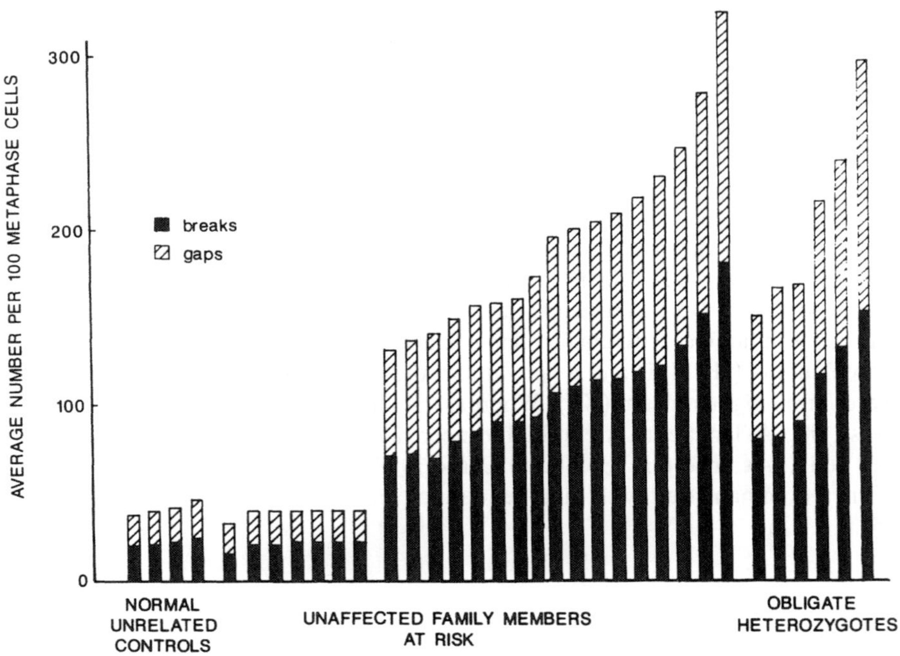

Fig. 1. Chromatid damage in skin fibroblasts from members of three Israeli ataxia telangiectasia (A-T) families after G_2 phase x-irradiation

identified by means of this assay in two previous blind studies (Parshad et al. 1985c; Shiloh et al. 1986).

Recently, in additional blind studies, we have applied this assay to peripheral blood lymphocytes. Seventy-two hour cultures of phytohemagglutinin-stimulated whole blood were X-irradiated (58 R in air in culture vessel) as a pellet, resuspended in culture medium and processed for chromosomes as described (Sanford et al. 1987). When collected with colcemid betweeen 0.5 to 1.5 h after X-irradiation, metaphase cells from nine obligate heterozygotes showed a significantly higher frequency of chromatid breaks and gaps than cells from a total of 33 of 35 normal controls (Fig. 2). The extent of damage in the heterozygotes was intermediate between that of the A-T patients (homozygotes) and normal controls. Of the four unaffected family members, three responded like normal and the remaining one like an A-T heterozygote. We conclude that carriers of the A-T gene(s) who are at high cancer risk and constitute ~ 2.8% of the general population (Swift et al. 1986) or ~ 7 million persons in the United States can be identified by this assay.

3 Detection of Cancer-Prone Individuals in Families with Hereditary DNS

In a collaborative study with M. H. Greene, formerly of the Environmental Epidemiology Branch of the National Cancer Institute, coded samples of skin fibroblasts from 25 members of 9 kindreds with hereditary DNS, 12 apparently normal spouses, and 11 additional unrelated normal individuals were tested for the extent of chromatid damage after G_2 phase X-irradiation. The cells from individuals with DNS or HCMM with DNS had significantly more chromatid breaks and gaps when entering metaphase 0.5 to 1.5 h after G_2 phase X-irradiation (68 R in air in culture vessel) than those

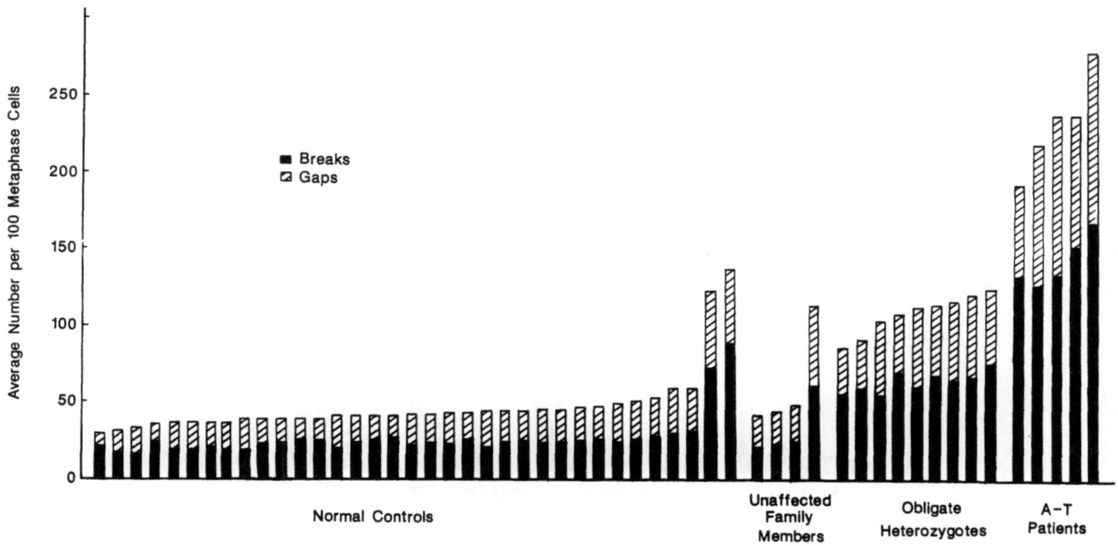

Fig. 2. Chromatid damage in stimulated peripheral blood lymphocytes from members of ataxia telangiectasia families after G_2 phase X-irradiation

from unaffected controls (Fig. 3). In two cases, the test results positively identified individuals before the clinical diagnosis of DNS (Sanford et al. 1987). A clinically normal obligate carrier of the HCMM/DNS gene showed the enhanced chromatid damage. Similarly, stimulated peripheral blood lymphocytes from one proband showed the abnormal response (Sanford et al. 1987). Thus, enhanced chromatid damage in cells entering metaphase after G_2 phase X-irradiation appears to be a marker of genetic predisposition to cancer in individuals from families with hereditary DNS.

4 Carrier Detection in Xeroderma Pigmentosum

Xeroderma pigmentosum is a rare autosomal-recessive, progressive, degenerative disease associated with sun sensitivity, cutaneous pigmentation, neoplasia, abnormal DNA repair and, in some patients, neurologic degeneration (Kraemer and Slor 1985). XP heterozygotes [carriers of the XP gene(s)] are clinically normal individuals who cannot be recognized by any simple, reliable test known to date. However, their identification is important for genetic counseling, epidemiologic studies, and cancer control. In a collaborative study with K. H. Kraemer of the Laboratory of Molecular Carcinogenesis, National Cancer Institute, we were able to detect clinically normal carriers of XP genes from coded samples of either peripheral blood lymphocytes or

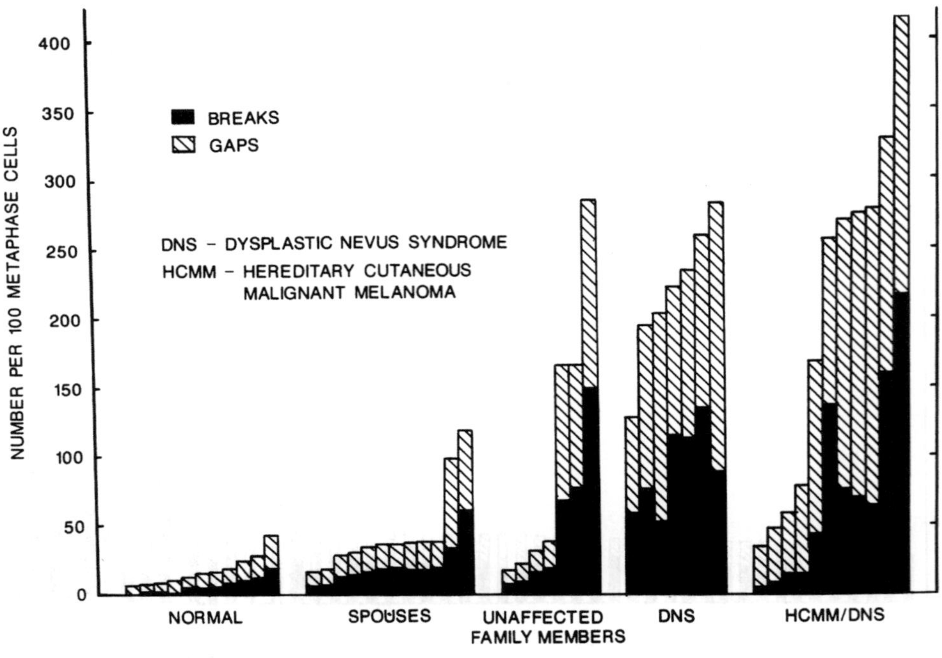

Fig. 3. Chromatid damage in skin fibroblasts from members of nine kindreds with hereditary dysplastic nevus syndrome (*DNS*) with or without hereditary cutaneous malignant melanoma (*HCMM*)

skin fibroblasts (Parshad et al. 1989). Metaphase cells of phytohemagglutinin-stimulated T-lymphocytes from eight obligate heterozygotes for XP were compared with those from nine normal controls at three intervals after X-irradiation (58 R in air in culture vessel) during G_2 phase of the cell cycle. Lymphocytes from the XP heterozygotes had two fold higher frequencies of chromatid breaks or chromatid gaps than normal when collected with colcemid from 0.5–1.5 h and 1.5–2.5 h after irradiation and fixed at 2.3 and 3.3 h postirradiation, respectively (Fig. 4). Lymphocytes from six XP homozygotes had frequencies of breaks and gaps three fold higher than normal.

Skin fibroblasts from an additional obligate XP heterozygote, when collected with colcemid from 0.5–1.5 h after X-irradiation (68 R in air in culture vessel), had a two fold higher frequency of chromatid breaks and a four fold higher frequency of gaps than fibroblasts from a normal control. The extent of damage in skin fibroblasts from this XP heterozygote was approximately half that observed in the XP homozygote (Parshad et al. 1989). These findings of elevated frequencies of chromatid breaks

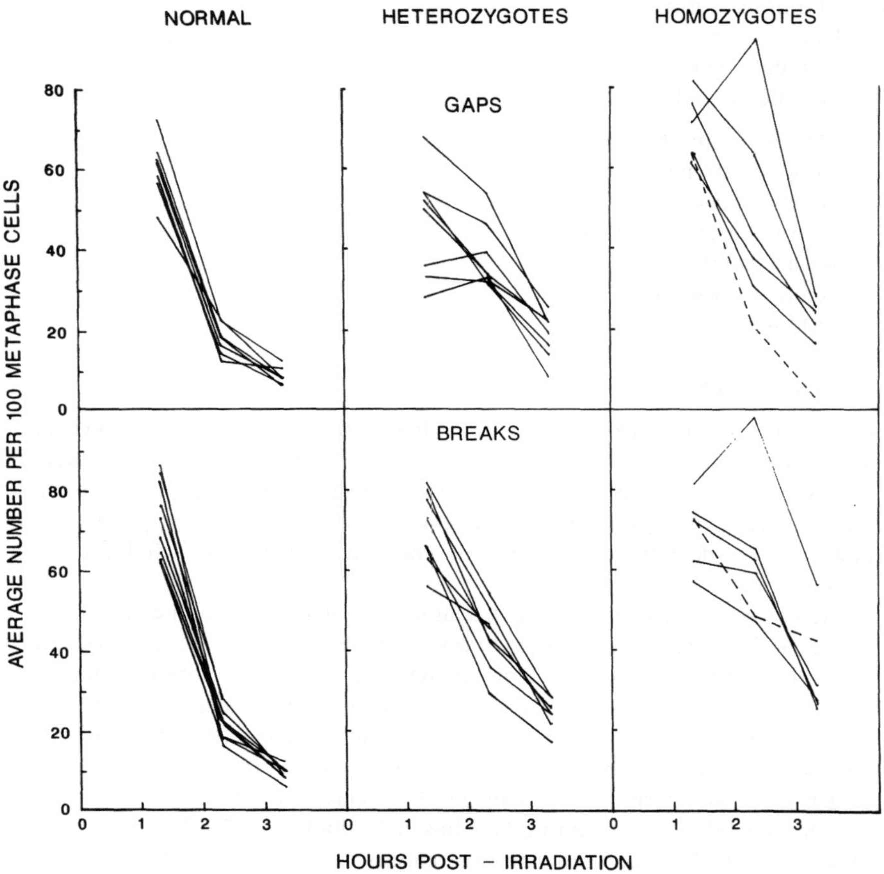

Fig. 4. Chromatid damage in stimulated peripheral blood lymphocytes from *normal* individuals, xeroderma pigmentosum (XP) *heterozygotes*, and XP patients (*homozygotes*). - - - Response of cells from XP-complementation group A-individual

and gaps following G_2 phase X-irradiation can be used to identify individuals carrying the XP gene(s) within known XP families.

5 Discussion

The present results show that individuals with genes for cancer proneness can be identified in families with any one of three genetic disorders, A-T, DNS, and XP. This identification is based on the abnormally high incidence of chromatid damage in metaphase cells collected by colcemid from 0.5–1.5 h after X-irradiation. This abnormal response has also been observed in skin fibroblasts and/or stimulated peripheral blood lymphocytes from individuals with the following cancer-prone genetic diseases or hereditary/familial cancers (Sanford et al. 1989; Parshad et al. 1985a):

A. *Cancer-prone genetic diseases*
   Bloom's syndrome
   Fanconi's anemia
   Familial polyposis
   Gardner's syndrome

B. *Hereditary/familial cancers*
   Retinoblastoma
   Wilms' tumor
   Nevoid basal cell carcinoma
   Breast cancer
   Acute myelogenous leukemia
   Osteosarcoma
   Liposarcoma
   Medullosarcoma

Of the 74 clinically normal control individuals tested to date (ages 1 to 96 years) only 4 (5%) have shown the abnormal response. The extent of chromatid damage following X-irradiation is a physiological response dependent on stringent culture conditions including control of pH, temperature, cell densitiy, and serum supplement (Sanford et al. 1989). Adequate normal controls, therefore, need to be included with the test cells.

It may seem surprising that such a large fraction (>90%) of the individuals tested with genetic diseases predisposing to cancer or hereditary neoplasms should have the same phenotype of persistent chromatid damage after X-irradiation during G_2 phase. Extensive literature proposes molecular mechanisms of chromosomal aberration formation in terms of DNA damage and repair (Kihlman 1971; Bender et al. 1974; Evens 1977; Taylor 1978; Natarajan et al. 1980; Preston 1980; Bryant 1984). It appears that this persistent chromatid damage after X-irradiation results from deficient or unbalanced DNA repair during G_2 phase (Parshad et al. 1983, 1985b; Gantt et al. 1986, 1987, 1989). The following two reasons can explain why this abnormal cytogenetic response is so commonly associated with cancer-proneness and hereditary cancers. First, DNA repair following initial strand incision is a multi-enzymatic process involving a substantial number of genes, mutations in any one of which could lead to the observed repair deficiency. Second, a DNA repair deficiency such as the G_2 phenotype, whether inherited or acquired, can provide genetic instability which appears to be essential for all neoplastic development.

References

Bender MA, Griggs HG, Bedford JS (1974) Mechanisms of chromosomal aberration production. III. Chemicals and ionizing radiation. Mutat Res 23: 197–212

Bryant PE (1984) Enzymatic restriction of mammalian cell DNA using *Pvu* II and *Bam* HI: evidence for the double strand break origin of chromosomal aberrations. Int J Radiat Biol 46: 57–65

Evans HJ (1977) Molecular mechanisms in the induction of chromosome aberrations. In: Scott D, Bridges BA, Sobels BA (eds) Progress in genetic toxicology. Elsevier, Amsterdam, pp 57–74

Gantt R, Parshad R, Price FM, Sanford KK (1986) Biochemical evidence for deficient DNA repair leading to enhanced G_2 chromatid radiosensitivity and suceptibility to cancer. Radiat Res 108: 117–126

Gantt R, Sanford KK, Parshad R, Price FM, Peterson WD Jr, Rhim JS (1987) Enhanced G_2 chromatid radiosensitivity, an early stage in the neoplastic transformation of human epidermal keratinocytes in culture. Cancer Res 47: 1390–1397

Gantt R, Sanford KK, Parshad R, Tarone RE (1989) Genetic predisposition to cancer and enhanced chromatid aberrations in human cells x-irradiated in G_2 phase. In: Park JF, Pelroy RA (eds) From molecules to man. Columbus, Battelle, pp 363–392

Greene MH, Clark WH Jr, Tucker MA, Kraemer KH, Elder DE, Fraser MC (1985) High risk of malignant melanoma in melanoma-prone families with dysplastic nevi. Am J Intern Med 102: 458–465

Kihlman BA (1971) Molecular mechanisms of chromosome breakage and rejoining. Adv Cell Mol Biol 1: 59–107

Kraemer KH, Slor H (1985) Xeroderma pigmentosum. Clin Dermatol 3: 33–69

Natarajan AT, Obe G, Zeeland AA van, Palitti F, Meyers M, Verdegaal-Immerzel EA (1980) Molecular mechanisms involved in the production of chromosomal aberrations. II. Utilization of *Neurospora* endonuclease for the study of aberration production by X-rays in G_1 and G_2 studies of cell cycle. Mutat Res 69: 293–305

Parshad R,Sanford KK, Jones GM (1983) Chromatid damage after G_2 phase X-irradiation of cells from cancer-prone individuals implicates deficiency in DNA repair. Proc Natl Acad Sci USA 80: 5612–5616

Parshad R, Sanford KK, Jones GM (1985a) Chromosomal radiosensitivity during the G_2 cell cycle period of skin fibroblasts from individuals with familial cancer. Proc Natl Acad Sci USA 82: 5400–5403

Parshad R, Sanford KK, Jones GM (1985b) Chromatid damage induced by fluorescent light during G_2 phase in normal and Gardner syndrome fibroblasts: interpretation in terms of deficient DNA repair. Mutat Res 151: 57–63

Parshad R, Sanford KK, Jones GM, Tarone RE (1985c) G_2 chromosomal radiosensitivity of ataxia-telangiectasia heterozygotes. Cancer Genet Cytogenet 14: 163–168

Parshad R, Sanford KK, Kraemer KH, Jones GM, Tarone RE (1990) Carrier detection in xeroderma pigmentosum. J Clin Invest 85: 135–138

Pippard EC, Hall AJ, Barker DJP, Bridges BA (1988) Cancer in homozygotes and heterozygotes of ataxia-telangiectasia and xeroderma pigmentosum in Britain. Cancer Res 48: 2929–2932

Preston RJ (1980) DNA repair and chromosome aberrations: the effect of cytosine arabinoside on the frequency of chromosome aberrations induced by radiation and chemicals. Teratog Carcinog Mutagen 1: 147–159

Sanford KK, Parshad R, Stanbridge EJ, Frost JK, Jones GM, Wilkinson JE, Tarone RE (1986) Chromosomal radiosensitivity during the G_2 cell cycle period and cytopathology of human normal x tumor cell hybrids. Cancer Res 46: 2045–2049

Sanford KK, Parshad R, Greene MH, Tarone RE, Tucker MA, Jones GM (1987) Hypersensitivity to G_2 chromatid radiation damage in familial dysplastic naevus syndrome. Lancet 1987 ii: 1111–1116

Sanford KK, Parshad R, Gantt R, Tarone RE, Jones GM, Price FM (1989) Factors affecting and significance of G_2 chromatin radiosensitivity in predisposition to cancer. Int J Radiol Biol 55: 963–981

Setlow RB (1978) Repair deficient human disorders and cancer. Nature 271: 713–717

Shiloh Y, Parshad R, Sanford KK, Jones GM (1986) Carrier detection in ataxia-telangiectasia. Lancet 1: 689–690

Swift M, Chase C (1979) Cancer in families with xeroderma pigmentosum. J Nat Cancer Inst 62: 1415–1421

Swift M, Morrell D, Cromartie E, Chamberlin AR, Skolnick MH, Bishop DT (1986) The incidence and gene frequency of ataxia-telangiectasia in the United States. Am J Hum Genet 39: 573–583

Taylor AMR (1978) Unrepaired DNA strand breaks in irradiated ataxia telangiectasia lymphocytes suggested from cytogenetic observations. Mutat Res 50: 407–418

Human Disorders with Increased Spontaneous and Induced Chromosomal Instability

T. M. Schroeder-Kurth, U. Cramer-Giraud, and U. Mannsperger[1]

Contents

1 Introduction

Wegner (1990) has reviewed in an exhaustive paper the cytogenetic characterization of human disorders with increased spontaneous and induced chromosome instability (194 references).

Today, the value of cytogenetic investigation of the classical syndromes: AT, BS, and FA still lies in the differential diagnosis of the suspected disease. The clinical pictures of the syndromes are different and clearly distinguishable from each other. However, there are clinically similar diseases of each condition which have to be identified as another genetic entity or as a sporadic disease with different implications for patient and family.

Each disease poses problems in confirming the clinical diagnosis for different reasons.

2 Ataxia Telangiectasia

The main clinical features of AT are progressive cerebellar ataxia and facial telangiectasia in early childhood associated with immunodeficiencies. Some patients have an elevated alpha fetoprotein level in serum. The patients usually die due to infections or develop cancer.

Differential clinical diagnosis is particularly difficult since the onset of the ataxia is slow and telangiectasia might develop later. Variants of the clinical manifestations as well as of the laboratory findings have been described.

Particularly the cytogenetic results cause problems in interpretation if typical chromosomal aberrations, like translocations involving chromosomes 7 and 14 (Fig. 1), are not present. Also, the sensitivity to certain mutagens, to γ- und X-rays are

[1] Institut für Humangenetik und Anthropologie der Universität, Im Neuenheimer Feld 328, 6900 Heidelberg 1, FRG

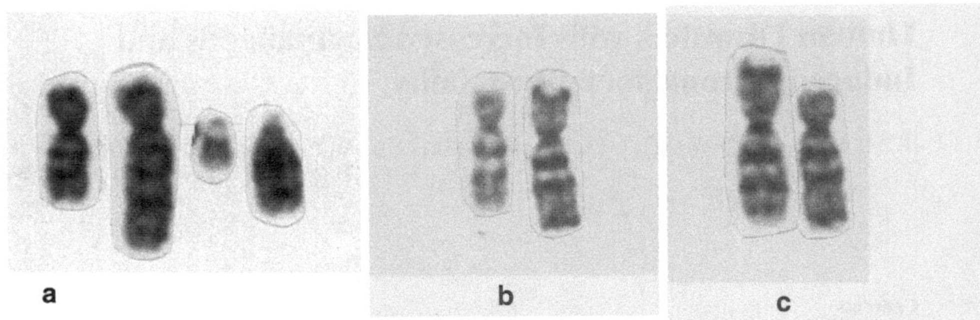

Fig. 1 a–c. Examples of typical inter- and intrachromosomal changes in AT. **a** t (7;14) (q32;q12); **b** t (7;7) (p13;q32); **c** inv (7) (p13q32). Clinical observations of R. Korinthenberg (Dept. of Neuropaediatrics, University of Heidelberg, Mannheim)

difficult to test in some cases, because not all the individuals with clinical AT show the same hypersensitivity, and tissues grow poorly (Taylor 1982).

Therefore, differential diagnosis in AT-like diseases requires a battery of chromosome instability tests in order to identify one or another AT-specific hypersensitivity. Wegner (1990) pointed to an important difference in the reaction of AT and non-AT (normal) cells towards ionizing radiation or radiomimetic drugs: in AT cells almost no inhibition of DNA synthesis occurs after treatment, while in non-AT cells there is significant inhibition measurable by scintillation counting of radioactivity incorporated by the whole cell population or by autoradiography on a single cell level. This striking difference can be used for diagnostic purposes. The clinical diagnosis of AT, cytogenetically or otherwise confirmed, does not represent one genetic entity but consists of five complementation groups.

AT variants, as described by Wegner (1990), may also show increased base line levels of chromosomal instability and an increase of breaks after mutagen treatment. A diffusable factor inducing chromosomal breakage in normal cells has been shown when cultured in the media of AT fibroblasts or if normal cells are cocultivated with AT fibroblasts (Shaham et al. 1980). Prenatal diagnosis is possible if the patient's cytogenetic characteristics are well known (Schroeder-Kurth 1987).

3 Bloom Syndrome

BS is characterized by the cardinal symptoms: intrauterine growth retardation and stunted growth after birth, skin sensitivity to UV-light, immunodeficiency and facial dysmorphies with prominent nose, a rectangular profile and a high susceptibility to cancer (German et al. 1985).

Cytogenetically the typical findings are well known: there is an increased rate of spontaneous chromatid aberrations with unique chromatid exchanges between homologous chromosomes forming a "mitotic chiasma" (Kuhn and Therman 1986; Schroeder and German 1974).

Since these configurations are relatively rare findings (0–2.4%, Ray and German 1981) the diagnosis should not rely solely on a chromosomal instability study but should always include actual proof of the ca. tenfold increased frequency of sister chromatid exchanges (SCEs; Chaganti et al. 1974). Prenatal diagnosis should be possible using SCE frequencies as the critical parameter, however, so far, no homozygote BS fetus has been identified prenatally.

The phenomenon of a spontaneous increase of SCE is thought to be unique for BS. Recently, Marx and Smith (1989) reported another condition showing compatible increases in SCEs which is not understood: parents of patients suffering from FA were found to have significantly increased SCE frequencies after diepoxybutan (DEB) treatment of the lymphocyte cultures. This group of FA patients possibly belongs to a unique cluster living around Bloemfontein, South Africa, since other FA heterozygotes from Europe do not exhibit this feature (Dallapiccola and Porfirion 1989).

However, for differential diagnosis of BS this observation is not relevant; it simply shows that there might be unexpected cellular disturbances leading to an increase of SCEs in other individuals and under different conditions. It is possible that within the limited population around Bloemfontein SCE frequencies after DEB treatment of the lymphocyte culture might serve as a method for FA-heterozygote detection.

4 Fanconi Anemia

There are still problems in diagnosing FA as well as in the cytogenetic confirmation procedures. FA cells are hypersensitive to DEB, mitomycin C (MMC), and other alkylating agents resulting in a significant increase in chromatid and chromosome aberrations in contrast to non-FA cells. DEB and MMC hypersensitive FA-cell lines have been tested in complementation experiments. So far, at least two complementation groups in FA are distinguishable (Buchwald et al. 1989; Digweed et al. 1989).

There are a few FA cases in which the lymphocyte cultures show only a low and therefore ambiguous hypersensitivity to DEB or MMC due probably to reverse mutations at the FA locus in the lymphocytes (Auerbach et al. 1989a; Arwert and Kwee 1989; Kwee et al. 1983). These cases might escape detection in a test performed by cocultivation of lymphocytes from the patient with those from a normal control individual in MMC-containing medium. There are also suspected cases of FA with aplastic anemia and/or malformations in which the lymphocytes are sensitive to MMC, but not to DEB, leading to a false positive diagnosis (Schroeder-Kurth et al. 1989). The use of DEB for FA diagnosis appears to be most reliable. It discriminates best between two groups of patients of which the DEB-positive group is considered to represent FA, the DEB-minus group to represent non-FA patients, and consisting of a variety of other genetic and non-genetic syndromes clinically similar to FA (Auerbach et al. 1989 a, b; Schroeder-Kurth et al. 1989).

Therefore, DEB testing is recommended as a prerequisite to complementation studies in FA in order to confirm the diagnosis and exclude DEB– (minus), MMC-sensitive non-FA cases which otherwise might appear as another complimentation group of FA.

The observations of a wide variability in the pattern of sensitivities of FA lymphocytes when tested with isonicotinic acid hydrazide (INH), MMC and DEB led to the conclusion that individual cellular DEB hypersensitivity follows the same pattern during the course of the disease (Fig. 2a) and that there is a constant pattern of hypersensitivities in families. This can be demonstrated in a family with two siblings affected with FA (Fig. 2b) and in another family with a hypersensitivity to MMC, not only in the two FA patients, but also in both parents (Fig. 2c). The identical pattern of sensitivity was also found in monozygotic twins suffering from FA; the onset of the aplastic anemia occurred at the same age (Table 1).

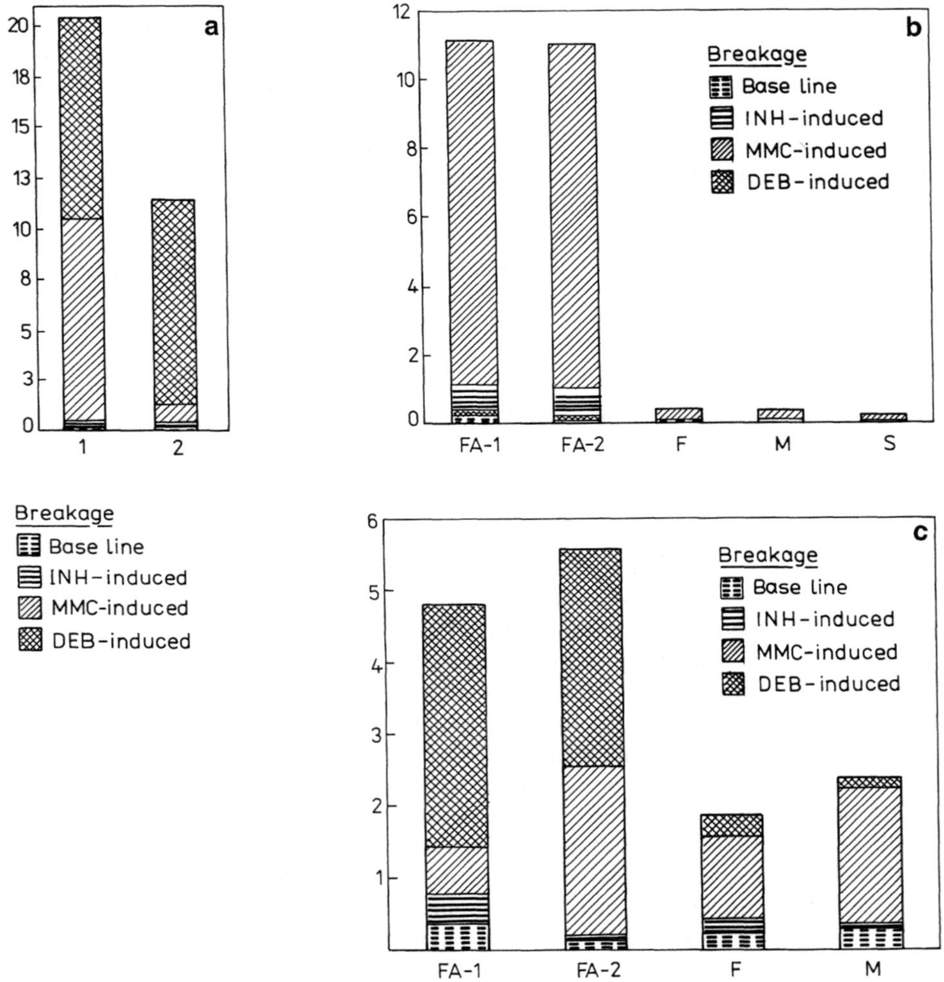

Fig. 2 a–c. Pattern of cellular sensitivity in FA families. The *columns* are a summary of the individual chromosomal instability (breaks/cell) from various tests: base line breakage, INH-, MMC-, and DEB-induced breakage (Schroeder-Kurth et al. 1989). **a** Constant DEB sensitivity and changing MMC sensitivity in two tests (1 and 2), 1 year apart (IFAR 1022/1). **b** Intrafamilial constancy of the pattern of sensitivity in two siblings with FA [*FA-1, FA-2,* father (*F*), mother (*M*), sister (*S*); IFAR 1052/1, 1052/2]. **c** Familial hypersensitivity to MMC in two siblings with FA (*FA-1, FA-2*) and their parents (*F, M*), however, specific DEB sensitivity to DEB only in the patients (IFAR 1124/1, 1124/2)

Table 1. Chromosomal instability[a] in identical twins with FA (clinical observation of M. Kabisch, Children's Hospital, University of Hamburg)

Monozygotic twins	Spontaneous			INH-induced			MMC-induced			DEB-induced		
	1	2	3	1	2	3	1	2	3	1	2	3
L.	20	0.29	1.4	19	0.28	1.4	75	>4.2	5.5	80	2.5	3.1
M.	22	0.37	1.5	37	0.75	1.7	52	>1.5	4.3	80	>3.0	3.6

[a] 1, Percent aberrant cells; 2, breaks/cell; 3, breaks/aberrant cell.

FA is one of the very few inherited diseases which can be cured by bone marrow transplantation (BMT). Most bone marrow donors are healthy, HLA identical siblings. However, sometimes other close relatives have donated their bone marrow in order to save the patient's life. It is of essential interest to observe continuously the cytogenetic changes in patients before and after BMT and to correlate these findings with their hematological and general health state. Five FA patients received bone marrow from siblings or relatives with opposite sex chromosomes. Thus, the donor lymphocytes carry their cytogenetic markers, as Blume et al. (1980) and Chaganti et al. (1980) already showed in their post-transplantation studies. Two other patients received bone marrow from siblings or relatives with identical sex chromosomes. The cytogenetic observations took place 1 to 46 months after BMT. The results or chromosomal breakage studies before and after BMT of three patients with donors of opposite sex and two patients with donors of identical sex are given in Table 2. For comparison, the values of the donor cells are included.

Base line breakage (N) became normal in three of five cases, in two cases breakage remained at high levels or increased in the donor cells.

MMC hypersensitivity clearly increased in two cases, decreased in two, and returned to normal in one case. DEB hypersensitivity decreased in all five cases significantly, however, it reached the range of control or donor cells only in three instances, while in two cases there was still considerable breakage present.

This shows that the donor lymphocytes, at least those identified by opposite sex chromosomes, became more sensitive than originally found in the donor cells in two of three cases, particularly when tested with MMC. Only one case (v.S. Sandra/Oliver) shows a normal cytogenetic finding which would actually exclude this FA patient from identification by DEB or MMC testing.

Using the XX or XY marker it was possible to identify in three cases donor metaphases growing exclusively in the lymphocyte cultures. Provided that donor cells in the two cases with identical sex chromosomes grow predominantly in the lymphocyte cultures, then these donor cells acquired a hypersensitivity to MMC, both of them showing also high endogenous breakage. Whether this is accomplished by a diffusable factor from the tissue surrounding the lymphocytes, or by cross-feeding or other mechanisms cannot be determined.

This finding of the in vivo cocultivation of FA cells with normal or FA-heterozygous cells contrasts to the first report on two FA patients who received BMT of the opposite sex by Shaham and Adler (1986). They compared spontaneous and DEB-induced chromosomal breakage in peripheral blood lymphocytes of the FA

Table 2. Base line breakage (N), MMC- and DEB-induced chromosomal instability before and after BMT in lymphocyte cultures (clinical report Friedrich et al. 1987)

Patient[a] (BM donor)	Break/cell culture condition N	MMC	DEB	Year of investigation	Intervall to BMT	Karyotype of lymphocytes	BM donor
M. Andrea	0.18	0.86	1.98	1985		46, XX	
	0.05	0.58	0.09	1987	1 year	46, XY	Brother
(Stefan)	(0.02)	(0.04)	(0.00)				
S. Muharem	1.20	–	> 4.82	1984		46, XY	
	0.04	0.94	> 2.18	1986			
	0.10	> 1.48	0.76	1986	7 months	46, XX	Sister
(Sidika)	(0.02)	(0.04)	(0.00)				
v. S. Sandra	0.42	–	> 10.0	1984		46, XX	
	0.47	2.36	> 10.0	1985			
	0.02	0.10	0.03	1987	19 months	46, XY	Brother
(Oliver)	(0.02)	(0.10)	(0.27)				
U. Ramasan	0.50	1.60	4.00	1985		46, XY	
	0.34	0.47	0.20	1986	8 months	46, XY	Brother
(Meshet)	(0.04)	(0.04)	(0.14)				
A. Yasmin	0.28	0.80	5.36	1984		46, XX	
	0.06	1.76	> 10.0	1985			
	0.52	3.12	1.68	1985	1 month	46, XX	Sister
(Soltan)	(0.02)	(0.02)	(0.16)				

[a] These results are part of the theories of U. Cramer-Giraud (in preparation). The IFAR No. of the FA patients are: M.A. 1055; S.M. 1047; S.S. 1030; U.R. 1073; A.Y. 1052.

patients following BMT without a full analysis of the sex chromosomes. Thus, their interpretation of the results – even with slightly increased chromosome breakage post-transplantation – was that FA cells do not produce any detectable clastogenic factor.

In vitro cocultivation of FA cells with CHO cells (Zakrzewski and Sperling 1980 a, b) resulted in an antagonistic response of increased chromosomal breakage in the CHO cells and a decrease of chromosomal breakage in the FA cells. Other authors (Nordensen et al. 1980; Shaham et al. 1980; Yoshida 1980) were not able to show a mutual interaction between FA cells and normal, cocultivated cells in vitro. Besides the basic interest in the clarification of any controversial finding concerning a clastogenic factor in AT, BS or FA, this must be of essential interest to the clinician who recommends BMT for FA patients. If there is a clastogenic factor in the FA tissue, inducing chromosomal breakage in the donor cells, then the bone marrow insufficiency might also occur after the bone marrow engrafting, followed by a new anemic stage of the disease. Therefore, we must be very precise in our posttransplantation investigation. Another approach is demonstrated in Table 3.

Two cases of BMT with cells of the opposite sex were investigated some months after the transplantation (Table 3). We used in situ hybridization (Schardin et al. 1985) of the X- and Y-chromosome in order to show how complete the acceptance of the transplanted bone marrow actually is. By this method it can be shown that not

Table 3. Results of in situ hybridization with Y- or X-probes after BMT

	Karyotype of FA patients	Karyotype of BM Donor	Months	After bone marrow transplantation		
				In situ hybridization of X or Y chromosomes		Remarks
				Metaphases	Interphases	
V. Elfriede	46, XX	46, XY	46	99 Y 1 –	0 X 979 Y 21 –	Small interphase nuclei without X or Y label
M. Yvonne	46, XX	46, XY	18	11 Y 0 –	994 Y 6 –	Due to deterioration of nuclei in FA cultures
Healthy control	46, XY	–	–	100 Y O –	999 Y 1 –	No such small nuclei, but one normal nucleus without Y

only metaphases can be evaluated for the presence of X- or Y-chromosomes, but also the interphase nuclei can be determined for the X- or Y-content: of 1000 interphase nuclei only 21 and 6 respectively, did not show X- or Y-label while only one nucleus of a male control did not contain a Y-signal. It is remarkable that the negative nuclei in the FA cases were particularly small and resemble broken cells or micronuclei, as often seen in FA-lymphocyte cultures (Schroeder-Kurth, unpubl. obs.) and also in first metaphases of PHA-stimulated lymphocytes (Obe et al. 1975).

These results once more demonstrate the successful BMT in two FA patients. A chromosomal breakage study will show whether the donor cells have acquired the phenomenon of endogenous chromosomal breakage and hypersensitivity to MMC camparable to the cases described above. Further observations are necessary to correlate these findings to the course of the disease, development of cancer, and the life expectancy (unpublished results).

These findings are even more interesting in view of the recent results of Rosselli and Moustacchi (1990) who did not find any increase of chromosomal breakage after cocultivation of human FA or "FA-like" mouse cells with normal human or mouse cells. They suggest that no detectable clastogenic factor is released by either human FA or FA-like mouse cells in their in vitro system. Our results show that normal lymphocytes growing in an FA patient acquire endogenous chromosomal breakage and MMC sensitivity. The in vivo system is more complex, or course, than the in vitro system used by Rosselli and Moustacchi. Thus although a comparison of the results is not possible, the results of the in vivo testing might prove important for the prognosis in each case.

Several unclassified cases with increased chromosomal instability and, in some cases, also with increased breakage after mutagen treatment, await differentiation (Wegner 1990). Some diseases might be discovered because of clinical similarities to well-known syndromes or eventually by careful cytogenetic studies.

Most cases today may escape detection and exact determination because elevated breakage usually will not be found by karyotyping, but only in an extended breakage study. Clarification of an unexpected, increased chromosome instability again is another task: unless peculiar, unique marker aberrations or cellular changes can be described, it remains difficult to classify single cases. Complementation experiments for determining different genetic defects are particularly valuable, if the clinical manifestations, together with cytogenetic properties at the cellular level, do not allow differentiation (Wegener 1990).

Acknowledgments. We are grateful to U. Böhlefeld and R. Schwinger for skillful technical assistance. This study is supported by the Deutsche Forschungsgemeinschaft.

References

Arwert F, Kwee ML (1989) Chromosomal breakage in response to cross-linking agents in the diagnosis of Fanconi anemia. In: Schroeder-Kurth TM, Auerbach AG, Obe G (eds), Faconi anemia, clinical, cytogenetic and experimental aspects. Springer, Berlin Heidelberg New York pp 83–92

Auerbach AD, Rogatko A, Schroeder-Kurth TM (1989 a) International Fanconi anemia registry: first report. In: Schroeder-Kurth TM, Auerbach AD, Obe G (eds) Fanconi anemia, clinical, cytogenetic and experimental aspects. Springer, Berlin Heidelberg New York pp 13–17

Auerbach AD, Rogatko A, Schroeder-Kurth TM (1989 b) International Fanconi anemia registry: relation of clinical symptoms to diepoxybutane sensitivity. Blood 73 (2): 391–396

Blume KG, Beutler E, Bross KJ, Schmitt GM, Spruce WE, Teplitz RL (1980) Genetic markers in human bone marrow transplantation. Am J Hum Genet 32: 414–419

Buchwald M, Clarke C, Ng J, Duckworth-Rysiecki G, Weksberg R (1989) Complementation and gene transfer studies in Fanconi anemia. In: Schroeder-Kurth TM, Auerbach AD, Obe G (eds) Fanconi anemia, clinical, cytogenetic and experimental aspects. Springer, Berlin Heidelberg New York pp 226–235

Chaganti RSK, Schonberg S, German J (1974) A manifold increase in sister chromatid exchanges in Bloom's syndrome lymphocytes. Proc Nat Acad Sci USA 71: 4508–4512

Chaganti RSK, Ratner LM, MC-KCC Bone Marrow Transplantation Group (1980) Use of chromosomal heteromorphisms in the analysis of hematologic reconstitution following hematogenic tissue transplantation. Am J Hum Genet 32: 65 A

Dallapiccola B, Porfirion B (1989) Chromosomal studies in Fanconi anemia heterozygotes. In: Schroeder-Kurth TM, Auerbach AD, Obe G (eds) Fanconi anemia, clinical, cytogenetic and experimental aspects. Springer, Berlin Heidelberg New York pp 145–158

Digweed M, Zakrzewski-Lüdcke S, Sperling K (1989) Complementation studies in Fanconia anemia using cell fusion and microinjection of mRNA. In: Schroeder-Kurth TM, Auerbach AD, Obe G (eds) Fanconi anemia, clinical, cytogenetic and experimental aspects. Springer, Berlin Heidelberg New York pp 236–254

Friedrich W, Ebell W, Blütters-Sawatzki R et al. (1987) Treatment of Fanconi anemia by bone marrow transplantation. Monatsschr Kinderheilkd 135: 253–258

German J, Bloom D, Passarge E (1985) The Bloom's syndrome registry: current status and incidence of cancer. Clin Genet 27: 310

Kuhn EM, Therman E (1986) Cytogenetics of Bloom's syndrome. Cancer Genet Cytogenet 22: 1–18

Kwee ML, Poll EHA, Kamp LLP van de, Koning H de, Eriksson AW, Loenje H (1983) Unusual response to bifunctional alkylating agents in a case of Fanconi anemia. Hum Genet 64: 384–387

Marx MP, Smith S (1989) Significance of cellular sensitivity in a group of parents of Fanconi anemia patients. In: Schroeder-Kurth TM, Auerbach AD, OIbe G (eds) Fanconi anemia, clinical, cytogenetic and experimental aspects. Springer, Berlin Heidelberg New York pp 137–144

Nordensen I, Björksten B, Lundh B (1980) Prevention of chromosomal breakage in Fanconi's anemia by cocultivation with normal cells. Hum Genet 56: 169–171

Obe G, Lüdcke BP, Waldenmaier K, Sperling K (1975) Premature chromosome condensation in a case of Fanconi's anemia. Hum Genet 28: 159–162

Ray JH, German J (1981) The chromosome changes in Bloom's syndrome, ataxia-teleangiectasia, and Fanconi's anemia. In: Arrighi FE, Rao PN, Stubblefield E (eds) Genes, chromosomes and neoplasia; Raven, New York pp 351–378

Rosselli F, Moustacchi E (1990) Cocultivation of Fanconi anemia cells and of mouse lymphoma mutants leads to interspecies complementation of chromosomal hypersensitivity to DNA-cross-linking agents. Hum Genet (in press)

Schardin M, Cremer T, Hager HD, Lang M (1985) Specific staining of human chromosomes in Chinese hamster × man hybrid cell lines demonstrates interphase chromosome territories. Hum Genet 71: 281–287

Schroeder TM, German J (1974) Bloom's syndrome and Fanconi's anemia: demonstration of two distinctive patterns of chromosome disruption and rearrangement. Humangenetik 25: 299–306

Schroeder-Kurth TM (1987) Pränatale Diagnostik von Sydromen mit erhöhter Chromosomenstabilität. In: Murken J (Ed) Pränatale Diagnostik und Therapie. Enke, Stuttgart pp 90–96

Schroeder-Kurth TM, Zhu TH, Hong Y, Westphal I (1989) Variation in cellular sensitivities among Fanconi anemia patients, non-Fanconi anemia patients, their parents and siblings, and control probands. In: Schroeder-Kurth TM, Auerbach AD, Obe G (eds) Fanconi anemia, clinical, cytogenetic and experimental aspects. Springer, Berlin Heidelberg New York pp 105–136

Shaham M, Adler B (1986) Chromosome breakage in Fanconi's anemia and normal cells following in vitro and in vivo cocultivation. Cancer Genet Cytogenet 23: 315–320

Shaham M, Bekcer Y, Cohen MM (1980) A diffusable clastogenic factor in ataxia teleangiectasia. Cytogenet Cell Genet 27: 155–161

Taylor AMR (1982) Cytogenetics of ataxia teleangiectasia. In: Bridges BA, Harnden DG (eds) Ataxia-teleangiectasia – a cellular and molecular link between cancer, neuropathology, and immune deficiency. Wiley, Chichester pp 53–81

Wegner RD (1990) Cytogenetic characterization of human disorders with increased spontaneous and induced chromosomal instability. In: Sabti RC, Obe G (eds) Enkaryotic chromosomes. Narosa, New Dehli (in press)

Yoshida MC (1980) Suppression of spontaneous and mitomycin C-induced chromosome aberrations in Fanconi's anemia by cell fusion with normal fibroblasts. Hum Genet 55: 223–226

Zakrzewski S, Sperling K (1980 a) Genetic heterogeneity of Fanconi's anemia demonstrated by somatic cell hybrids. Hum Genet 56: 81–84

Zakrzewski S, Sperling K (1980 b) Antagonistic effect of cocultivation on mitomycin C-induced aberration rate in cells of a patient with Fanconi's anemia and in Chinese hamster cells. Hum Genet 56: 85–88

Possible Causes of Variability of the Adaptive Response in Human Lymphocytes

G. Olivieri and A. Bosi[1]

Contents

1 Introduction

Mutagenesis research has been focused for a long period on the interaction between mutagens (mostly X-rays) and the cell structure modified by them. Subsequently, the emphasis was shifted to mutation as a cellular process in which the primary lesion was extensively modified by enzymatic DNA-repair systems constitutively present in the cell (Auerbach 1976). Finally, over the past 15 years, considerable interest has been focused on inducible cellular functions (possibly DNA repair activities) occurring in the cell after exposure to very low doses of mutagens, probably through DNA lesions (for reviews, see Walker 1985; Elespuru 1987). DNA-damage-inducible processes are probably part of a set of general biological phenomena by means of which the cells and the organism respond to various types of stress (Luckey 1980; McClintock 1984; Lindquist 1986; Morgan et al. 1986). When stress is represented by a mutagenic agent one of the most interesting responses is the adaptive response (AR; Samson and Cairns 1977). The mechanisms underlying the AR are reasonably well known in prokaryotes in which they were discovered (Sekiguchi and Nakabeppu 1987; Rebeck et al. 1988). In plant (Rieger et al. 1982; Heindorff et al. 1987) and animal cells (Samson and Schwartz 1980, 1983) investigation is focused mainly on the reduction of the yield of induced chromatid aberrations by various pretreatment procedures (clastogenic adaptation).

This phenomenon tends to be considered as an adaptive response even though its molecular basis is still unknown. We have therefore called adaptive response the reduction of clastostogenic damage observed in human lymphocytes cultivated in tritiated thymidine (^{3}HdThd) and subsequently exposed to high doses of X-rays (Olivieri et al. 1984; Wiencke et al. 1986).

Even though other research had already pointed to the existence of similar phenomena in other animal and plant systems, considerable efforts have been con-

[1] Dipartimento di Genetica e Biologia Molecolare, Università di Roma "La Sapienza", P. le Aldo Moro, 5, 00185 Rome, Italy

centrated over the past 5 years on this type of AR in human lymphocytes. The most important characteristics that have emerged are:

1. Very low doses of X-rays (0.5–1 cGy) act as ^{3}Hdthd (Shadley and Wolff 1987).
2. G_1 cells irradiated as early as 4 h after stimulation did show the AR, which persists for at least three cell cycles; a 4-h interval was enough for expression of the AR (Shadley et al. 1987).
3. The AR is dependent on both the total dose of the pretreatment and on the rate at which the dose is given (Shadley and Wiencke 1989).
4. A pretreatment with low doses of radiation can affect various types of clastogenic lesions induced by chemical mutagens (Wolff et al. 1988).
5. The AR is inhibited by 3-aminobenzamide (Wiencke et al. 1986; Shadley and Wolff 1987; Vijayalaxmi and Bukart 1988) and by cycloheximide (Wolff et al. 1989); furthermore, nicotinamide deficiency mimics the effects of 3AB (Wiencke 1987).
6. Exposure of lymphocytes to low doses of radiation induces several proteins (repair proteins?) (Wolff et al. 1989).
7. The AR can also be induced by low doses of bleomycin (Vijayalaxmi and Burkart 1989) H_2O_2, MNNG, and MMS (Wolff et al. 1988, 1989).

An AR involving also other end points has been observed in human lymphocytes and Chinese hamster cells (Ikushima 1987; Sanderson and Morley 1986; Morimoto et al. 1986; Seu Gupta and Bhattacharjee 1988). It is noteworthy that the dose-effect curves obtained with low doses of ionizing radiation could be indicative of AR in human lymphocytes (Pohl-Rüling and Fisher 1979; Pohl-Rüling et al. 1986; Luchnik and Sevankaev 1976). Finally, no adaptive response was observed by Schmid et al. (1989) in three donors.

Recently, it has been shown that there are variations between individuals with respect to the induction of AR in their lymphocytes (Sankaranarayanan et al. 1989; Bosi and Olivieri 1989). In this chapter we will present data of experiments done in order to understand the causes of such a variability.

2 Materials and Methods

Experiments were carried out using cultures of blood from donors which, in previous experiments (Bosi and Olivieri 1989), had displayed an adaptive response (donors 4, 11, 16) or not (donors 1, 3, 6, 18). In the present report AR+ will be used to indicate donors having displayed an adaptive response and AR-those that have not. Whole blood (0.5 ml) was added to 4.5 ml of RPMI 1640 medium without fetal calf serum (Wolff et al. 1984), 2 mM glutamine, 100 units/ml penicillin, 100 μg/ml streptomycin, and 2% phytohemagglutinin M (Gibco). The experiments consisted first of exposing cultured human lymphocytes to adapting treatments and subsequently challenging the cells with high doses of X-rays. The cells were scored to see whether the prior exposure reduced the number of chromatid and isochromatid breaks induced by the challenging doses. Few chromatid exchanges (less than 10%) were recorded but not included in the analysis. Five experiments were carried out as follows: Exp. I with donors 6, 16, 18; Exp. II with donors 1, 11; Exp. III with donors 3, 4; Exp. IV with donor 3, from whom the blood sample was taken at 9 a.m. and 5 p.m.; Exp. V with donors 1, 3.

In all the experiments the conditioning pretreatment consisted of 0.02 Gy of X-rays administered 26 (Exps. I and V) or 30 h (Exps. II, III and IV) after stimulation with PHA; in Exp. I also low doses of MMS were used as conditioning pretreatment. The cells were subsequently challenged with 0.75 or 0.40 Gy of X-rays and fixed 2 h later or 1.5 Gy of X-rays followed by fixing 6 h later. The challenge treatment was carried out after 72 (Exp. I), 50 (Exps. II and IV), or 54 h (Exps. III and V). In all the experiments except the first, several compounds, whose effect on the adaptive response was to be studied, were added to some of the cultures. The following substances were used: adrenal cortex extracts (Maxicortex „2000" by „Manetti e Roberts") in Exps. II, III, IV, and V; insulin (Actrapid HM by Novo Farm.) in Exp. III; thymus extracts (Leucotrofina by Ellem) in Exp. III; hydrocortisone (Sigma) in Exp. V; interleukin-2 and interferon-gamma (Boehringer) in Exp. V. Irradiation was carried out with 200 kVp X-rays (Gilardoni MGL 200/8D, 0,2 mm Cu added filtration, 8 mA, 0.60 Gy/min). Two hours before fixation 0.1 ml of colcemid (final concentration 2×10^{-7} M) was added to each culture, and fixation was performed according to standard cytological procedures; for each point examined two parallel cultures were set up. A diagram of general experimental protocols is shown in Table 1.

3 Results

Experiments were carried out using human lymphocytes from donors analyzed previously (Bosi and Olivieri 1989). Lymphocytes were used: four donors were AR− and three AR+. The results of the first experiment, in which the AR of two previous AR-donors (donors 6 and 18) and one previous AR+ donor (16) was studied, are shown in Table 2. Conditioning pretreatment consisted of 0.2 Gy of X-rays or three different low doses of MMS. The lymphocytes were challenged with 0.75 or 1.5 Gy of X-rays. The cultures were fixed 2 h after treatment with 0.75 Gy or 6 h after treatment with 1.5 Gy. To our surprise the two AR− donors displayed a distinct AR: conditioning pretreatment with both X-rays and an optimal range of low doses of

Table 1. General experimental protocols

Experiment No.	Donor No.	Compound added to the cultures	Conditioning pretreatment	Challenge treatment
I	16; 6; 18	–	0.02 Gy (X-rays) or MMS	0.75 or 1.5 Gy (X-rays)
II	11; 1	Adrenal cortex extracts	0.02 Gy (X-rays)	0.40 Gy (X-rays)
III	4; 3	Adrenal cortex extracts Insulin; thymus extracts	0.02 Gy (X-rays)	0.40 or 1.5 Gy (X-rays)
IV	3	Adrenal cortex extracts	0.02 Gy (X-rays)	0.40 or 1.5 Gy (X-rays)
V	3; 1	Adrenal cortex extracts Hydrocortisone; IL-2; IFN-γ	0.02 Gy (X-rays)	0.40 Gy (X-rays)

Table 2. (Exp. 1). The effect of different types of conditioning pretreatment on the frequency of chromatid aberrations induced in human lymphocytes challenged with 0.75 or 1.5 Gy of X-rays (100 cells were scored for each point)

Donor	No. of chromatid and isochromatid deletions for differnt types of treatment									
No./ Sex	None	0.02 Gy	0.036 mM MMS	0.75 Gy	0.02 + 0.75 Gy	0.0048 mM MMS + 0.75 Gy	0.018 mM MMS + 0.75 Gy	0.036 mM MMS + 0.75 Gy	1.5 Gy	0.02 + 1.5 Gy
16; M	0	3	1	180	229[b]	242[c]	228[b]	152	30	29
6; F	3	1	2	482[a]	412[a,b]	135[c]	167[c]	213	–	–
18; M	2	3	3	356	301[d]	299[d]	321	339	32	18[d]

[a] Two hundred cells were scored for these points.
[b] $P < 0.02$.
[c] $P < 0.01$.
[d] $P < 0.05$ (Student's t-test).

MMS led to the onset of a smaller number of aberrations both after 0.75 and 1.5 Gy of X-rays (only one of the two donors was tested at this dose). The reverse response was obtained using lymphocytes from donor 16, who had been AR+ in preceding experiments. Overall, the results of Exp. I indicate that the absence of AR does not depend on stable constitutional traits but rather on some transient physiological parameters. In order to identify these parameters, in the subsequent four experiments we added to the cultures compounds which are present in variable quantities in blood and that could possibly interfere with lymphocyte metabolism. The results of Exp. II, in which the lymphocytes of two previous donors, one AR+ (11) and one AR− (1), were used, are shown in Table 3. In this experiment, under standard culture conditions, neither donor displayed an AR. However, if adrenal cortex extracts are added to the cultures at the time of conditioning pretreatment, the AR occurs: for donor 11 this is true at both doses, while for donor 1 it is true only for the higher dose. The presence of adrenal cortex extracts in the cultures immediately after stimulation with PHA has no effect on the presence or absence of AR. In Exp. III (Table 4) we further tested the action of adrenal cortex extracts added to the cultures 30 min earlier, at the same time as, or 1.5–3 h after conditioning pretreatment. We also tested the action of insulin and thymus extracts. The experiment was carried out using lymphocytes from two donors who had previously proved to be AR+ (4) and AR− (3). Under standard culture conditions, the two donors confirmed the presence (donor 4) or absence (donor 3) of AR. The results in Table 4 also show that the different treatment with adrenal cortex extracts, although not modifying the Ar in donor 4, lead to its appearance in donor 3, in cells fixed either 2 or 6 h after irradiation. The same effect is exerted by thymus extracts; insulin, on the other hand, has no effect. Experiments II and III thus show that adrenal cortex extracts can have an effect on the presence of AR in cultures in which it is absent under standard conditions. Since circadian variations in adrenal corticosteroids are well known, in Exp. IV a test was made for possible variations in AR in lymphocytes taken from donor 3 (AR−) at 9 a.m. and 5 p.m. Adrenal cortex extracts at doses equal to or greater

Table 3. (Exp II). Effect of adrenal cortex extracts on the adaptive response of human lymphocytes. A "conditioning" pretreatment of 0.02 Gy of X-rays at 30 h and a "challenge" treatment of 0.40 Gy of X-rays at 50 h were used (200 cells were scored for each point; 100 for the controls)

Donor	Adrenal Cortex extracts concentration[a]	Number of chromatid and isochromatid deletions for various types of treatment			
No./sex		None	0.02 Gy	0.40 Gy	0.02 + 0.40 Gy
11; F	–	3	1	178	191
	0.32 μg/ml (1 h)	–	–	152	148
	0.32 μg/ml (30 h)	–	–	229	160[b]
	3.2 μg/ml (1 h)	1	2	216	211
	3.2 μg/ml (30 h)	1	2	277	159[c]
1; F	–	1	0	122	124
	0.32 μg/ml (1 h)	–	–	168	199
	0.32 μg/ml (30 h)	–	–	162	180
	3.2 μg/ml (1 h)	3	1	186	190
	3.2 μg/ml (30 h)	1	0	168	121[c]

[a] Expressed as hydrocorticone concentration; the number in parentheses are the times after PHA stimulation at which the extracts were added to the cultures.
[b] $P < 0.02$. (Student's t-test)
[c] $P < 0.01$. (Student's t-test)

than those used in previous experiments were added to some cultures. The results in Table 5 show that, under standard conditions, donor 3 displayed no AR in either sample. The adrenal cortex extracts allow AR to be obtained when they are used at the same dose as in the previous experiments (not at the higher dose) and if they are added to the culture at the time of conditioning pretreatment, although not when they are added 4 h earlier. Using the blood sample taken at 9 a.m. we also tested several cultures by fixing them 6 h after irradiation with 1.5 Gy of X-rays. The results were similar to those obtained by fixation 2 h after irradiation. Hydrocortisone is an important component of the adrenal cortex extracts; for this reason, in Exp. V we tested it at two different doses. In this experiment we also tested both the adrenal cortex extracts and two compounds which could interfere with the metabolism of cultured lymphocytes, namely interleukin-2 (IL-2) or during infections, namely, interferon-gamma (IFN-γ). The two donors tested in this experiment were previously AR−. Also, in this experiment (Table 6) in standard cultures, the two donors displayed no AR. Addition to the cultures of adrenal cortex extracts and hydrocortisone at the lower dose led to the onset of AR (in donor 1 the results with the lower dose of hydrocortisone display the same trend as in donor 3, although they are not statistically significant). IL-2 effectively induced AR only in donor 3 (however, donor 1 displayed the same trend). On the other hand, IFN-γ did not modify the standard situation in either donor.

Table 4. (Exp. III). Effect of adrenal cortex extracts, insulin and thymus extracts, added to the cultures at different times after PHA stimulation, on the adaptive response of human lymphocytes. A "conditioning" pretreatment of 0.02 Gy of X-rays at 30 h and a "challenge" treatment of 0.40 or 1.5 Gy of X-rays were used (100 cells were scored for each point)

Donor	Treatment	Number of chromatid and isochromatid deletions for various types of treatment					
No./ sex		None	0.02 Gy	0.40 Gy	0.02 + 0.40 Gy	1.5 Gy	0.02 + 1.5 Gy
	–	5	3	84	59[c]	–	–
	Ad. cort. extracts[a] 2.4 μg/ml (29.5 h)	3	2	71	43[e]	–	–
	Ad. cort. extracts 2.4 μg/ml (30 h)	–	–	–	48[c]	–	–
4: M	Ad. cort. extracts 2.4 μg/ml (31.5 h)	–	–	–	52	–	–
	Ad. cort. extracts 2.4 μg/ml (33 h)	–	–	–	62	–	–
	Insulin 0.04 U/ml (30 h)	–	–	–	–	–	–
	Insulin 0.4 U/ml (30 h)	3	2	129	117	–	–
	Thymus extracts 30 μg/ml (30 h)	1	2	124	87[d]	–	–
	–	1	0	65	93[c]	24	27
	Ad. cort. extracts 2.4 μg/ml (29.5 h)	1	3	105	71[d]	–	–
	Ad. cort. extracts 2.4 μg/ml (30 h)	–	–	–	70[c]	–	–
3; M	Ad. cort. extracts 2.4 μg/ml (31.5 h)	–	–	–	84	47[b]	18[b, d]
	Ad. cort. extracts 2.4 μg/ml (33 h)	–	–	–	90	–	–
	Insulin 0.04 U/ml (30 h)	–	–	101	117	–	–
	Insulin 0.4 U/ml (30 h)	3	3	106	108	–	–
	Thymus extracts 30 μg/ml (30 h)	0	1	155[b]	120[b, c]	–	–

[a] Expressed as hydrocortisone concentration; the numbers in parentheses are the times after PHA stimulation at which the extracts were added to the cultures.
[b] Two hundred cells were scored for these points.
[c] $P < 0.05$ (Student's t-test).
[d] $P < 0.02$.
[e] $P < 0.01$.

4 Discussion

If AR would occur in somatic cells other than lymphocytes and in germinal cells, this would modify the evaluation of the genetic risk by low doses of chemical mutagens and ionizing radiation (Sobels 1989). Indeed, in view of the AR, the low doses should be studied not only in view of the probability with which they induce mutations but also because of the modifications they produce in the cells which in turn are capable of substantially modifying any subsequent mutagenic treatment. The induced enzyme system appears to be extremely complex (Walker 1985; Wolff et al. 1989; Rebeck et al. 1988; Boothman et al. 1989; Akaboshi and Howard-Flanders 1989). Only the investigation at the molecular level will elucidate all the effects that can derive from it and which probably affect various end points.

As far as the AR in human lymphocytes is concerned, the results of the experiments described here indicate that the variablility of AR found in various donors is not linked to their genetic constitution but depends on some transient physiological parameters. Two donors, who were AR+ in previous experiments, displayed no adaptive response. On the other hand, two AR− donors displayed a clear-cut AR

Table 5. (Exp. IV). Effect of adrenal cortex extracts on the adaptive response of human lymphocytes in two samples of blood drawn at 9 a.m. and 5 p.m. from the same donor No. 3. A "conditioning" pretreatment of 0.02 Gy of X-rays at 30 h and a "challenge" treatment of 0.40 or 1.5 Gy were used (100 cells were scored for each point)

Time	Adrenal cortex extracts concentration[a]	Number of chromatid and isochromatid deletions for various types of treatment					
		None	0.02 Gy	0.40 Gy	0.02 + 0.40 Gy	1.5 Gy	0.02 + 1.5 Gy
9 a.m.	–	5	1	212[b]	184[b]	14	27
	2.4 μg/ml (26 h)	2	1	199[b]	179[b]	18	18
	2.4 μg/ml (30 h)	–	–	–	157[b, c]	–	8[d]
	12 μg/ml (29.5 h)	7	3	149[b]	182[b]	16	24
5 p.m.	–	2	2	67	81	–	–
	2.4 μg/ml (26 h)	1	2	70	70	–	–
	2.4 μg/ml (30 h)	–	–	–	47[c]	–	–
	12 μg/ml (29.5 h)	4	3	72	62	–	–

[a] Expresed as hydrocortisone concentration; the numbers in parentheses are the times after PHA stimulation at which the extracts were added to the cultures.
[b] Two hundred cells were scored for these points.
[c] $P < 0.05$ (Student's t-test).
[d] $P < 0.02$.

Table 6. (Exp. V). Effect of adrenal cortex extracts, hydrocortisone, interleukin-2 (IL-2), interferon-gamma (IFN-γ) added to the cultures 26 h after PHA stimulation on the adaptive response of human lymphocytes. A "conditioning" pretreatment of 0.02 Gy of X-rays at 26 h and a "challenge" treatment of 0.40 Gy of X-rays at 54 h were used (100 cells were scored for each point)

Donor No./sex	Treatment	Number of chromatid and isochromatid deletions for various types of treatment			
		None	0.02 Gy	0.40 Gy	0.02 + 0.40 Gy
3; M	–	4	1	107[a]	105[a]
	Ad. cort. extracts 2.4 μg/ml	1	1	85	60[b]
	Hydrocort. 2 μg/ml	1	0	72	49[b]
	Hydrocort. 6 μg/ml	8	3	78	79
	IL-2 60 U/ml	2	0	77	51[b]
	IFN-γ 600 U/ml	1	1	62	54
	IFN-γ 1200 U/ml	3	2	–	64
1; F	–	1	0	95[a]	105[a]
	Ad. cort. extracts 2.4 μg/ml	1	1	115[a]	87[a, b]
	Hydrocort. 2 μg/ml	0	2	57	42
	Hydrocort. 6 μg/ml	4	1	51	58
	IL-2 60 U/ml	1	0	75	60
	IFN-γ 600 U/ml	0	2	62	61
	IFN-γ 1200 U/ml	1	2	–	82

[a] Two hundred cells were scored for these points.
[b] $P < 0.05$ (Student's t-test).

in Exp. I. Furthermore, during the experiments, cases were observed in which, under standard culture conditions, the donor displayed no AR, and yet an AR could be evidenced by modifying the culture conditions. Therefore, our evidence seems to point to the great importance for AR of the metabolic state of cells during conditioning. We do not know exactly what modifications are produced in the metabolism of the lymophocytes by the compounds which also elicited an AR in donors who displayed none under standard conditions. However, in all probability, the compounds used by us all have the power to stimulate protein synthesis in cells. Considerable evidence points to the fact that the presence of an AR in lymphocytes or in other biological systems needs a certain level of cellular metabolic activity and an unimpaired protein symthesis (Heindorff et al. 1987). Indeed, it is true that: (1) AR cannot be induced in unstimulated lymphocytes (Moquet et al. 1987; Shadley et al. 1987; Afzal et al., unpubl. results) and dry seeds (Nicoloff et al. 1985); (2) protein synthesis inhibitors such as cyclohexamide in lymphocytes (Wolff et al. 1989) and cyclohexamide and chloranphenicol in plant cells (Heindorff et al. 1987) prevent AR; (3) AR can be prevented by blocking the poly(ADP-ribosyl) action (Wiencke et al. 1986; Wiencke 1987; Vijayalaxmi and Bukart 1988); (4) new proteins are produced by lymphocytes stimulated using low doses (Wolff et al. 1989).

Further research is needed in order to understand the true action mechanism of the compounds used by us. It is interesting to note that, whatever the action mechanism of these compounds, they are physiologically present in the plasma and can therefore cause the AR to vary. In this sense it is of special interest to point out that the presence in the plasma of cortico-suprarenal hormones is regulated among other things both by life-style and various stress situations.

As in previous experiments (Bosi and Olivieri 1989) in Exp. I donor 18 displayed a high G2 sensitivity. In the lymphocyte sample fixed 6 h after irradiation the frequency of aberrations in donor 18 was not, however, higher than in the other donors. At both fixation times (i.e., 2 and 6 h after irradiation, respectively) donor 18 displayed an AR, while donor 16 did not. This confirms that AR is not linked to variations in individual sensitivity, which are often found in G2 (Natarajan et al. 1982; Obe and Beek 1984), nor to the type of protocol used (i.e., fixation 2 or 6 h after irradiation, respectively).

It should be noted that, in Exp. I instead of an AR, donor 16 showed a significant increase in the number aberrations after conditioning treatment. This finding has already been reported previously (Bosi and Olivieri 1989) and usually occurred after challenge treatments with MMS (Wolff et al. 1988). This synergism is difficult to interpret. Conceivably, the constitutive and induced DNA repair systems are very complex and, as a result of the induction of several of these enzymes, is possibly liable to imbalances affecting the final result of damage repair.

Acknowledgments. This work was supported by a grant from the Consiglio Nazionale delle Ricerche (Centro Genetica Evoluzionistica) and by Euratom Grant B16-0186-I.

References

Akaboshi E, Howard-Flanders P (1989) Proteins induced by DNA-damaging agents in cultured *Drosophila* cells. Mutat Res 227: 1–6

Auerbach C (1976) Mutation research. Chapman & Hall, London

Boothman DA, Bouvard I, Hughes EN (1989) Identification and characterization of X-ray-induced proteins in human cell. Cancer Res 49: 2871–2878

Bosi A, Olivieri G (1989) Variability of the adaptative response to ionizing radiations in humans. Mutat Res 211: 13–71

Elespuru RK (1987) Inducible responses to DNA damage in bacteria and mammalian cells. Environ Mol Mutagen 10: 97–116

Heindorff KR, Rieger I, Schubert I, Michaelis A, Aurich O (1987) Clastogenic adaptation of plant cells – reduction of the yield of clastogen induced chromatid aberrations by various pretreatment procedures. Mutat Res 181: 157–171

Ikushima T (1987) Chromosomal responses to ionizing radiation reminiscent of an adaptive response in cultured Chinese hamster cells. Mutat Res 180: 215–221

Lindquist S (1986) The heat shock response. Annu Rev Biochem 55: 1151–1191

Luchnik NV, Sevankaev AV (1976) Radiation-induced chromosomal aberrations in human lymphocytes. 1. Dependence on the dose of gamm-rays and an anomaly at low doses. Mutat Res 36: 363–378

Luckey TD (1980) Hormesis with ionizing radiation. (CCR Press, Boca Raton, Fla)

McClintock B (1984) The significance of responses of the genome to challenge. Science 226: 792–801

Moquet JE, Lloyd DC, Prosser JS, Edwards AA (1987) Sister-chromatid exchanges induced by mitomycin C after exposure of human lymphocytes in G_0 to a low dose of X-radiation. Mutat Res 176: 143

Morgan RW, Christman MF, Jacobson FS, Storz G, Ames BN (1986) Hydrogen peroxide-inducible proteins in *Salmonella typhimurium* overlap with heat shock and other stress protein. Proc Natl Acad Sci USA 83: 8059–8063

Morimoto KM, Sato-Mizuno, Koizumi A (1986) Adaptation-like response to the chemical induction of sister chromatid exchanges in human lymphocytes. Hum Genet 73: 81–85

Natarajan AT, Meijers M, Rijn JLS van (1982) Individual variability of human cells in induction of chromosomal alterations by mutagens. In: Sorsa M, Vainio H (eds) Mutagens in our environment. Liss, New York, pp. 75–88

Nicoloff HK, Gecheff R, Rieger, Michaelis A (1985) Clastogenic adaptation in barley: differential response of presoaked and dry seeds. Mutat Res 143: 83–85

Obe G, Beek B (1984) Human peripheral lymphocytes in mutation research. In: Obe G (ed) Mutation in man. Springer, Berlin Heidelberg New York, pp 177–197

Olivieri G, Bodycote J, Wolff S (1984) Adaptive response of human lymphocytes to low concentration of radioactive thymidine. Science 223: 594–597

Pohl-Rüling J, Fisher P (1979) The dose-effect relationship of chromosome aberration to and irradiation in a population subjected to an increased burden of natural radioactivity. Radiat Res 80: 61–81

Pohl-Rüling J, Fischer P et al. (1986) Chromosomal damage induced in human lymphocytes by low doses of D-T neutrons. Mutat Res 173: 267–272

Rebeck GW, Coons S, Carroll P, Samson L (1988) A second DNA methyltransferase repair enzyme in *Escherichia coli*. Proc Natl Acad Sci 85: 3039–3043

Rieger RA, Michaelis A, Nicoloff H (1982) Inducible repair processes in plant root tip meristems? "Below additivity effects" of unequally fractionated clastogen concentrations. Biol Zentralbl 101: 125–138

Samson L, Cairns J (1977) A new pathway for DNA repair in *Escherichia coli*. Nature 267: 281–283

Samson L, Schwartz JL (1980) Evidence for an adaptive DNA repair pathway in CHO and human skin fibroblast cell lines. Nature 287: 861–863

Samson L, Schwartz JL (1983) The induction of resistance to alkylating damage in mammalian cells. In: Lawrence CW (ed) Induced mutagenesis: molecular mechanisms and their implications for environmental protection. Plenum, New York, pp 291–309

Sanderson BJS, Morley AA (1986) Exposure of human lyphocytes to ionizing radiation reduces mutagenesis by subsequent ionizing radiation. Mutat Res 164: 347–351

Sankaranarayanan K, Duyn A von, Loos MJ, Natarajan AT (1989) Adaptative response of human lymphocytes to low-level radiation from radioisotopes or X-rays Mutat Res 211: 7–12

Schmid E, Bauchinger M, Nahrstedt U (1989) Adaptive response after X-irradition of human lymphocytes? Mutagenesis 4: 87–89

Sekiguchi M, Nakabeppu Y (1987) Adaptive response: induced synthesis of DNA repair enzymes by alkylating agents. Trends Genet 3: 51–54

Seu Gupta S, Bhattacharjee SB (1988) Induction of repair functions by hydrogen peroxide in Chinese hamster cells. Int J Radiat Biol 53: 935–942

Shadley JD, Wiencke JK (1989) Induction of the adaptive response by X-rays is dependent on radiation intensity. Int J Radiat Biol 56: 107–118

Shadley J, Wolff S (1987) Very low doses of X-rays can cause human lymphocytes to become less susceptible to ionizing radiation. Mutagenesis 2: 95–96

Shadley JD, Afzal V, Wolff S (1987) Characterization of the adaptive response to ionizing radiation induced by low doses of X-rays to human lymphocytes. Radiat Res 111: 511–517

Sobels FH (1989) Models and assumptions underlying genetic risk assessment. Mutat Res 212: 77–89

Vijayalaxmi, Bukart W (1988) Effect of 3-aminobenzamide on chromosome damage in human blood lymphocytes adapted to bleomycin. Mutagenesis 4: 187–189

Vijayalaxmi, Bukart W (1989) Resistance and cross-resistance to chromosome damae in human blood lymphocytes adapted to bleomycin. Mutat Res 211: 1–5

Walker GC (1985) Inducible DNA repair systems. Annu Rev Biochem 54: 425–457

Wiencke JK (1987) Nicotinamide deficiency in human lymphocytes prevents the [^{3}H]thymidine-induced adaptive response for the repair of X-ray-induced chromosomal damage. Exp Cell Res 171: 518–523

Wiencke JK, Afzal V, Olivieri G, Wolff S (1986) Evidence that the [^{3}H]thymidine-induced adaptive response of human lymphocytes to subsequent doses of X-rays involves the induction of a chromosomal repair mechanism. Mutatgenesis 1: 375–380

Wolff S, Afzal V, Lindquist PB (1984) Cultured human lymphocytes proliferate faster in medium lacking fetal calf serum and antibiotics. Mutat Res 129: 207–213

Wolff S, Afzal V, Wiencke JK, Olivieri G, Micheli A (1988) Human lymphocytes exposed to low doses of ionizing radiation become refractory to high doses of radiation as well as chemical mutagens that induce double-strand breaks in DNA. Int J Radiat Biol 53: 39–48

Wolff S, Wiencke JK, Afzal V, Youngblom J, Cortes F (1989) The adaptive response of human lymphocytes to very low doses of ionizing radiation: a case of induced chromosmal repair with the induction of specific proteins. In: *Low dose radiation: biological bases of risk assessment.* Taylor, London (in press).

Adaptation of Human Lymphocytes to Radiation or Chemical Mutagens: Differences in Cytogenetic Repair

S. WOLFF[1], G. OLIVIERI[1, 2], and V. AFZAL[1]

Contents

1 Introduction

It is now known that the production of DNA damage by a variety of insults from ultraviolet light, alkylating agents, heat, or oxidizing agents can induce a series of responses that are often related to the repair of the initial damage (see Boothman et al. 1989). One type of induced repair, which was first found in *Escherichia coli*, has been called the adaptive reponse (Samson and Cairns 1977). In these experiments with bacteria, and in later experiments with mammalian cells (Samson and Schwartz 1980, 1983), it was found that cells that had been pre-exposed to low levels of alkylating agents became less susceptible to the damaging effects of higher doses of the same or other alkylating agents. Adaptation after low-level exposures has also been found to occur in the plant *Vicia faba* exposed to chemical clastogens (Rieger et al. 1982) and in *E. coli* exposed to hydrogen peroxide (Demple and Halbrook 1983). The adaptation after exposure to alkylating agents was related to the induction of an alkyltransferase (Karran et al. 1979; Olsson and Lindahl 1980). Ionizing radiation, however, brings about its damage mainly by the induction of chromosome aberrations, which are the result of DNA double-strand breaks (Wolff 1972, 1978; Evans 1977; Morgan et al. 1988) that can be induced directly by the X-rays or indirectly by the conversion of X-ray-induced single-strand breaks to double-strand breaks by endogenous single-strand nucleases (Natarajan et al. 1980). Thus, with ionizing radiation, any repair mechanism would be independent of the induction of an alkyltransferase.

---

[1] Laboratory of Radiobiology and Environmental Health, University of California, San Francisco, San Francisco, California 94143-0750, USA
[2] Dipartimento di Genetica e Biologia Moleculare, Universitá Degli Studi di Roma, Italy

2 Human Lymphocytes

2.1 Tritiated Thymidine and X-Rays

In human lymphocytes, pre-exposure of the cells to low doses of tritiated thymidine, [^{3}H]dThd, or X-rays renders the cells resistant to the induction of chromosomal damage by subsequent high doses of X-rays. In the original experiments in which the phenomenon of adaptation to ionizing radiation was discovered (Olivieri et al. 1984), human lymphocytes were exposed to [^{3}H]dThd throughout the culture period, and then were X-irradiated with 150 cGy 6 h before fixation. The results obtained showed that pre-exposure of lymphocytes to very low doses of [^{3}H]dThd could decrease the number of chromatid deletions induced by the 150 cGy challenge (Table 1). The results obtained when [^{3}H]dThd was added to blood cultures throughout the entire culture period, however, were quite variable, largely because, in blood, the amount of [^{3}H]dThd incorporated into the cells is highly dependent upon the catabolism of dThd to an unincorporable form (Bodycote and Wolff 1986). This variability was markedly reduced by pulse treating the cells with [^{3}H]dThd only when they were in S-phase of culture. Under such conditions, catabolism is minimized and the amount of [^{3}H]dThd incorporated is controlled to a far greater extent than possible if the dThd is added to the medium for the entire culture period (Wiencke et al. 1986).

In a series of experiments in which the cells were exposed to diminishing concentrations of [^{3}H]dThd in an attempt to find how much was necessary to induce the effect, it was found that exposures to 18–37 Bq/ml (0.0005–0.001 μCi/ml) of [^{3}H]dThd, which in themselves did not induce any chromatid breaks observable in 200 cells, did reduce the number of breaks induced by a subsequent dose of 150 cGy of X-rays by a factor of two. Calculation of the number of disintegrations that could have occurred between the time of administration of [^{3}H]dThd and the challenge dose of 150 cGy indicated that this adaptive response could be brought about by as little as one disintegration, showing that the response was not caused by a chronic exposure to the material, as was the case in the original experiments of Samson and Cairns (1977) with alkylating agents, but was brought about by a very low acute dose. When experiments were carried out in which very low doses of X-rays themselves were used as the adapting dose (Shadley and Wolff 1987), it was found that exposures to 0.5–1 cGy (0.5–1 rad) alone could cause adaptation and lead to a reduction in the number of chromatid breaks produced by 150 cGy given at a later time (Table 2).

In the early experiments with [^{3}H]dThd, questions arose regarding whether or not the response could be attributed to a sampling artifact related to radiation-induced delays in cell progression and variations in the radiosensitivity of cells throughout the G_2 phase of the cell cycle. The initial experiments had indeed shown that G_2 cells that reached metaphase 3 h after the 150 cGy challenge were very much more sensitive than were S-cells that reached metaphase 6 h after the challenge. It was also found, however, that the cells showed uniform sensitivity when fixed anywhere from 5–11 h after the 150 cGy challenge, indicating that the adaptation brought about by [^{3}H]dThd could not be attributed to stage sensitivity phenomena and changes in the progression of treated cells to mitosis, which would lead to the scoring of different populations of cells at metaphase.

Table 1. Effect of low concentrations of [^{3}H]dThd on the number of chromatid deletions induced by 150 cGy of X-rays (100 cells per point). Data from Olivieri et al. (1984)

Treatment	Deletions[a]
None	0
[^{3}H]dThd (0.01 μCi/ml)	2
[^{3}H]dThd (0.1 μCi/ml)	5
150 cGy	36
[^{3}H]dThd (0.01 μCi/ml)plus 150 cGy	23 (38)
[^{3}H]dThd (0.1 μCi/ml) plus 150 cGy	13 (41)

[a] Expected values in parentheses.

Table 2. Frequency of chromatid deletions in human lymphocytes treated with 0.5 or 1 cGy of X-rays followed by 150 cGy of X-rays (200 cells per point). Data from Shadley and Wolff (1987)

Treatment	Deletions[a]
Control	4
0.5 cGy (34 h)	4
1 cGy (34 h)	6
150 cGy (48 h)	81
0.5 cGy (34 h) plus 150 cGy (48 h)	45 (81)
1 cGy (34 h) plus 150 cGy (48 h)	48 (83)

[a] Expected values in parentheses.

Similar experiments carried out after exposure to very low doses of X-rays (Table 3) have now shown that after an adapting dose of 1 cGy, the response occurs at all fixation times from 6–12 h, although in this series of experiments the yield of aberrations induced by 150 cGy alone increased with fixation time.

This adaptive response to very low doses of radiation can be inhibited by 3-aminobenzamide (3AB), which inhibits poly(ADP-ribose) polymerase (Benjamin and Gill 1980), an enzyme induced in response to the production of strand breaks within the cell, and which has been suggested to be involved in the repair of chromosomal breaks (Wiencke et al. 1986). It was postulated that low-dose exposures of X-rays induced a repair mechanism that efficiently repaired the initial breaks induced by subsequent high doses of radiation, and thus resulted in less total damage being left in the cell. In an attempt to see if inhibition of poly(ADP-ribose) polymerase by 3AB could actually prevent chromosomal repair, an experiment was carried out in G_1 lymphocytes in which the total dose of X-rays was fractionated into two 100-cGy doses, separated by various periods of time. Under such conditions, dicentrics and rings, which are brought about by the interaction of two independently produced chromosome breaks, increase approximately as the square of the total dose if it is not fractionated. When, however, the dose is fractionated with sufficient time between the two fractions for breaks induced by the first dose to become repaired

Table 3. Effect of 1 cGy pretreatment (at 24 h) on the frequency of chromatid deletions obtained in human lymphocytes challenged with 150 cGy of X-rays 48 h after stimulation and fixed 6–12 h later (200 cells were scored for each point)

Treatment (cGy)	Fixation time after challenge (h)	Chromatid and isochromatid deletions	Expected
None	6	0	
1	6	4	
150	6	78	
1 + 150	6	48	82
150	8	88	
1 + 150	8	62	92
150	10	100	
1 + 150	10	52	104
150	12	117	
1 + 150	12	90	121

before breaks are induced by the second dose, no interaction can occur between the breaks induced by the first and second doses. Under these conditions, the yield of aberrations is simply the sum of the yields of dicentrics and rings induced by the first and second dose. In control experiments, it had been found that breaks from the first dose of 100 cGy were repaired before 3 h had elapsed, and thus were unable to interact with breaks induced by a second dose given 3 h later. When 3AB was present between the two doses, however, the breaks from the first dose remained unrepaired and capable of interacting with breaks from a second dose given as much as 7 h after the first (Table 4). These experiments showed that 3AB actually did prevent the repair of chromosomal breaks, even in G_1 cells.

2.2 Oxidizing Agents

It had already been determined in *E. coli* that cells pretreated with nontoxic concentrations of hydrogen peroxide, H_2O_2, became less susceptible to the killing effects of γ-rays (Demple and Halbrook 1983). It was not known whether this H_2O_2-induced repair mechanism that affected cell killing in *E. coli* was the same as the radiation-induced repair mechanism for chromosomal damage in human lymphocytes. It was known, however, that damage induced by both H_2O_2 and X-rays was mediated by reactive oxygen species. Experiments testing whether or not chemical agents that induce reactive oxygen species could induce this chromosomal repair mechanism in human lymphocytes were carried out in which the lymphocytes were pretreated with low concentrations of H_2O_2 (1.0–50 μM) from 24–28 of culture and then irradiated with 150 cGy of X-rays at 48 h (Wolff et al. 1989). When the cells were fixed 6 h later and the number of induced chromatid aberrations was scored, it was found that prior exposure of cells to any of the H_2O_2 concentrations, which in themselves did not induce aberrations, reduced the number of X-ray-induced chromatid breaks by approximately one-half (Table 5). Thus, it appears that chemical agents that produce

Table 4. 3AB between two 100-cGy doses of X-rays inhibits repair of breaks from the first dose for at least 7 h (200 cells per point). Data from Wiencke et al. (1986)

Dose 1	Interval (h)	3-AB	Dose 2	Aberrations
100	7	−	100	98
100	7	+	100	173

Table 5. Pre-exposure of human lymphocytes to low doses of H_2O_2 decreases the amount of cytogenetic damage induced by X-rays. Data from Wolff et al. (1989)

Treatment	Deletions / No. of cells
Control	1/100
H_2O_2	
1 μM	0/100
10 μM	0/100
25 μM	1/100
50 μM	1/100
X-rays (150 cGy)	70/200
H_2O_2 + X-rays	
1 μM + 150 cGy	51/250
10 μM + 150 cGy	43/200
25 μM + 150 cGy	41/200
50 μM + 150 cGy	49/200

DNA strand breaks via oxygen species, like-X-rays, are capable of inducing a chromosomal repair mechanism in human lymphocytes. The experiments also raised the possibility that the repair mechanism induced by low doses of X-rays is the result of the induction of oxygen radicals.

2.3 Radiomimetics

Because ionizing radiations are efficient inducers of DNA double-strand breaks, and because chromosome aberrations are the result of such DNA double-strand breaks, it is thought that these breaks might also be the lesions responsible for the induction of the repair brought about by exposure to 1 cGy of X-rays, $[^3H]$dThd, or H_2O_2. Further support for the role of double-strand breaks came from the observation that the adaptive response is modulated by poly(ADP-ribose) polymerase (Wiencke et al. 1986), an enzyme stimulated in response to such breaks (Benjamin and Gill 1980). To test whether or not DNA double-strand breaks are indeed capable of inducing the repair mechanism, experiments were carried out with the radiomimetic chemical bleomycin, which itself produces double-strand breaks in DNA. Human lymphocytes were pretreated with low concentrations of bleomycin ranging from 2.5 to 100 ng/ml at 24–28 h of culture (Wolff et al. 1989). The cells were exposed to 150 cGy of X-rays at 48 h, and the number of induced chromatid aberrations was scored in cells fixed 6 h later. When cells were exposed to 25–50 ng/ml of bleomycin from 24–28 h of culture, the cells became refractory to the subsequent exposure to a high dose of X-rays. As with the adaptation induced by low doses of X-rays, the response occurred only over a narrow range of adapting doses. Doses less than 25 or greater than 50 ng/ml did not induce the response. It has been proposed that for the response to be turned on with X-rays, a given number of lesions must be produced within a limited period of time (Shadley and Wiencke 1989), but that an excess number of lesions results in saturation that inhibits the response. The same now seems to be the case for the induction of adaptation by bleomycin. At concentrations of bleomycin greater than 50 ng/ml, or with acute X-ray doses greater than 20 cGy, adaptation does not occur (Shadley and Wolff 1987). In similar experiments in which bleomycin was added to the cells at 4 h of culture and left in until the time of fixation at 54 h, doses as low as 10 ng/ml did cause adaptation (Vijayalaxmi and Burkart 1989a, b). In plants, too, bleomycin can induce adaptation (Heindorff et al. 1987).

3 Induction of Proteins (Enzymes)

It had been shown previously (Shadley et al. 1987) that a 1-cGy pretreatment does not significantly decrease the yield of deletions induced by 150 cGy until 4 h after the pretreatment. To characterize the putative repair mechanism at the molecular level, experiments were carried out to see if any enzymes (proteins) responsible for the response were being synthesized at the times when the pretreatments became effective (Youngblom et al. 1989). Cells were first exposed to 1 cGy of X-rays at 44 h of culture and then exposed to 10 μg/ml of the protein synthesis inhibitor cycloheximide at 1–2, 2–4, or 4–6 h after the 1-cGy dose. Only when protein synthesis was inhibited 4–6 h after the adapting dose was the adaptive response to 1 cGy prevented (Table 6). This corresponds to the elapsed time after exposure required for maximum adaptation. These results indicate that, 4–6 h after exposure to low doses of radiation, a protein is being synthesized that could take part in the adaptation.

Table 6. Inhibition of the adaptive response of human lymphocytes to 1 cGy of X-rays by cycloheximide (CHM) administered 4–6 h after the 1-cGy exposure (200 cells per point). Data from Youngblom et al. (1989)

Treatment	Deletions
Control	2
X-rays (1 cGy)	4
CHM (10 μg/ml)	0
1 cGy + CHM	2
X-rays (150 cGy)	74
1 cGy + 150 cGy	39
1 cGy + CHM (0–2 h) + 150 cGy	29
1 cGy + CHM (2–4 h) + 150 cGy	41
1 cGy + CHM (4–6 h) + 150 cGy	75

The results with cycloheximide indicated that protein synthesis is necessary for the adaptive response to be induced by 1 cGy of X-rays. To help characterize the repair mechanism at a molecular level, two-dimensional gel electrophoresis was carried out on human lymphocytes exposed to 1 cGy to see if, indeed, proteins were being induced. Four separate experiments were carried out in which proteins were extracted from unirradiated human lymphocytes and from lymphocytes exposed to 1 cGy (Wolff et al. 1989). It was possible to detect several proteins that were reproducibly present in all four of the samples exposed to 1 cGy but were not present in the controls. These induced proteins are considered to be good candidates for being the induced repair enzyme(s).

Such inducible proteins that are involved in repair mechanisms have been implicated in a variety of systems. Usually, however, this occurs only after exposure to high doses. For instance, Howard and Cowie (1976) had found that exposure of the alga *Closterium moniliferum* to 10 MeV electrons changed the shoulder of survival curves obtained after exposure of cells to a series of doses 4 h after the initial exposure. They suggested that potentially lethal damage (PLD) was being repaired. Similarly, just as we have found that cycloheximide, a protein synthesis inhibitor, can prevent the adaptive response if it is applied in the critical period 4–6 h after the initial adapting dose, two-dimensional gel electrophoresis of X-ray-induced polypeptides in human malignant melanoma cells has shown that the addition of cycloheximide inhibits the formation of these polypeptides and also inhibits PLD repair (Boothman et al. 1989).

4 Alkylating Agents

4.1 MMS

In another series of experiments to see if the repair mechanism induced by low doses of radiation from either [^{3}H]dThd or 1 cGy of X-rays could affect various types of clastogenic lesions induced in DNA by subsequent exposure to chemical mutagens and carcinogens, it was found that low doses of radiation caused adaptation that affected the response to chemical agents, such as bleomycin, that produce double-strand breaks in DNA, but did not cause adaptation to methyl methanesulfonate (MMS), an S-dependent agent that alkylates DNA and then leads to single-strand breaks (Wolff et al. 1988). In fact, in these experiments, it was found that prior exposures to low doses of ionizing radiation led to a synergistic response with subsequent high-dose exposures to MMS.

4.2 MNNG

Because MMS is unlike N-methyl-N′-nitro-N-nitrosoguanidine (MNNG), in that the two alkylating agents give a different spectrum of methylated bases (Laval and Laval 1984), with MMS producing a far higher ratio of N^7-methylguanine to O^6-methylguanine than does MNNG, experiments were carried out with MNNG to see if the synergism obtained between X-rays and a subsequent exposure to the simple alkylating agent MMS was obtained with another simple methylating agent that produces a different spectrum of lesions. The experiments showed, as had been found previously (Wolff et al. 1988), that a conditioning treatment with 1 cGy of X-rays was synergistic with a subsequent high dose of MMS (Table 7). When 0.018 mM MMS or 600 ng/ml MNNG was used as a conditioning treatment, they, too, were found to be synergistic with a subsequent exposure to a high dose of MMS. When, however, the challenge treatment was X-rays, then a pretreatment or conditioning with 1 cGy of X-rays, 0.018 mM MMS, or 600 ng/ml of MNNG decreased the numbers of aberrations induced by the subsequent challenge with 150 cGy of X-rays (Table 8), showing that conditioning with simple alkylating agents, as with the conditioning by H_2O_2 or by bleomycin, can reduce the yield of aberrations induced subsequently by a high dose of X-rays. It was also found that X-rays, MMS, or MNNG can reduce the numbers of aberrations induced by subsequent exposures to a high dose of MNNG (Table 9).

5 Conclusion

These results indicate that conditioning with a variety of DNA-damaging agents, such as ionizing radiation, alkylating agents, radiomimetic compounds, and H_2O_2, can induce the repair mechanism that makes the cells less susceptible to cytogenetic damage induced by a variety of agents used as a challenge. This cross-adaptivity is

Table 7. Low doses of X-rays, MMS, or MNNG are synergistic with a subsequent high dose of MMS

Conditioning treatment	Challenge treatment[a] (MMS, 0.42 mM)	No. of aberrations in 200 cells[b]
1 cGy (1 rad) X-rays		
−	−	3
+	−	3
−	+	114
+	+	169*
0.018 mM MMS		
−	−	3
+	−	2
−	+	97
+	+	133*
600 ng/ml MNNG		
−	−	3
+	−	1
−	+	97
+	+	147**

[a] − No treatment; + treatment.
[b] Significantly higher than the sum of conditioning and challenge treatments given separately: * $p < 0.01$, ** $p < 0.001$ (one-tailed t-test).

Table 8. Low doses of X-rays, MMS, or MNNG cause human lymphocytes to be less susceptible to the induction of cytogenetic damage

Conditioning treatment	Challenge treatment[a] (1.5 Gy X-rays)	No. of aberrations / No. of cells[b]
1 cGy (1 rad) X-rays		
–	–	4/100
+	–	3/100
–	+	31/100
+	+	14/100*
0.018 mM MMS		
–	–	4/100
+	–	3/100
–	+	31/100
+	+	16/100*
600 ng/ml MNNG		
–	–	3/200
+	–	2/100
–	+	58/200
+	+	32/200**

[a] – No treatment; + treatment.
[b] Significantly lower than the sum of conditioning and challenge treatments given separately: * $p < 0.01$, ** $p < 0.005$ (one-tailed test).

Table 9. Low doses of X-rays, MMS, or MNNG cause human lymphocytes to become less susceptible to an exposure to a high dose of MNNG

Conditioning treatment	Challenge treatment[a] (MNNG, 30 μg/ml)	No. of aberrations / No. of cells[b]
1 cGy (1 rad) X-rays		
–	–	3/100
+	–	1/100
–	+	111/200
+	+	65/200*
–	+	115/200
+	+	81/200*
–	+	120/200
+	+	80/200*
0.018 mM MMS		
–	–	1/100
+	–	3/100
–	+	40/100
+	+	11/100*
600 ng/ml MNNG		
–	–	1/100
+	–	2/100
–	+	40/100
+	+	18/100*

[a] – No treatment; + treatment.
[b] * Significantly lower than the sum of conditioning and challenge treatments given separately ($p < 0.005$, one-tailed t-test).

similar in some ways to that observed by Rieger et al. (1982) in *Vicia faba*, although the synergistic response found when cells are challenged with MMS is unusual.

To date, the results indicate that all of the agents, including MMS, can induce an adaptive response in human lymphocytes that makes the cells less susceptible to damage induced by higher doses of some of the agents. The only exception to this occurs when the challenge is with MMS. Here, it is found that pre-exposure either to X-rays or to MMS itself can render the cells more susceptible to damage induced by a higher dose. These results, therefore, indicate that MMS, with its different spectrum of lesions, produces enough damage in DNA of the type necessary to induce the adaptive response, but that the predominant lesion that leads to the production of chromosomal breaks after a challenge with MMS differs from that found with the other agents.

Acknowledgments. Work supported by the Office of Health and Environmental Research, U. S. Dept. of Energy, contract DE-AC03-76-SF01012.

References

Benjamin RC, Gill DM (1980) Poly(ADP-ribose) synthesis in vitro programmed by damaged DNA. A comparison of DNA molecules containing different types of strand breaks. J Biol Chem 255: 10502–10508

Bodycote J, Wolff S (1986) Metabolic breakdown of [^{3}H]thymidine and the inability to measure human lymphocyte proliferation by incorporation of radioactivity. Proc Natl Acad Sci USA 83: 4749–4753

Boothman DA, Bouvard I, Hughes EN (1989) Identification and characterization of X-ray-induced proteins in human cells. Cancer Res 49: 2871–2878

Demple B, Halbrook J (1983) Inducible repair of oxidative DNA damage in *Escherichia coli.* Nature 304: 466–468

Evans HJ (1977) Molecular mechanisms in the induction of chromosome aberrations. In: Scott D, Bridges BA, Sobels FH (eds) Progress in genetic toxicology. Elsevier Amsterdam, pp 57–74

Heindorff K, Rieger R, Michaelis A, Takehisa S (1987) Clastogenic adaptation triggered by S-phase-independent clastogens in *Vicia faba.* Mutat Res 190: 131–135

Howard A, Cowie FG (1976) Induced resistance in a desmid *Closterium moniliferum.* Radiat Res 65: 540–549

Karran P, Lindahl T, Griffin B (1979) Adaptive response to alkylating agents involves alteration in situ of O^6-methylguanine residues in DNA. Nature 280: 76–77

Laval F, Laval J (1984) Adaptive response in mammalian cells: crossreactivity of different pretreatments on cytotoxicity as contrasted to mutagenicity. Proc Natl Acad Sci USA 81: 1062–1066

Morgan WF, Fero ML, Land MC, Winegar RA (1988) Inducible expression and cytogenetic effects of the *Eco*RI restriction endonuclease in Chinese hamster ovary cells. Mol Cell Biol 8:4204–4211

Natarajan AT, Obe G, Zeeland AA Van, Palitti F, Meijers M, Verdegaal-Immerzeel EAM (1980) Molecular mechanisms involved in the production of chromosomal aberrations. II. Utilization of *Neurospora* endonuclease for the study of aberration production by X-rays in G_1 and G_2 stages of the cell cycle. Mutat Res 69: 293–305

Olivieri G, Bodycote J, Wolff S (1984) Adaptive response of human lymphocytes to low concentrations of radioactive thymidine. Science 223: 594–597

Olsson M, Lindahl T (1980) Repair of alkylated DNA in *Escherichia coli:* methyl group transfer from O^6-methylguanine to a protein cysteine residue. J Biol Chem 255: 10569–10571

Rieger R, Michaelis A, Nicoloff H (1982) Inducible repair processes in plant root tip meristems? "Below-additivity effects" of unequally fractionated clastogen concentrations. Biol Zentralbl 101: 125–138

Samson L, Cairns J (1977) A new pathway for DNA repair in *Escherichia coli.* Nature 267: 281–283

Samson L, Schwartz JL (1980) Evidence for an adaptive DNA repair pathway in CHO and human skin fibroblast cell lines. Nature 287: 861–863

Samson L, Schwartz JL (1983) The induction of resistance to alkylating damage in mammalian cells. In: Lawrence CW (ed) Induced mutagenesis: molecular mechanisms and their implications for environmental protection. Plenum, New York, pp 291–309

Shadley JD, Wiencke JK (1989) Induction of the adaptive response by X-rays is dependent on radiation intensity. Int J Radiat Biol 56: 107–118

Shadley JD, Wolff S (1987) Very low doses of X-rays can cause human lymphocytes to become less susceptible to ionizing radiation. Mutagenesis 2: 95–96

Shadley JD, Afzal V, Wolff S (1987) Characterization of the adaptive response to ionizing radiation induced by low doses of X-rays to human lymphocytes. Radiat Res 111: 511–517

Vijayalaxmi, Burkart W (1989a) Effect of 3-aminobenzamide on chromosome damage in human blood lymphocytes adapted to bleomycin. Mutagenesis 4: 187–189

Vijayalaxmi, Burkart W (1989b) Resistance and cross-resistance to chromosome damage in human blood lymphocytes adapted to bleomycin. Mutat Res 211: 1–5

Wiencke JK, Afzal V, Olivieri G, Wolff S (1986) Evidence that the [^{3}H]thymidine-induced adaptive response of human lymphocytes to subsequent doses of X-rays involves the induction of a chromosomal repair mechanism. Mutagenesis 1: 375–380

Wolff S (1972) The repair of X-ray-induced chromosome aberrations in stimulated and unstimulated human lymphocytes. Mutat Res 15: 435–444

Wolff S (1978) Relation between DNA repair, chromosome aberrations, and sister chromatid exchanges. In: Hanawalt PC, Friedberg EC, Fox CF (eds) DNA repair mechanisms. Academic Press, London, pp 751–760

Wolff S, Afzal V, Wiencke JK, Olivieri G, Michaeli A (1988) Human lymphocytes exposed to low doses of ionizing radiations become refractory to high doses of radiation as well as to chemical mutagens that induce double-strand breaks in DNA. Int J Radiat Biol 53: 39–48

Wolff S, Wiencke JK, Afzal V, Youngblom J, Cortés F (1989) The adaptive response of human lymphocytes to very low doses of ionizing radiation: a case of induced chromosomal repair with the induction of specific proteins. In: Low dose radiation: biological bases of risk assessment. Taylor & Francis, London, pp 446–454

Youngblom JH, Wiencke JK, Wolff S (1989) Inhibition of the adaptive response of human lymphocytes to very low doses of ionizing radiation by the protein synthesis inhibitor cycloheximide. Mutat Res 227: 257–261

Radio-Adaptive Response: A Novel Chromosomal Response in Chinese Hamster Cells in Vitro

T. IKUSHIMA[1]

Contents

1 Introduction

When exposed to various DNA damaging stresses, a variety of prokaryotic and eukaryotic cells exhibit induction of an adaptation mechanism. The adaptive response was first reported by Samson and Cairns (1977) in *Escherichia coli* as an inducibile form of DNA repair acting on alkylation damage, in which the bacterial cells that had been exposed to low levels of alkylating agents became less susceptible to the subsequent high doses of the same and other compounds. In this adaptive response, the induction of adaptation proteins, DNA glycosylase and DNA alkyltransferase are involved (Karran et al. 1979; Evensen and Seeberg 1982), which is distinct from the SOS response, another inducible effect resulting from DNA damage (Jeggo et al. 1977). An analogous adaptive response has also been observed in some higher mammalian systems (e.g., Samson and Schwartz 1980; Laval and Laval 1984) and in higher plants (Rieger et al. 1982). Interestingly, the adaptive response has been recently found to occur in human lymphocytes exposed to low doses of ionizing radiation (Olivieri et al. 1984; Shadley and Wolff, 1987; Sankaranarayanan et al. 1989). A pretreatment with "conditioning" or "adapting" low doses of either tritium β-rays or X-rays resulted in a reduction in the yield of chromosomal aberrations induced by a subsequent ("challenge") higher dose. Such a clastogenic radio-adaptive response has also been observed in the induction of micronuclei and sister chromatid exchanges (SCEs) in proliferative Chinese hamster V79 cells (Ikushima 1987, 1989), and of chromosomal aberrations in *Vicia faba* root tip cells (Heindorff et al. 1987). It is now clear that the radio-adaptive response can occur in cells from a rather universal range of eukaryotes. While the network of the adaptive response to alkylating agents in bacteria is well understood at the molecular level (for a review, see Lindahl et al. 1988), very little is known about the mechanisms responsible for the radio-adaptive response in higher eukaryotes.

---

[1] Research Reactor Institute, Kyoto University, Kumatori-cho, Sennan-gun, Osaka, 590-04 Japan

The work done to ascertain the radio-adaptive response in V79 Chinese hamster cells in vitro is briefly reviewed, and proposed mechanisms of the radio-adaptive response are discussed in the light of an inducible genetic network of adaptation. The term "radio-adaptive response" will be used here as the adaptive response induced by low doses of ionizing radiation, in contrast to the "chemo-adaptive response" which is induced by low levels of chemicals such as alkylating agents.

2 Materials and Methods

Cell Culture. Chinese hamster cells (a cloned derivative of the V79-B310H, modal chromosome number 2n = 22) were cultured as monolayers in plastic tissue culture flasks (Falcon) containing α-modified Eagle MEM (Sigma), 10% fetal bovine serum (M. A. Bioproducts), 100 units/ml penicillin and 100 μg/ml streptomycin at 37 °C in an atmosphere of 95% air/5% CO_2. The cell generation time is 12 h (G_1:1 h, S:8 h, G_2:2 h, M:1 h) under these culture conditions.

Irradiation with ^3H β-Rays, ^{60}Co γ-Rays, Reactor Radiations and UV-B. For irradiation with β-rays from incorporated tritium, the ells were cultured for 16 h in the presence of either [Me-^{3}H] thymidine ($3.7 \times 10^{-3}-7.4$ kBq/ml; spec. act. 1517 GBq/mM, NEN) or tritiated water (370, 740 kBq/ml). The cells were also cultured in medium containing 1 μM of nonradioactive thymidine as a control. After rinsing with the medium, the cells were irradiated with ^{60}Co γ-rays (0.4 Gy/min) or reactor radiations from the D_2O thermal column of the Kyoto University Research Reactor (KUR) operated at 5 MW in our institute (Ikushima 1972). Irradiation with a conditioning low-dose of ^{60}Co γ-rays was performed at a dose rate of 5 cGy/min. For irradiation with near-ultraviolet light (UV-B, 313 nm), the cells plated on Falcon polystyrene Petri dishes were washed with PBS and exposed to the light of a health lamp (Toshiba FL20SE) which penetrated the Petri dish covers at 97.5 J m^{-2} min^{-1} (measured by a black-ray ultraviolet meter, Ultra-Violet Product, Inc.).

Treatment with Chemical Mutagens. The adapted cells were cultured in medium containing the given concentrations of MMC, EMS and cisplatin from the fourth to sixth hour after the adapting γ-ray irradiation. The treated cells were washed in medium three times and cultured again under the initial conditions.

Scoring of Micronuclei and SCEs. For scoring micronuclei cytochalasin B (5 μg/ml) induced binucleate cells were harvested 6 or 10 h (5 μ/ml) after the challenging with high doses and fixed in a mixture of methanol and acetic acid (3:1 V/V). The slides were stained with Giemsa or acridine orange. The frequency of micronuclei was determined by scoring 1000 cells for each point. In the SCE experiments, 5-bromodeoxyuridine (5 μM final concentration) was added to the cultures immediately after the treatments with challenging agents, and then after two cell cycles, colcemid (final conc., 2×10^{-7} M) was added 2 h before fixation. SCEs where scored in 50 second-division metaphases stained by a modified fluorescent-plus-Giemsa method (Ikushima and Wolff 1974; Perry and Wolff 1974).

Others. For monitoring the uptake of ^{3}H-dThd into DNA, the cells cultured in the presence of ^{3}H-dThd for 16 h were harvested onto glass fiber filters (Whatman GF/C) and washed with cold TCA and ethanol. The radioactivity was counted with a liquid-scintillation counter.

Radiations and chemicals, and their combination used for conditioning and challenging are outlined in Table 1.

3 Results

An Optimal Dose Range of Tritium β-Rays for Induction of Radio-Adaptive Response. Initial experiments showed that the yield of micronuclei induced by 1 Gy of γ-rays in cells pretreated with a low concentration of ^{3}H-dThd (0.37–1.85 kBq/ml) was less than the sum of the yields by ^{3}H-dThd and γ-rays separately, and that the magnitude of resistance induced by ^{3}H-dThd pretreatments seemed to be higher with decreasing conditioning dose, and high conditioning doses did not induce adaptive resistance (Ikushima 1987). This type of experiment has been expanded to determine whether conditioning with much lower doses of ^{3}H β-rays can still induce the radio-adaptive response; a wide range of lower concentrations down to 3.7×10^{-3} kBq/ml of ^{3}H-dThd was used. The results are given in Fig. 1, compiled together with the previous results. A concentration of 0.185 kBq/ml of ^{3}H-dThd still led to a significantly lower yield of micronuclei than the sum of the damage induced by the conditioning ^{3}H-dThd and challenging γ-rays doses separately. That is, the S-value (index of the interaction) is significantly lower than unity. The yield of micronuclei is approximately at the additivity level in the cells conditioned with 3.7×10^{-2} or 3.7×10^{-3} kBq/ml of ^{3}HdThd. These results indicate that a narrow range of conditioning ^{3}H-doses is effective for induction of the radio-adaptive response.

Radio-Adaptive Response for SCE Induction. With respect to the induction of SCEs, the cells conditioned with a low dose of ^{3}H-dThd (0.74 kBq/ml) became obviously less susceptible to challenging doses of γ-rays (Fig. 2). In cells pretreated with β-rays from ^{3}H-dThd, the frequency of SCEs induced by γ-rays was clearly less than that induced by each radiation administered separately (protective effects), showing that the induced DNA repair mechanisms copes with the SCE lesions more efficiently

Table 1. Radiations and chemicals, and their combinations used for conditioning and challenging

Conditioning	Challenging
β-Rays;	
^{3}H-thymidine (18.6 keV)	^{60}Co γ-Rays
tritiated water (HTO)	Thermal neutrons
γ-Rays;	MMC
^{60}Co (1.17 and 1.33 MeV)	UV-B (313 nm)
Neutrons;	EMS
thermal [H(n,r), N(n,p)]	Cisplatin

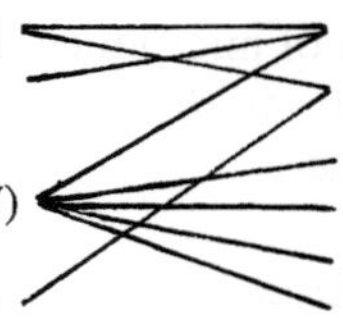

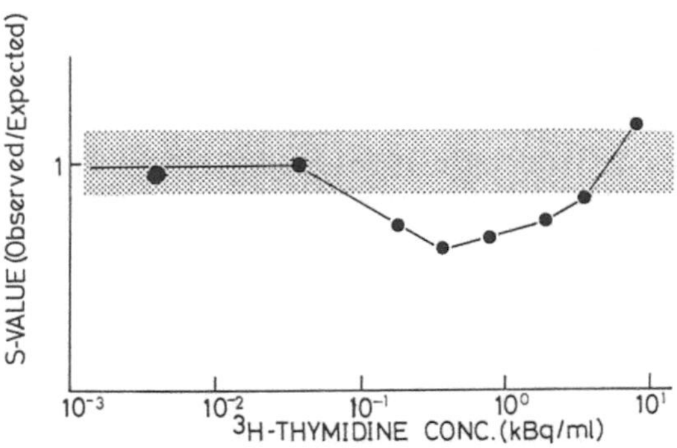

Fig. 1. Dependence of induction of radio-adaptive response on the adapting dose of ^{3}H β-rays for micronucleus induction. The *S-value* (index of interaction) was determined as ratio of the observed yield to the expected, the latter being the sum of the yield induced by both treatments separately. The *shaded* area shows the control level

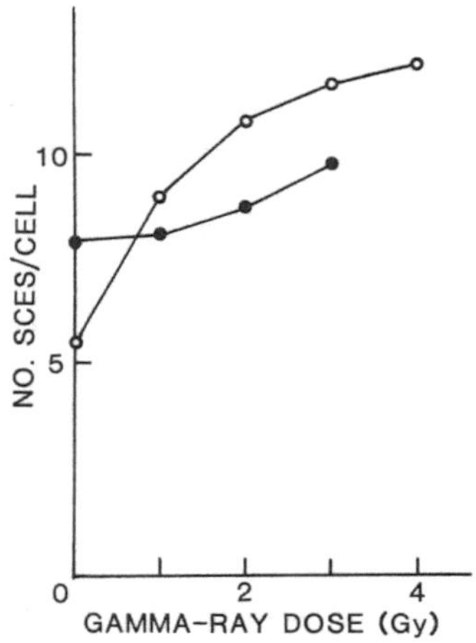

Fig. 2. Effects of ^{3}H-dThd pretreatment on the induction of SCEs by subsequent irradiation with γ-rays. Pretreated (●) and nontreated control (○) cells

than with the lesions leading to micronuclei induced by challenging with high radiation doses. The radio-adaptive response induced by low-level tritium exposure is different for two different types of chromosomal damage.

Induction of Radio-Adaptive Response by Low Doses of Tritiated Water β-Rays. When cells are cultured in the presence of HTO, it is expected that the cells will be homogeneously exposed intracellularly to β-rays different to the exposure to ^{3}H-dThd. To determine whether doses of β-rays from HTO can also trigger the radio-adaptive response, the yield of micronuclei induced by γ-rays was measured in cells pre-exposed to low concentrations of HTO. In this experiment, micronuclei were assayed in 500 binucleated cells induced by cytochalasin B for each point. As shown in Table 2, the significant reduction of the yield of micronuclei was seen in any case. These results indicate that low-level exposure to β-rays from HTO as well as ^{3}H-dThd can induce the radio-adaptive response.

Induction of Radio-Adaptive Response by Low Doses of γ-Rays. To test the possibility of induction of a radio-adaptive response by external, acute exposure to low doses of γ-rays, the cells exposed first to conditioning doses of either 1 or 5 cGy of γ-rays were subsequently challenged with 1 Gy of γ-rays at intervals of 2 or 4 h. Then the yield of micronuclei was measured in the cells fixed 6 h after the challenging dose. As shown in Fig. 3, a reduction in the yield of micronuclei induced by the high challenging doses was found in the cells adapted with either 1 or 5 cGy followed by 4-h culture. Such a reduced number of micronuclei, however, was not seen in the cells challenged only 2 h after the conditioning dose. These results indicate that the external exposure to low-level γ-rays can induce the radio-adaptive response to γ-ray induced micronuclei, and that the radio-adaptive response requires a period of time (4 h) for full development.

No Induction of Radio Adaptive Response by Reactor-Radiations. As shown in Fig. 4, the pre-exposure to a low dose of high LET reactor radiation cannot induce the radio-adaptive response for induction of micronuclei by the subsequent challenge doses of the same type of radiation. The thermal neutron flux and the dose rate of contaminating γ-rays of the reactor radiations were 3×10^9 n cm^{-2} s^{-1} and 2.0 Gy h^{-1} at the site in the heavy water facility of KUR used in this experiment. They were monitored by Au-foil activation and Li-F coated thermoluminescent dosimetry. The results suggest that the radio-adaptive response might be dependent on the type of damage initially induced.

Decay of Adaptive Response, and Inhibition of Its Induction by 3-Aminobenzamide (3AB). When the ^{3}H-dThd adapted cells were incubated in complete medium for 15 h prior to the challenging dose, or treated with 2 mM 3AB, an inhibitor of poly (ADP-ribose) polymerase, for the final 10 h during the adaptation, the below-additivity effects of the ^{3}H-pretreatment on micronecleus induction by the subsequent challenge with 1 Gy of γ-rays was no longer observed. The yield of micronuclei was at the same level as that of non-pretreated cells (Fig. 5).

Table 2. Induction of micronuclei (MNs) by γ-rays in Chinese hamster V79 cells pre-exposed to low-level HTO β-rays

Adapting HTO (kBq/ml)	*Challenge* γ-rays (Gy)	*No. MNs/500 binucleated cells* Observed	Expected
–	–	5	
–	1	35	
–	2	68	
370[a]	–	6	
370	1	17[b]	39
370	2	48[b]	73
740	–	11	
740	1	24[b]	44
740	2	52[b]	79

[a] The dose absorbed by the cells cultured in the presence of 370 kBq/ml HTO was determined to be 16.3 cGy assuming a water content of 84%.

[b] Significantly different from expected, $p < 0.01$ (one-tailed Student's t-test).

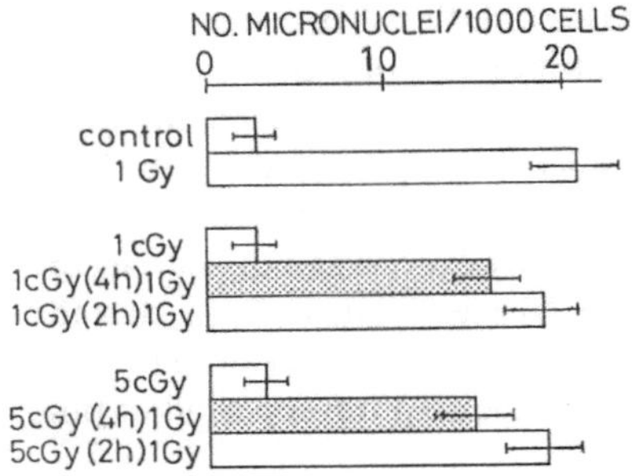

Fig. 3. Induction of micronuclei in cells pre-exposed to 1 or 5 cGy of γ-rays. The adapted cells were challenged with 1 Gy of γ-rays 2 or 4 h after the adapting dose. *Bars* represent SE

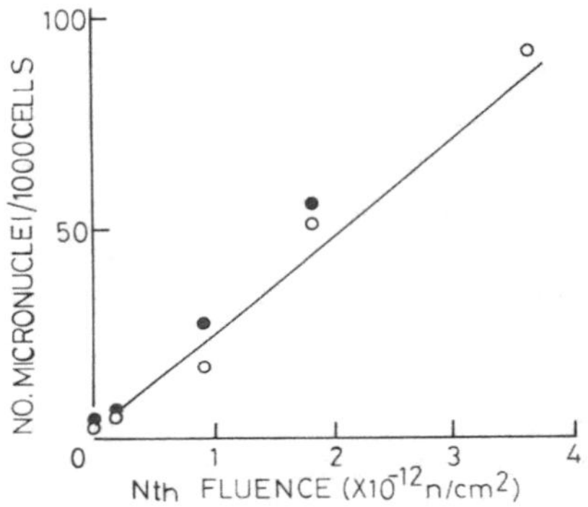

Fig. 4. Effects of pre-exposure to a low dose of reactor radiation on the induction of micronuclei by subsequent high doses of reactor radiation. Pre-exposed (●) and nonexposed control (○) cells

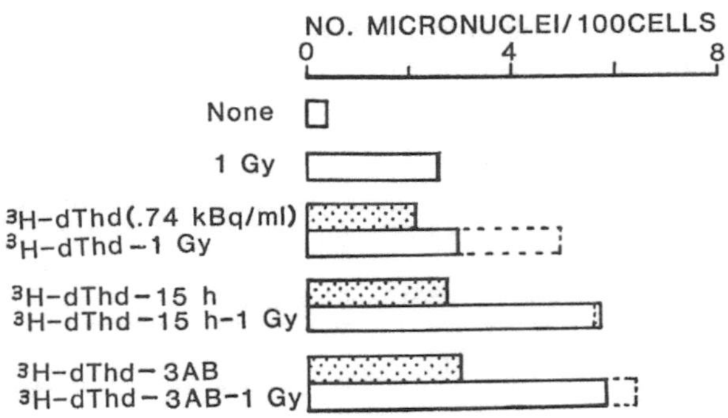

Fig. 5. Effects of pretreatment with ^{3}H-dThd on the induction of micronuclei by γ-rays. Cells treated with ^{3}H-dThd were grown for 15 h or treated with 2 mM of 3AB for 10 h before challenge with γ-irradiation. The *dotted lines* show the expected yields

Cross-Adaptation of ^{3}H-Induced Radio-Adaptive Response to Clastogenic Effects of Reactor Radiation. The cells conditioned with a low dose of β-rays from ^{3}H-dThd were challenged by high doses of high LET reactor radiations. As shown in Fig. 6, the cross-adaptation was not observed for induction of micronuclei, but of SCEs. A synergistic response of an increased, rather than decreased, number of micronuclei was found. For SCEs, a protective effect was observed following challenge with γ-rays (Fig. 2). This response might reflect different mechanisms of induction of micronuclei and SCEs.

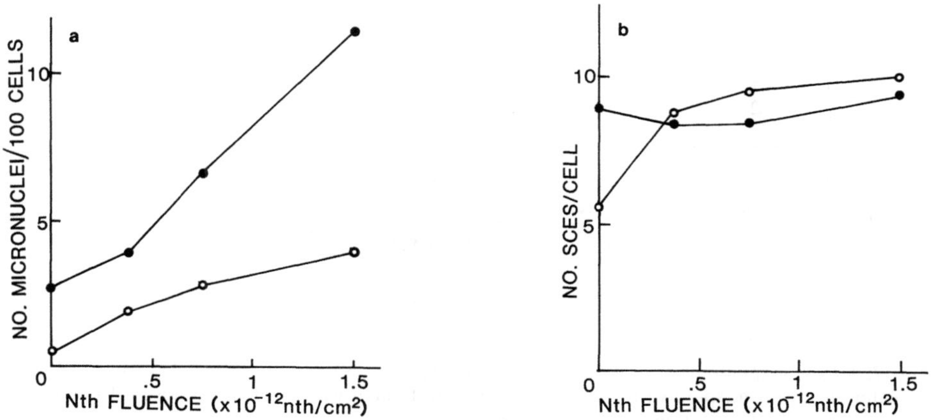

Fig. 6a, b. Effects of the pretreatment with ^{3}H-dThd on the induction of micronuclei (*a*) and SCEs (*b*) by reactor radiations. Pretreated (●) and nontreated control (○) cells

Cross-Adaptation of γ-Ray Induced Radio-Adaptive Response to SCE Induction by Other Chemical or Physical Mutagens. Experiments have been carried out to investigate whether the radio-adaptive response can affect various types of chromosomal lesions leading to SCEs produced by other chemical or physical mutagens. UV-B, MMC, EMS, and cisplatin were used as challenging agents. UV-B is a major component of sunlight and produces (6–4) photoproducts in DNA (Haseltine 1983). MMC is a cross-linking agent (Tomasz et al. 1974), EMS alkylates DNA (Singer 1975), and cisplatin induces intra-DNA strand cross-links (Eastman 1983). The cells were exposed to UV-B, or treated with these chemical agents for 2 h after another 4-h culture following the first low dose of γ-rays. In the cells challenged with MMC or UV-B, significant suppressed induction of SCEs was found, but no cross-adaptation was seen with EMS or cisplatin (Fig. 7).

4 Discussion

It has been found that pre-exposure to low-level chronic β-rays from incorporated ^{3}H-dThd or HTO protected growing Chinese hamster V79 cells against the additional induction of chromosomal damage by subsequent exposure to higher acute doses of ^{60}Co γ-rays (Ikushima 1987). The pre-exposure exhibited quantitatively different effects on the two types of chromosomal damage, micronuclei, and SCEs. A below-additivity effect was seen for micronucleus induction but this effect was rather weak compared to SCE induction, for which a protective effect was observed. This differential response might reflect a different process in the formation of micronuclei and SCEs (Ikushima 1977). A radio-adaptive response is evident for chromatid aberrations (Olivieri et al. 1984) and mutations (Sanderson and Morley 1986) in human lymphocytes, and for clastogenic effects in *Vicia faba* root tip cells (Heindorff et al. 1987), and it is found for SCEs and micronuclei in proliferating Chinese hamster cells in vitro.

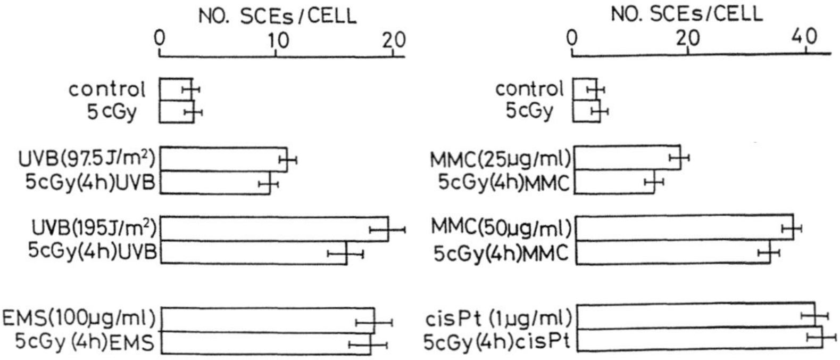

Fig. 7. Cross adaptation of γ-ray induced radio-adaptive response to MMC, UV-B, EMS and cisplatin. The cells adapted with 5 cGy of γ-rays were treated with challenging doses of mutagens for 2 h as described in Section 2. *Bars* represent SE

The data presented here indicate that there is an optimal dose range of the adapting ^{3}H β-rays for an induction of a radio-adaptive response in cultured Chinese hamster cells (Fig. 1; Ikushima 1989). An optimal adapting dose has also been found in the adaptive response induced by alkylating agents (Samson and Schwartz 1980), X-rays (Shadley and Wolff 1987; Shadley and Wiencke 1989), or bleomycin (Vijayalaxmi and Burkart 1989). When ^{3}H-dThd at concentrations much lower or much higher than this range was applied, the radio-adaptive response manifested as a micronucleus-reducing effect was not observed. The optimal ^{3}H-dose range for the induction of a radio-adaptive response was estimated to be between 0.1 to 10 cGy on the basis of measurements of the ^{3}H radioactivity incorporated into DNA (Ikushima 1987, 1989). Acute exposure to low doses (1 or 5 cGy) of γ-rays also induced a radio-adaptive response, but low-level exposure to high LET reactor radiations did not (Figs. 1–4). These results suggest that the radio-adaptive response occurs in cells pre-exposed to low doses of low LET radiations such as β-, γ- and X-rays from any sources either internally or externally under restricted conditions, but not of high LET radiations (Olivieri et al. 1984; Ikushima 1987; Shadley and Wolff 1987; Sankaranarayanan et al. 1989). This dependency on the type of radiation might reflect the quality and quantity of chromosomal lesions which trigger the radio-adaptive response.

The radio-adaptive response in Chinese hamster cells can be inhibited by 3-aminobenzamide, an inhibitor of poly(ADP-ribose) polymerase and is no longer observed after one cell division following the adapting dose (Fig. 5; Ikushima, 1987). In human lymphocytes the induced adaptive response persists for three cell cycles (Shadley et al. 1987; Wiencke et al. 1986). The full development of the radio-adaptive response did not occur until 4 h after the conditioning dose (Fig. 3; Ikushima 1987), as reported in X-irradiated human lymphocytes (Shadley and Wolff 1987; Shadley et al. 1987). These results indicate that the radio-adaptive response is attributed to the induction of a chromosomal repair mechanism which repairs chromosomal lesions induced by the high challenge doses, and thus results in the reduction of total chromosomal damage. Radiation induced repair processes in mutagenesis have also been suggested to occur in X-irradiated human lymphocytes and higher plants (Rieger et al. 1982; Leenhouts et al. 1982; Sanderson and Morley 1986), and in UV-irradiated V79 cells (Chatterjee et al. 1984).

The cells adapted with low doses of γ-rays displayed a cross-adaptation to SCE lesions produced by UV-B and MMC, but not to those by EMS and cisplatin (Fig. 7) as summarized in Table 3. Human lymphocytes pre-exposed to low doses of X-rays exhibited cross-resistance to the induction of chromatid aberrations by MMC, bleomycin, H_2O_2, and MNNG (N-methyl-N'-nitro-N-nitrosoguanidine) but not to that by MMS (methyl methanesulfonate) (Wolff et al. 1988, 1989). Such a cross-adaptability has been observed in X-irradiated *Vicia faba* which cross-adapted to bleomycin (Heindorff et al. 1987).The cross-adaptation might be dependent on the induction of an adaptive repair process that copes efficiently with the DNA lesions caused by similar DNA-damaging agents. It is not well known whether the cross-adaptation, positive mutagens induce the same spectrum of DNA lesions, or whether the same repair processes operate to repair these mutagen-induced chromosomal lesions. It appears, however, that the adaptive response by agents which induce DNA lesions mediated by active oxygen species such as hydroxy radicals exhibits cross-adaptation to each other, since it is well known that active oxygen species are produced by ionizing radiation and near-UV irradiation as well as H_2O_2 (Demple and Halbrook 1983). UV-B produces mutagenic DNA lesions such as (6–4) photo-products (Haseltine 1983) and MMC induces cross-links in DNA (Tomasz et al. 1974), while EMS alkylates DNA (Singer 1975) and cisplatin preferentially forms DNA-intrastrand adducts between the N-7 atoms of adjacent purines (Eastman 1983). All these agents are efficient SCE inducers and produce different spectra of DNA lesions compared with ionizing radiations (e.g., Ward 1988). It is, therefore, understandable that radiation-inducible repair processes differ from the adaptive response to alkylating agents, which in *Escherichia coli* involves the synthesis of O^6-alkylguanine-DNA transferase and, DNA glycosylase (Waldstein et al. 1982; Lindahl et al.

Table 3. Cross-adaption of radio-adaptive responses in mammalian cells

Adapting	Challenging	Effect
Chinese hamster	V79 cells	
β-Rays	γ-Rays	+
	N_{th}-neutrons	−(+)[a]
γ-Rays	γ-Rays	+
	UV-B	+
	MMC	+
	EMS	−
	Cisplatin	−
Human lymphocytes[b]		
X-rays	Bleomycin	+
	H_2O_2	+
	MMC	+
	MNNG	+
	MMS	−

[a] − for SCEs and + for micronuclei.
[b] Data from Wolff and co-workers (see References).

1988). Poly(ADP-ribose) polymerase is probably one of the potential candidate for the adaptive enzymes in mammalian cells (Wiencke et al. 1986; Ikushima 1987). The activity of poly(ADP-ribose) synthetase is stimulated by DNA strand breaks and poly(ADP-ribose) activates a DNA ligase required for DNA repair (Benjamin and Gill 1980; Criessen and Shall 1982). Though the regulation of a repair network for ionizing-radiation induced DNA damage is beyond our present knowledge, the activation or promotion of the expression of genes coding for adaptation or protection proteins such as ligase, DNA glycosylase, topoisomerase, superoxide dismutase, catalase, peroxidase, and others in addition to poly(ADP-riose) synthetase may be responsible for the radio-adaptive response. It has been found that low doses of ionizing radiations can induce several proteins not found in nonirradiated human lymphocytes (Wolff et al. 1989). Recent findings of the variability of the radio-adaptive response among individuals suggest genetic determinations (Sangakranarayanan et al. 1989; Bosi and Olivieri 1989). Many questions, however, remain unsolved to date: what is the direct trigger for the induction of the radio-adaptive response, and what molecular mechanism is responsible for the radio-adaptive response? A working hypothesis of the process of the radio-adaptive response is depicted in Fig. 8. The identification of inducible proteins and the isolation of the genes coding for the enzymes which are involved in the radio-adaptive response are necessary for a full understanding of radiation-inducible cytogenetic repair as one of the multiple defense mechanisms against a variety of environmental genotoxic impacts.

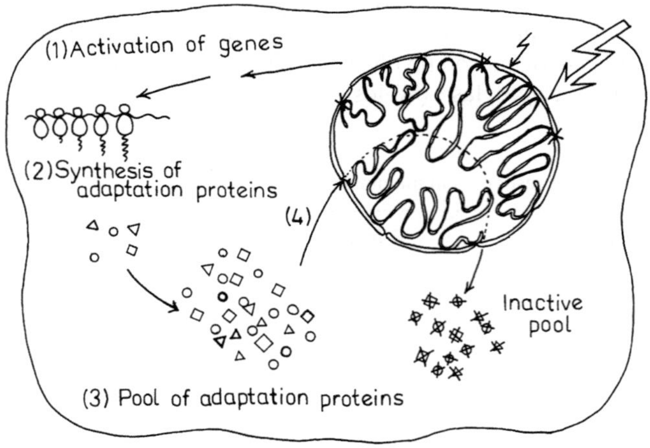

Fig. 8. Proposed mechanism for the radio-adaptive response. *1* A defined amount of chromosomal lesions produced by low doses of ionizing radiation trigger activation of transcription of genes coding for adaptation or protection proteins. *2* Adaptation or protection proteins are synthesized, and *3* a large amount of these proteins accumulate within the cells. *4* The pooled proteins cope efficiently with new chromosomal lesions induced by subsequent high doses of ionizing radiations

Acknowledgements. Part of the work was supported by a Grand-in-Aid from the Ministry of Education, Science and Culture, Japan.

References

Benjamin RC, Gill DM (1980) Poly(ADP-ribose) synthesis in vitro programmed by damaged DNA. A comparison of DNA molecules containing different types of strand breaks. J Biol Chem 255: 10502–10508

Bosi A, Olilvieri G (1989) Variability of the adaptive response to ionizing radiations in humans. Mutat Res 211: 13–17

Chatterjee S, Bhattacharjee SB, Bhattacharya N, Bhaunk G (1984) UV non-mutable mutant from V-79 Chinese hamster cells. Mol Gen Genet 193: 251–254

Criessen D, Shall S (1982) Regulation of DNA ligase activity by poly(ADP-ribose). Nature 196: 271–272

Demple B, Halbrook J (1983) Inducible repair of oxidative DNA damage in *Escherichia coli*. Nature 304: 466–468

Eastman A (1983) Characterization of the aducts produced by *cis*-diamine dichloroplatinum(II) and *cis*-dichloro(ethylene-diamine) platinum(II). Biochemistry 22: 3927–3933

Evensen G, Seeberg E (1982) Adaptation to alkylation resistance involves the induction of DNA glycosylase. Nature 296: 773–775

Haseltine WA (1983) Ultraviolet light repair and mutagenesis revisited. Cell 33: 13–17

Heindorff K, Rieger R, Michaelis A, Takehisa S (1987) Clastogenic adaptation triggered by S-phase-independent clastogens in *Vicia faba*. Mutat Res 190: 131–135

Ikushima T (1972) Relative biological effectiveness of heavy particles from the reaction ^{10}B (n, α)^{7}Li for cytogenetic effects in a ploidy series of the genus *Triticum*. Jpn J Genet 47: 401–410

Ikushima T (1977) Role of sister chromatid exchanges in chromatid aberration formation. Nature 268: 325–236

Ikushima T (1987) Chromosomal response to ionizing radiation reminiscent of an adaptive response in cultured Chinese hamster cells. Mutat Res 180: 215–221

Ikushima T (1989) Radio-adaptive response: characterization of a cytogenetic repair induced by low-level ionizing radiation in cultured Chinese hamster cells. Mutat Res 227: 241–246

Ikushima T, Wolff S (1974) Sister chromatid exchanges induced by light flashes to 5-bromodeoxy-uridine and 5-iodo-deoxyuridine substituted Chinese hamster chromosomes. Exp Cell Res 87: 15–19

Jeggo P, Defais M, Samson L, Schendel P (1977) An adaptive response of *E. coli* to low levels of alkylating agent: comparison with previously characterized DNA repair pathways. Mol Gen Genet 157: 1–9

Karran P, Lindahl T, Griffin B (1979) Adaptive response to alkylating agents involves alteration in situ of O^6-methylguanine residues in DNA. Nature 280: 76–77

Laval F, Laval J (1984) Adaptive response in mammalian cells: crossreactivity of different pretreatments on cytotoxicity as contrasted to mutagenicity. Proc Natl Acad Sci USA 81: 1062–1066

Leenhouts HP, Broertjes C, Sijsma MS, Chadwick KH (1982) Radiation stimulated repair in *Saintpaulia*: its cellular basis and effect on mutation frequency. Environ Exp Bot 22: 301–306

Lindahl T, Sedgwick B, Sekiguchi M, Nakabeppu Y (1988) Regulation and expression of the adaptive response to alkylating agents. Annu Rev Biochem 57: 133–157

Olivieri G, Bodycote J, Wolff S (1984) Adaptive response of human lymphocytes to low concentrations of radioactive thymidine. Science 223: 594–597

Perry P, Wolff S (1974) New Giemsa methods for the differential staining of sister chromatids. Nature 251: 156–158

Rieger R, Michaelis A, Nicoloff H (1982) Inducible repair processes in plant root tip meristems. "Below additivity effects" of unequally fractionated clastogen concentrations, Biol 101: 125–138

Samson L, Cairns J (1977) A new pathway for DNA repair in *Escherichia coli*. Nature (Lond) 267: 281–283

Samson L, Schwartz JL (1980) Evidence for an adaptive DNA repair pathway in CHO and human skin fibrobrast cell lines. Nature 287: 861–863

Sanderson BJS, Morley AA (1986) Exposure of human lymphocytes to ionizing radiation reduces mutagenesis by subsequent ionizing radiation. Mutat Res 164: 347–351

Sankaranarayanan K, Duyn AV, Loos MJ, Natarajan AT (1989) Adaptive response of human lymphocytes to low-level radiation from radioisotopes or X-rays. Mutat Res 211: 7–12

Shadley JD, Wiencke JK (1989) Induction of the adaptive response by X-rays is dependent on radiation intensity. Int J Radiat Biol 56: 107–118

Shadley JD, Wolff S (1987) Very low doses of X-rays can cause human lymphocytes to become less susceptible to ionizing radiation. Mutagenesis 2: 95–96

Shadley JD, Afzal V, Wolff S (1987) Characterization of the adaptive response to ionizing radiation induced by low doses of X-rays to human lymphocytes. Radiat Res 111: 511–517

Singer B (1975) The chemical effects of nucleic acid alkylation and their relation to mutagenesis and carcinogenesis.Prog Nucl Acid Res Mol Biol 15: 219–284

Tomasz M, Mercado CM, Olson J, Chatterjee N(1974) The mode of interaction of mitomycin with deoxyribonucleic acid and polynucleotides in vitro. Biochemistry 13: 4878–4887

Vijayalaxmi, Burkart W (1989) Resistance and crossresistance to chromosome damage in human blood lymphocytes adapted to bleomycin. Mutat Res 211: 1–5

Waldstein F, Cao AE, Setlow RB (1982) Adaptive resynthesis of O^6-methylguanine-accepting protein can explain the difference between mammalian cells proficient and deficient in methyl excision repair. Proc Natl Acad Sci USA 79: 5117–5121

Ward JF (1988) DNA damage produced by ionizing radiation in mammalian cells: identities, mechanisms of formation and repairability. Prog Nucl Acid Res Mol Biol 35: 95–125

Wiencke JK, Afzal V, Olivieri G, Wolff S (1986) Evidence that the [^{3}H]thymidine-induced adaptive response of human lymphocytes to subsequent doses of X-rays involves the induction of a chromosomal repair mechanism. Mutagenesis 1: 375–380

Wolff S, Afzal V, Wiencke JK, Olivieri G, Michaeli A (1988) Human lymphocytes exposed to low doses of ionizing radiations become refractory to high doses of radiation as well as to chemical mutagens that induce double-strand breaks in DNA. Int J Radiat Biol 53: 39–48

Wolff S, Wiencke JK, Afzal V, Youngblom J, Cortes F (1989) The adaptive response of human lymphocytes to very low doses of ionizing radiation: a case of induced chromosomal repair with the induction of specific proteins. In: Baverstock KF, Stather JW (eds) Low dose radiation: biological basis of risk assessment. Taylor & Francis, London, pp 446–454

On Adaptive Responses of Plant Meristem Cells in Vivo – Protection Against Induction of Chromatid Aberrations

R. RIEGER[1], A. MICHAELIS[1], and S. TAKEHISA[2]

Contents

1 Introduction

When cells are confronted with environmental stress they may react by adaptive responses which eventually enhance their resistance to the damaging impact and provide additional capacities to survive the crisis.

The best documented of such inducible mechanisms is the adaptive response of bacteria to alkylating (Samson and Cairns 1977; for review see Yarosh 1985; Teo et al. 1986) and oxidative agents (Demple and Halbrook 1983). Analogous responses are known for animal cells cultivated in vitro (for review see Frosina and Abbondandolo 1985).

Stimulated by these observations and by some earlier reports on inducible protective functions of higher plant cells (Rieger and Michaelis 1965; Broertjes 1972; Sybenga and Kleijer 1976), which remained largely unrecognized, we decided to study in more detail the potential adaptive responses of plant cells that may serve to alleviate the effect of genotoxic agents.

2 Procedure and Rationale

In vivo experiments with Vicia faba (broad bean) root tip meristems provided evidence (Rieger et al. 1982) that plant cells which are stressed by various agents

[1] Zentralinstitut für Genetik und Kulturpflanzenforschung Gatersleben der AdW der DDR, Gatersleben, 4325, GDR
[2] Department of Biology, Keio University, Yokohama, Japan

(conditioning treatment) prior to treatment with clastogens inducing chromatid aberrations (challenge treatment), may respond to the conditioning treatments by inducible protective functions. These eventually reduce the yield of metaphases with chromatid aberrations induced by the challenge treatment, i.e., the conditioning treatments resulted in adaptive responses ultimately serving to reduce the genotoxicity of clastogen impact (for review, see Heindorff et al. 1987b; Rieger and Michaelis 1988a; Neumann et al. 1989).

Protective functions triggered by appropriate conditioning treatments were inferred when the yield of induced chromatid structural changes induced by subsequent challenge treatment was significantly lower than that obtained by challenge treatment alone. Figure 1 shows the major types of chromatid aberrations evaluated in metaphases of the first cell cycle after challenge treatment: chromatid breaks, isochromatid breaks, duplication deletions, intercalary deletions, and chromatid translocations. The experimental procedure used to study adaptive responses in Vicia faba root tip meristems (reconstructed karyotype ACB) and the influence of various inhibitors on these is outlined in Fig. 2. In principle, main root meristems were conditioned by using small clastogen concentrations, various metabolic inhibitors, heat shocks and heavy metals and, after various intertreatment times (IT), were challenged with higher concentrations of various clastogens (see Tables 1–4). Subsequent to challenge

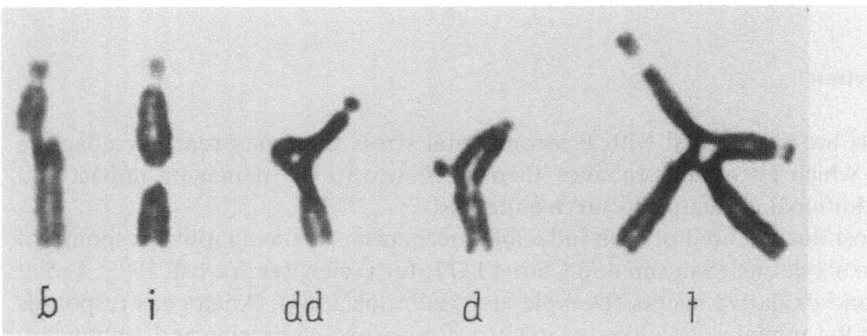

Fig. 1. The main types of chromatid aberrations as observed in Vicia faba (broad bean) root tip meristem cells: *b* chromatid break; *i* isochromatid break; *dd* duplication deletion; *d* intercalary deletion; *t* reciprocal chromatid translocation

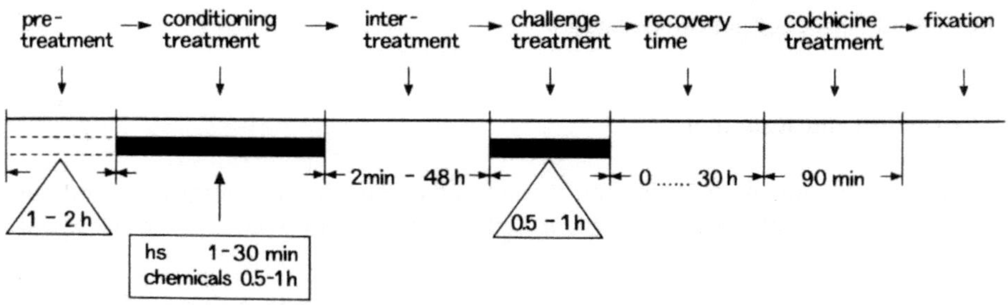

Fig. 2. The scheme according to which the experiments on adaptive response in Vicia faba were performed

Table 1. Clastogens tested with respect to their capability to trigger protective functions against induction of chromatid aberrations[a]

Types of clastogens tested	Clastogen	Conditioning treatment	Inter-treatment time (h)	Challenge treatment	Protection[b]	References
S-phase-dependent agents	MNU[c]	1 h 10^{-4} M	2	1 h 10^{-3} M	+	Rieger et al. (1982)
	Trenimon	0.5 h 10^{-7} M	2	0.5 h 2.5×10^{-6} M	+	Unpubl.
Alkylating	TEM	0.5 h 10^{-5} M	2	0.5 h 10^{-4} M	+	Rieger et al. (1982)
Non-alkylating	MH	0.5 h 5×10^{-5} M	2	0.5 h 4×10^{-4} M	+	Rieger et al. (1982)
	Ethanol	2 h 2×10^{-1} M	2	2 h 4×10^{-1} M	–	Rieger et al. (1982)
S-phase independent agents	Bleomycin	1 h 0.1 mg/l	2	1 h 10 mg/l	+	Heindorff et al. (1987c)
	X-rays	0.06 Gy	1	0.6 Gy	+	Heindorff et al. (1987c)

[a] For experimental dctails, see references.
[b] + Significant reduction (by more than 50 %) of aberration yield due to the conditioning treatment (protection observed); – no protection.
[c] For abbreviations, see Table 2.

Table 2. Combinations of agents tested with respect to the occurrence of cross-protection

Combination of agents	Conditioning treatment	Intertreatment time	Challenge treatment	Cross-protection[a]	Reference
Alkylating vs alkylating agent	1 h 10^{-4} M MNU[b]	2	0.5 h 10^{-4} M TEM	+	Rieger et al. (1982)
	0.5 h 10^{-5} M TEM	2	1 h 10^{-3} M MNU	+	Rieger et al. (1982)
	1 h 10^{-2} EMS	2	0.5 h 10^{-4} M TEM	+	Unpubl.
Alkylating vs nonalkylating agent	1 h 10^{-2} M EMS	2	0.5 h 4×10^{-4} M MH	−	Unpubl.
	1 h 10^{-4} M MNU	2	0.5 h 4×10^{-4} M MH	−	Rieger et al. (1982)
	0.5 h 5×10^{-5} M MH	2	1 h 10^{-3} M MNU	−	Rieger et al. (1982)
	1 h 0.1 mg/l BM	2	0.5 h 2.5×10^{-6} M TRE	+	Heindorff et al. (1987c)
	1 h 0.1 mg/l BM	2	0.5 h 10^{-4} M TEM	+	Heindorff et al. (1987c)
	0.5 h 10^{-5} M TEM	2	1 h 10 mg/l BM	−	Heindorff et al. (1987c)
Nonalkylating vs nonalkylating agent	1 h 0.1 mg/l BM	2	0.5 h 4×10^{-4} M MH	+	Heindorff et al. (1987c)
	0.5 h 5×10^{-5} M MH	2	1 h 10 mg/l BM	+	Heindorff et al. (1987c)
	1 h 0.1 mg/l BM	2	0.6 Gy X-rays	+	Heindorff et al. (1987a)
	0.06 Gy X-rays	2	0.5 h 4×10^{-4} M MH	+	Heindorff et al. (1987c)
Peroxides or hydrazines vs nonalkylating agent	0.5 h 5×10^{-4} M HP	2	0.5 h 5×10^{-4} M MH	+	Heindorff et al. (1985b)
	0.5 h 10^{-3} M BHP	2	0.5 h 5×10^{-4} M MH	+	Heindorff et al. (1985b)
	1 h 10^{-4} M HYD	2	0.5 h 5×10^{-4} M MH	+	Heindorff et al. (1985a)
Peroxides or hydrazines vs alkylating agent	0.5 h 5×10^{-4} M HP	2	0.5 h 10^{-4} M TEM	−	Heindorff et al. (1985b)
	0.5 h 10^{-3} BHP	2	0.5 h 10^{-4} M TEM	−	Heindorff et al. (1985a)
	1 h 10^{-3} HYD	2	0.5 h 10^{-4} M TEM	−	Heindorff et al. (1985a)

[a] + Cross-protection observed; − no cross-protection.
[b] Abbreviations: EMS, ethyl methanesulfonate; MH, maleic hydrazide (1,2-dihydropyridazine-3,6-dione); MNU, N-methyl-N-nitrosourea; TEM, triethyleneiminotriazine; BM, bleomycin; TRE, trenimon (2,3,5-tris(ethyleneimino)-1,4-benzoquinone); HP, hydrogen peroxide; BHP, tert. butyl hydroperoxide; HYD, N,N'-diformylhydrazine.

Table 3. Protective functions against aberration induction by TEM or MH, respectively, triggered by conditioning treatment with various metabolic inhibitors

Conditioning treatment	Inter-treatment time (h)	Challenge treatment 0.5 h 10^{-4} M TEM	0.5 h 5×10^{-4} M MH	References
1 h 2.5×10^{-5} M EB[a]	2	–[b]	+	Michaelis et al. (1988)
1 h 1.9×10^{-4} M NA	2	–	+	Michaelis et al. (1988)
1 h 5×10^{-3} M BA	2.5	+	–	Rieger et al. (1987)
1 h 10^{-4} M BA	2.5	–	n.t.	Rieger et al. (1987)
1 h 10^{-3} M BSI	2	+	+	Unpubl.
2 h 10^{-3} M BSI	24	–	+	Unpubl.

[a] EB, Ethidium bromide; NA, nalidixic acid; BA, benzamide; BSI buthionine sulfoximine.
[b] – No protection observed; + protection observed; n.t. not tested.

treatment and recovery times (RT from 0 to 30 h depending on the use of S-phase-independent or S-phase-dependent agents) the yield of metaphases with chromatid aberrations was established and compared to the yield obtained after challenge treatment alone. Significant differences as to the two yields compared were taken to provide evidence of protective functions triggered by conditioning treatments. Various metabolic inhibitors were used to study their influence on adaptive response (Table 5). They were adminstered prior to conditioning treatments. Effects of conditioning treatments in the presence or absence of inhibitor pretreatment on clastogen challenging provided the basis for conclusions with respect to the influences of inhibitors on adaptive responses triggered by conditioning.

3 Results

The main results obtained by using the experimental schedule outlined above are summarized in the following sections. They provide evidence that a variety of agents used for conditioning treatment trigger adaptive responses to clastogen challenging, that the protective effects decay with intertreatment time and depend on the conditioning parameters, and that inhibitor pretreatment prior to conditioning exerts differential effects depending on the agents used for conditioning and challenging, respectively. Though the molecular mechanisms underlying the inducible protective responses of plant cells are presently obscure, the data at hand allow some preliminary conclusions that may pave the way for biochemical studies.

3.1 Protective Effects Against Clastogen Challenging are Triggered by a Variety of Agents Used for Conditioning Treatment

This conclusion is documented by the results summarized in Tables 1–4: Small clastogen concentrations, various metabolic inhibitors, heat shocks, and inorganic

Table 4. Protection against aberration induction by various clastogens triggered by heat shock and various inorganic salts

Conditioning treatment	Inter-treatment time (h)	Challenge treatment 0.5 h 10^{-4} M TEM[a]	0.5 h 5×10^{-4} M MH[a]	0.5 h 2.5×10^{-6} M Trenimon[b]	1 h 10 mg/l Bleomycin[b]	60 rad (0.6 Gy) X-rays[b]	References
1–30 min 40 °C HS[c]	1	+[d]	+	+	+	+	Rieger et al. (1985) Rieger and Michaelis (1988b)
1 h 10^{-2} M NH_4Cl	2	+	+				Michaelis et al. (1986)
1 h 10^{-3} M $MnCl_2$	2	+	+				Rieger and Michaelis (1988a)
1 h 10^{-4} M $NiCl_2$	24	+	+				Unpubl.
1 h 10^{-5} M $HgCl_2$	2	+	+				Rieger and Michaelis (1988a)
1 h 10^{-2} M KCl	2	+	+				Unpubl.
1 h 10^{-2} M $(NH_4)_2SO_4$	2	–	n.t.				Michaelis et al. (1986)
1 h 10^{-3} M $ZnSO_4$	2	+	+				Michaelis et al. (1986)
1 h 10^{-2} M NH_4NO_3	2	+	+				Unpubl.
1 h 10^{-5} M Cd $(NO_3)_2$	2	+	+				Rieger and Michaelis (1988)
1 h 10^{-3} M Ca $(NO_3)_2$	24	–	–				Unpubl.

[a] Rieger et al. (1985)
[b] Rieger and Michaelis (1988b)
[c] Heat shock.
[d] + Protection observed; – no protection; n.t. not tested.

Table 5. Differential effects of pretreatment with various metabolic inhibitors on protection against TEM and MH triggered by various conditioning treatments

Pretreatment	Conditioning treatment	Inter-treatment time (h)	Challenge treatment 0.5 h 10^{-4} M TEM	Challenge treatment 0.5 h 5×10^{-4} M MH	References
2 h 10^{-6} M CH[a]	0.5 h 10^{-5} M TEM	2	–[c]	–	Rieger et al. (1984)
2 h 10^{-6} M CH	10 min. 40 °C HS[b]	1	–	n.t.	Rieger et al. (1986)
2 h 10^{-6} M CH	1 h 10^{-5} M $Cd(NO_3)_2$	24	+	–	Unpubl.
1 h 5×10^{-3} M BA	0.5 h 10^{-5} M TEM	2	–	–	Rieger et al. (1987)
1 h 5×10^{-3} M BA	0.5 h 10^{-5} M MH	2	–	–	Rieger et al. (1987)
1 h 5×10^{-3} M BA	10 min. 40 °C HS	1	–	+	Rieger et al. (1985)
1 h 5×10^{-3} M BA	1 h 10^{-2} M NH_4Cl	2	–	+	Rieger and Michaelis (1988a)
1 h 5×10^{-3} M BA	1 h 10^{-5} M $HgCl_2$	2	+	+	Rieger and Michaelis (1988a)
1 h 5×10^{-3} M BA	1 h 10^{-3} M $ZnSO_4$	2	–	+	Rieger and Michaelis (1988a)
1 h 10^{-3} M BSI	1 h 10^{-5} M $Cd(NO_3)_2$	2	–	–	Unpubl.
2 h 10^{-3} M BSI	1 h 10^{-5} M $Cd(NO_3)_2$	24	–	+	Unpubl.
2 h 10^{-3} M BSI	1 h 10^{-4} M $NiCl_2$	24	–	–	Unpubl.
1 h 10^{-3} M BSI	1 h 10^{-5} M $HgCl_2$	2	+	+	Unpubl.
1 h 10^{-3} M BSI	1 h 10^{-2} M NH_4Cl	2	+	+	Unpubl.
1 h 10^{-3} M BSI	1 h 10^{-3} M $ZnSO_4$	2	–	–	Unpubl.

[a] Cycloheximide.
[b] Heat shock.
[c] – No protection observed; + protection observed; n.t. not tested.

salts used for conditioning treatment were found to exert protection against challenge treatment with various clastogens.

In a first series of experiments the same agents (clastogens) were used for conditioning (at low concentration) and, after 2 h intertreatment time, for challenge treatment (high concentration). The results are summarized in Table 1 and show that only in the case of ethanol treatment no protective activity against challenge treatment was observed (ethanol is an inducer of chromatid aberrations in Vicia faba; see Michaelis et al. 1959). Conditioning treatments with all other clastogens triggered protective effects against challenging by the same agents, i.e., reduced the yield of metaphases with induced chromatid aberrations.

The next question was to determine whether or not different agents used for conditioning and challenge treatment, respectively, can trigger cross-protection. The results (Tables 2–4) provide evidence for differential effects depending on the agents used for conditioning and challenging. All alkylating agents tested were found to exert cross-protection but maleic hydrazide (MH) conditioning did not trigger protection against challenge treatment with alkylating agents and vice versa. In some cases cross-protection was dependent on the sequence of treatment with various clastogens. Bleomycin (BM) triggered protection against triethylenemelamine (TEM), but TEM proved unable to protect against BM. When, instead of TEM, MH and BM were used for conditioning or challenge treatment, protection occurred with both treatment sequences. The differences are thus not due to the simple fact that one agent is S-phase-dependent (TEM and MH) the other S-phase-independent (BM) but may be connected with the different spectra of lesions induced by alkylating and nonalkylating agents. This inference is supported by the observation that differential effects on TEM and MH challenging, respectively, were also observed when peroxides or hydrazines were used for conditioning treatment. Protective effects were triggered by these agents against aberration induction by MH; no protection occurred when TEM was used for challenge treatment.

Table 3 summarizes the observations made when various metabolic inhibitors were used for conditioning treatment before TEM or MH challenging. Conditioning by ethidium bromide (EB), which inhibits DNA as well as RNA synthesis, and, 2 h later, challenging by MH or TEM, resulted in protection against MH but not against TEM. The same result was obtained when instead of EB nalidixic acid (NA) was used for conditioning treatment (Michaelis et al. 1988). NA inhibits primarily DNA replication.

Opposite results were obtained when benzamide (BA) was used for conditioning. BA is an inhibitor of ADP-ribosyltransferase (ADPRT) and ADP-ribosylation, which may play a role in the response of cells to DNA damaging agents (for review, see Shall 1984; Gaal and Pearson 1985). With BA conditioning, concentration-dependent protection occurred against TEM but not against MH.

When buthionine sulfoximine (BSI) was used for conditioning treatment, protection occurred against MH and TEM challenging when the intertreatment time was 2 h. With an intertreatment time of 24 h, protection still occurred against MH but no longer against TEM, i.e., for the latter challenger the protective effect decayed quicker than for MH. BSI is a specific inhibitor of γ-glutamyl-cysteine synthetase and inhibits the synthesis of phytochelatins, a class of heavy metal-binding peptides which are apparently involved in the accumulation, detoxification, and metabolism of metal ions in plant cells (Grill et al. 1987). The data obtained by using these inhibitors once more demonstrate that the combination of agents used for conditioning and challenging is decisive for the occurrence of protective effects. They provide additional evidence of agent-specific adaptive responses of plant cells.

Table 4 provides a survey as to the effect of heat shock (hs) or salt conditioning on clastogen challenge treatment of Vicia faba root tip meristem cells with the alkylating agent TEM, on the one hand, and with MH, on the other hand. The data show that in addition to conditioning by low clastogen concentrations and certain metabolic inhibitors, these pretreatments may also trigger protective effects against TEM and MH.

Heat shocks (hs) are well known to result in the synthesis of heat shock (or stress) proteins. Their synthesis is usually correlated with increased thermotolerance (for review, see Neumann et al. 1989). Various inorganic salts may also result in induced synthesis of at least a fraction of the hs proteins. As evident from Table 4, both heat shocks (40 °C for 1–30 min) and at least some of the salts tested triggered protective effects against aberration induction by TEM and MH when these clastogens were administered after an appropriate intertreatment time. No protective effects were observed when calcium nitrate (unpubl.) or ammonium sulfate (Michaelis et al. 1986) were used for conditioning treatment prior to TEM or MH challenging.

The data summarized in Tables 1–4 support the conclusion that a variety of agents used for conditioning may trigger specific cellular functions which eventually alleviate the genotoxic effects of challenge treatments resulting in chromatid structural changes.

3.2 The Protective Effect of Conditioning Treatment Decays with Time and is Dependent on the Conditioning Parameters

Variations of the time-span between conditioning and challenge treatment have been used to investigate how long protective functions may last and how quickly they may become effective. Onset and duration of protective effects triggered by conditioning treatments were found to be dependent on the agent being used for conditioning. In the case of conditioning by MH or TEM, no protection to the same agents was observed when the intertreatment time was longer than 6 h (Rieger et al. 1984). When, however, conditioning was by cadmium nitrate or nickel-II-chloride the protective effect against TEM and MH challenging was still observable after 24 h intertreatment time (unpubl. results).

Differential onset and decay of heat shock (10' at 40 °C) protection against aberration induction by TEM or MH, respectively, is illustrated in Figs. 3 and 4 for hs conditioning. Heat shock protection was effective after less than 10 min and lasted up to 210 min in the case of challenge treatment with the nonalkylating agent MH. Heat shock-induced protective responses against the alkylating agent TEM needed more than 10 min for expression and decayed more quickly. After an intertreatment time longer than 180 min, hs protection against TEM was no longer observed. This clastogen-specific difference in hs response indicates that hs triggers at least two protective functions, one specific for MH, the other for TEM.

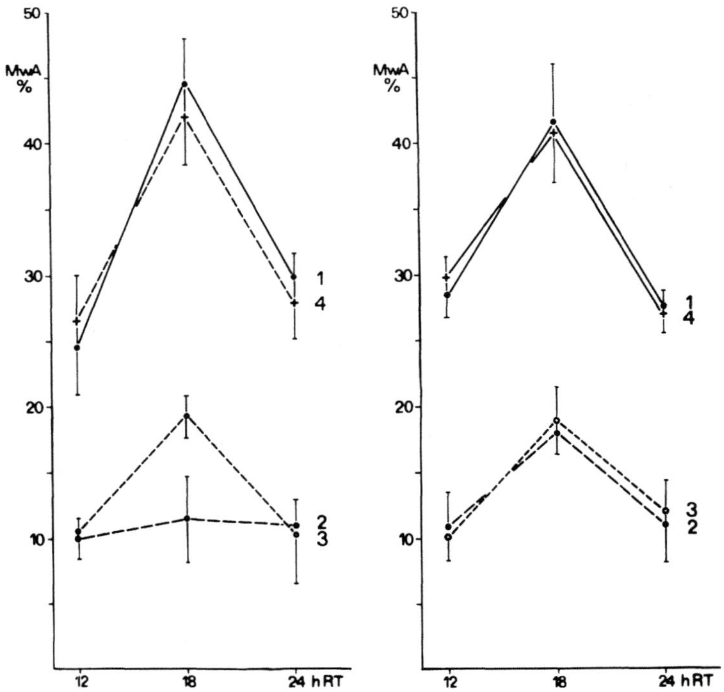

Fig. 3. Effects of long intertreatment times (IT) between heat shock (hs) and MH (*left*) or TEM challenge treatment (*right*) on the yield of metaphases with chromatid aberrations (*MwA %*) at recovery times (*RT*) of 12, 18, and 24 h: hs protection is absent when IT is longer than 210 min in the case of MH challenging; protective functions disappear when IT is longer than 180 min in the case of TEM challenging. *Left* Curve *1*: 0.5 h 5×10^{-4} M MH; curve *2*: 10 min 40 °C hs – 1 h IT – 0.5 h 5×10^{-4} M MH; curve *3*: 10 min 40 °C hs – 210 min IT – 0.5 h 5×10^{-4} M MH; curve *4*: 10 min 40 °C hs – 240 min IT – 0.5 h 5×10^{-4}M MH. *Right* Curve *1*: 0.5 h 10^{-4} M TEM; curve *2*: 10 min 40 °C hs – 1 h IT – 0.5 h 10^{-4} M TEM; curve *3*: 10 min 40 °C hs – 180 min IT – 0.5 h 10^{-4} M TEM; curve *4*: 10 min 40 °C hs – 210 min IT – 0.5 h 10^{-4} M TEM

Whether or not protective effects are being triggered was found to be dependent on the concentration of the conditioning agents used (Rieger et al. 1984) or, in the case of heat shocks, on the temperature and duration of heat shock (Rieger et al. 1986; Rieger and Michaelis 1988a). Heat shocks (10 min) at 38 or 40 °C, but not at 36 °C, were able to reduce the yield of chromatid aberrations induced by challenge treatment with either MH (Fig. 5) or TEM. Reduction of heat shock duration to 1 min at 38 an 40 °C, respectively, triggered protection against MH when the hs temperature was 40 °C. Only partial protection occurred when the heat shock was at 38 °C as shown in Fig. 6.

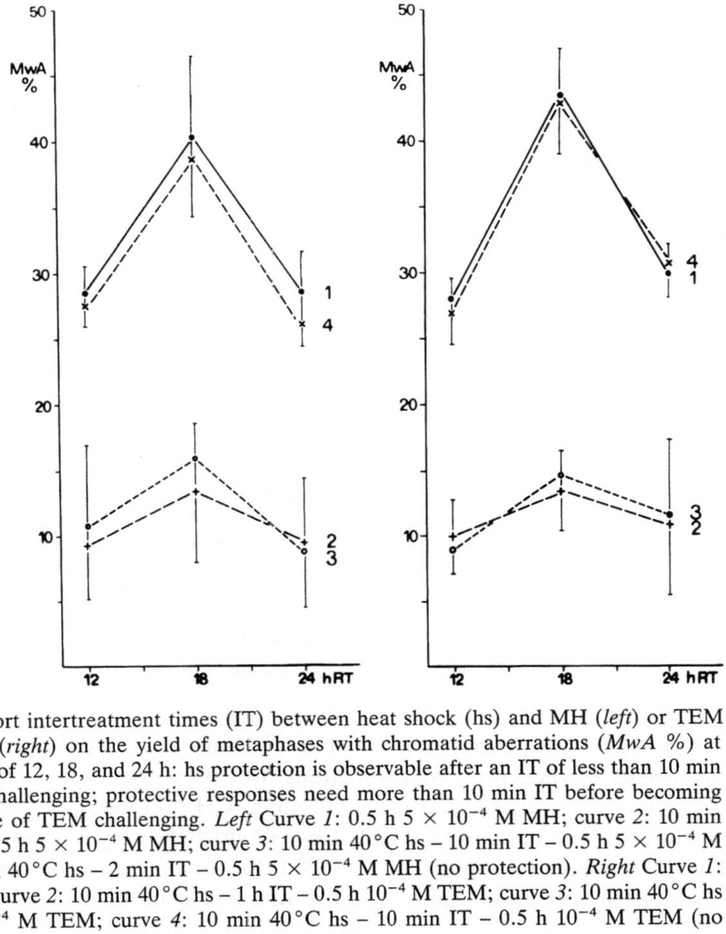

Fig. 4. Effects of short intertreatment times (IT) between heat shock (hs) and MH (*left*) or TEM challenge treatment (*right*) on the yield of metaphases with chromatid aberrations (*MwA* %) at recovery times (*RT*) of 12, 18, and 24 h: hs protection is observable after an IT of less than 10 min in the case of MH challenging; protective responses need more than 10 min IT before becoming expressed in the case of TEM challenging. *Left* Curve *1*: 0.5 h 5 × 10^{-4} M MH; curve *2*: 10 min 40 °C hs – 1 h IT – 0.5 h 5 × 10^{-4} M MH; curve *3*: 10 min 40 °C hs – 10 min IT – 0.5 h 5 × 10^{-4} M MH; curve *4*: 10 min 40 °C hs – 2 min IT – 0.5 h 5 × 10^{-4} M MH (no protection). *Right* Curve *1*: 0.5 h 10^{-4} M TEM; curve *2*: 10 min 40 °C hs – 1 h IT – 0.5 h 10^{-4} M TEM; curve *3*: 10 min 40 °C hs – 0.5 IT – 0.5 h 10^{-4} M TEM; curve *4*: 10 min 40 °C hs – 10 min IT – 0.5 h 10^{-4} M TEM (no protection)

3.3 Effects of Pretreatment with Various Metabolic Inhibitors Before Conditioning Treatment on Protection Against TEM and MH Challenging

It was shown in Section 3.1 (Table 3) that some metabolic inhibitors may trigger protective functions against TEM and/or MH when they were used for conditioning treatments. Three specific metabolic inhibitors: cycloheximide (CH), benzamide (BA), and buthionine sulfoximine (BSI) have additionally been administered prior to various other conditioning treatments in order to study their effects on conditioning-induced protective functions against TEM and MH challenge treatments, respectively. The hope was that their effect might provide additional insights into the mechanisms underlying conditioning-induced protective functions. The results are summarized in Table 5.

Cycloheximide (CH) was used to inhibit cytoplasmic protein synthesis prior to conditioning by TEM, MH, or heat shock. Under such conditions protection against

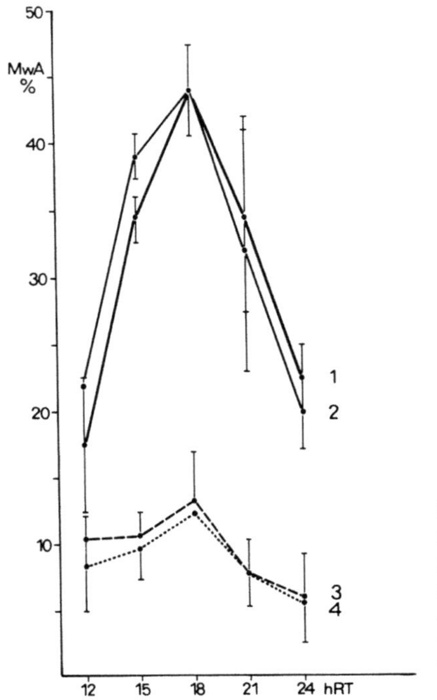

Fig. 5. The yield of metaphases with chromatid aberrations (*MwA* %) as affected by heat shock (hs) treatment (10 min) at different temperatures followed, after 1 h (IT), by challenge treatment with MH. Curve *1*: 0.5 h 5 × 10^{-4} M MH; curve *2*: 10 min hs at 36 °C → 1 h IT → 0.5 h 5 × 10^{-4} M MH; curve *3*: 10 min hs at 38 °C → 1 h IT → 0.5 h 5 × 10^{-4} M MH; curve *4*: 10 min hs at 40 °C → 1 h IT → 0.5 h 5 × 10^{-4} M MH

TEM or MH was no longer observable. This indicates that de novo protein synthesis is necessary when protective functions are to be triggered in these cases. When cadmium nitrate was used for conditioning, CH pretreatment prevented protection against MH but not against TEM. Since phytochelatin synthesis cannot be inhibited by CH treatment (Scheller et al. 1987), these γ-glutamyl peptides may in fact be involved in protection against TEM. Protection against MH is obviously caused by a different mechanism since pretreatment with CH prevented it (Heindorff et al. 1987b and unpubl.). This again underlines the conclusion that different protective functions which are specific for the agents being used for challenge treatment can be triggered by conditioning treatments. Phytochelatins may be involved in Cd^{2+}-triggered protection against TEM (no effect of CH on the effect) but they are possibly without importance for Cd^{2+}-triggered protection against MH (CH-prevention of protection).

Benzamide (BA), which inhibits ADP-ribosyltransferase (ADPRT), was administered prior to conditioning treatments with TEM, MH, heat shock, ammonium chloride (NH_4Cl), zinc sulfate ($ZnSO_4$) and mercury chloride ($HgCl_2$). In the absence of BA these agents triggered protective effects against TEM and MH, provided the appropriate treatment sequences were chosen (see Table 4). BA pretreatment prevented protection by TEM conditioning against TEM challenging and protection by MH conditioning against MH challenging suggesting that ADP-ribosylation is somehow involved in both cases (Rieger et al. 1987). When, however, BA was administered prior to conditioning by heat shock, the occurrence of NH_4Cl and $ZnSO_4$ protective effects against TEM were prevented but those against MH were upheld (Rieger and Michaelis 1988a). In the case of $HgCl_2$ conditioning, BA pretreatment was without

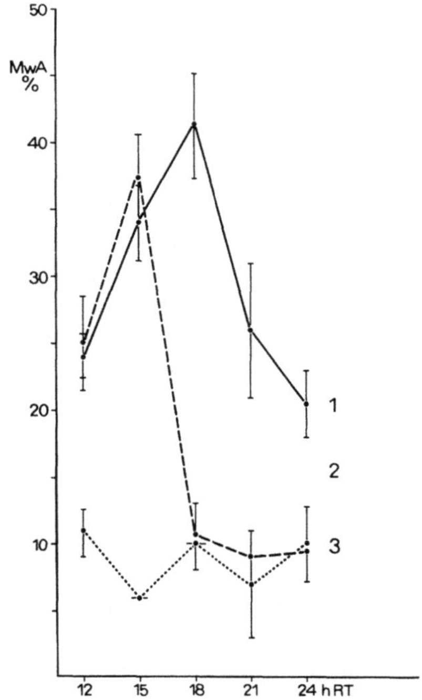

Fig. 6. The yield of metaphases with chromatid aberrations (*MwA* %) as affected by heat shock (hs) pretreatment for 1 min at 38 or 40 °C followed, 1 h (IT) later, by challenge treatment with MH. Protective effects observed only after recovery times (RT) of 18, 21, and 24 h for 38 °C hs. Curve *1*: 0.5 h 5×10^{-4} M MH; curve *2*: 1 min hs at 38 °C → 1 h IT → 0.5 h 5×10^{-4} M MH; curve *3*: 1 min hs at 40 °C → 1 h IT → 0.5 h 5×10^{-4} M MH

any effect on protection against TEM and MH (Table 5). These observations together with the finding that BA conditioning treatment triggered protection against TEM but not against MH challenging (Table 3) illustrate the complexity of the situation with respect to the inducible protective functions in plant cells. Some of the results indicate that ADP-ribosylation may play a role in the protective responses under certain, not yet sufficiently defined conditions.

Again, a rather complex situation was noted when the glutathione synthetase inhibitor buthionine sulfoximine (BSI) was adminstered prior to conditioning with various salts and challenge treatment with TEM or MH (Table 5). BSI used for conditioning prior to challenging with TEM or MH resulted in protection against both agents when the intertreatment time was 2 h; with 24 h intertreatment time protection occurred only against MH (see Table 3).

BSI is the most effective inhibitor of glutathione biosynthesis. The tripeptide glutathione is the major free thiol in living cells which participates in processes such as detoxification of xenobiotics, removal of hydroperoxides, protection against effects of ionizing radiation, and maintenance of the sulfhydryl status of proteins. Since various salts were found to trigger protective functions in plant cells (Table 4) which alleviate the clastogenic effectivity of TEM or MH, we presumed that phytochelatins (γ-glutamyl peptides) might possibly be involved in metal salt-induced protective functions observed in plant cells. Inhibition of their synthesis by BSI pretreatment was expected to give an answer to this question.

BSI pretreatment prior to conditioning by ammonium chloride and mercury chloride did not prevent protection against TEM and MH challenge treatment. When, however, conditioning was by cadmium nitrate, nickel-II-chloride or zink sulfate, BSI pretreatment prevented protection against both TEM and MH. Prolongation of intertreatment time from 2 to 24 h resulted in BSI prevention of Cd^{2+} protection against TEM but not against MH, i.e., the modifying effect of BSI on MH challenging decayed much quicker than in the case of TEM challenging. With Ni^{2+} conditioning, BSI treatment abolished Ni^{2+}-triggered protection against both TEM and MH even in the case of 24-h intertreatment time.

The results presently at hand allow the conclusion that at least some of the adaptive responses can be specifically inhibited by inhibitor treatment before conditioning, others evidently not. This inference is once more indicative of the fact that plant cells may react to conditioning treatments by a variety of responses that eventually reduce the clastogenic effectivity of challenge treatments.

4 Discussion and Conclusions

Though the experimental results summarized in this report raise more questions than they solve, they add to those findings which document that cells developed mechanisms against environmental insults. All the agents used for conditioning treatments and resulting in adaptive responses may be viewed as factors which the cells sence as stress and to which they respond by inducible functions counteracting, at least in part, the injurious strain. The data reviewed here show that plant cells are capable of reducing the damaging impact of S-phase-dependent and S-phase-independent clastogens by various inducible mechanisms. The molecular mechanisms responsible for these modifications remain as yet obscure but several conclusions can be drawn:

1. Nontoxic conditioning pretreatment with various agents (such as low clastogen concentrations, certain metabolic inhibitors, heat shocks and various salts) may trigger protective cellular functions which reduce the yield of chromatid aberrations induced by challenge treatment with various clastogens (Tables 1–4), i.e., result in adaptive responses to clastogens.
2. Adaptive responses are in many cases dependent on de novo protein synthesis and are prevented from occurring when protein synthesis is inhibited by cycloheximide (Table 5).
3. Onset and decay of inducible functions that counteract the formation of chromatid aberrations may vary and are dependent on the agents used for conditioning and challenge treatment.
4. Presence or absence of cross-protection after use of different agents for contitioning and challenging depend on the agents used for conditioning and challenging, i.e., the cells are obviously capable of triggering a number of adaptive responses which eventually exert protection against clastogens (Tables 2, 3, and 4).
   This conclusion is supported by the fact that protective effects were observed after conditioning and challenge treatment with alkylating agents, on the one hand, and after conditioning and challenge treatment with MH, on the other hand (Table 1). Cross-protection was, however, absent when alkylating agents were used for conditioning and MH for challenge treatment (and vice versa; see Table 2). Thus, two different protective mechanisms are being triggered, one by low concentra-

tions of alkylating agents and specific for aberration production by alkylating agents, the other by low concentrations of MH, peroxides and hydrazines and specific for aberration production by MH.

Heat shocks and some salts used for conditioning treatment resulted in protection against both alkylating agents and MH. These treatments thus either trigger two protective functions at the same time or a third one. Since benzamide (BA) pretreatment before conditioning by hs, NH_4Cl or $ZnSO_4$ abolished protection against TEM but not against MH, two different mechanisms are more probable (Table 5). ADP-ribosylation may be involved in hs- or salt-induced protection against TEM but not in protection against MH. Since BA pretreatment abolished the protective effects against TEM and MH when these agents were used, instead of hs or salts, for conditioning, multiple protective functions are indicated.

5. Buthionine sulfoximine (BSI) pretreatment before conditioning treatment with $Cd(NO_3)_2$ abolished protection against both TEM and MH when the intertreatment time (IT) was 2 h, only against TEM, and not against MH, when IT was 24 h. When conditioning was with $NiCl_2$, BSI prevented protection against TEM and MH when the IT was 24 h. The same was true in the case of conditioning treatment by $ZnSO_4$ and an IT of 2 h. In these cases inhibition by BSI of phytochelatin synthesis may be responsible for the prevention of protective effects.

   When $Cd(NO_3)_2$, $NiCl_2$, or BSI were used separately for conditioning treatment 2 h prior to TEM or MH challenging, decreased yields of metaphases with chromatid aberrations were observed after challenging treatment with both these clastogens (Tables 3 and 4). Thus, all three types of conditioning treatment triggered protective functions. Adaptive responses were prevented when BSI and $Cd(NO_3)_2$ or $NiCl_2$ were consecutively administered before TEM or MH challenging (Table 5). This means that the protective effects triggered by BSI conditioning, on the one hand, and $Cd(NO_3)_2$ or $NiCl_2$ conditioning on the other hand, became completely compensated when the agents were used for consecutive pretreatments. We can presently only conclude that the underlying mechanisms are obscure.
6. The multitude of reactions observed when some specific metabolic inhibitors were used for pretreatment prior to conditioning treatment (Table 5) or when intertreatment time was appropriately modified allows the conclusion that, in addition to de novo protein synthesis, other enzymatic functions are involved in the various adaptive responses of plant cells.

Taken together, the results now at hand show that conditioning pretreatment may, in many cases at least, render the Vicia faba root tip meristem cells less sensitive to DNA damaging agents, as evidenced by lower yields of induced chromatid aberrations. Basically similar results have been reported for other plants (e.g., Veleminsky et al. 1983; Stepanyan et al. 1983; Nicoloff et al. 1985; Dubinina et al. 1986; Dubinina and Krushova 1989a, b). Together with the results reported for bacteria and animal cells cultivated in vitro, these findings show that adaptive responses which may transiently protect against genotoxic effects are a widespread phenomenon. Whether or not inducible DNA repair pathways are involved in the protective effects observed in higher plant cells is presently an open question.

Acknowledgments. We are grateful to Drs. I. Schubert and O. Aurich for critical reading of the manuscript and helpful suggestions. Our thanks are due to Mrs. I. Siegel and Mrs. H. Trippner for typing the manuscript and to Mrs. H. Ernst and Mrs. A. Lauterwald for their help with the figures.

References

Broertjes C (1972) Use in plant breeding of acute, chronic or fractionated doses of X-rays or fast neutrons as illustrated with leaves of *Saintpaulia*. Agric Res Rep 7761: 74

Demple B, Helbrook J (1983) Inducible repair of oxidative DNA damage in *E. coli*. Nature 304: 466–468

Dubinina LG, Krushova ZI (1989a) Effect of nitrosoguanidine in *Crepis capillaris* cells and adaptive response. Genetika (Moscow) 25: 269–276

Dubinina LG, Krushova ZI (1989b) Effects of chloramphenicol and cycloheximide on adaptive response in *Crepis capillaris* cells. Genetika (Moscow) 25: 448–453.

Dubinina LG, Krushova ZI, Sergievskaya (1986) The influence of pretreatment with nonmutagenic doses of ethyleneimine on mutagenesis in *Crepis capillaris* cells. Genetika 12: 2805–2812

Frosina G, Abbondandolo A (1985) The current evidence for an adaptive response to alkylating agents in mammalian cells with special reference to experiments with in vitro cell cultures. Mutat Res 154: 85–100

Gaal JC, Pearson CK (1985) Eukaryotic nuclear ADP-ribosylation reactions. Biochem J 230: 1–18

Grill E, Winnacker EL, Zenk MH (1987) Phytochelatins, a class of heavy metal binding peptides from plants are functionally analogous to metallothioneins. Proc Natl Acad Sci USA 84: 439–443

Heindorff K, Aurich O, Rieger R, Michaelis A (1985a) Pretreatment of *Vicia faba* root tip meristems with hydrazines results in "clastogenic adaptation" to maleic hydrazide. Mutat Res 142: 183–186

Heindorff K, Michaelis A, Aurich O, Rieger R (1985b) Peroxide pretreatment of *Vicia faba* root tip meristems results in "clastogenic adaptation" to maleic hydrazide but not to TEM. Mutat Res 142: 23–27

Heindorff K, Rieger R, Michaelis A, Takehisa S (1987a) Clastogenic adaptation triggered by S-phase independent clastogens in *Vicia faba*. Mutat Res 190: 131–135

Heindorff K, Rieger R, Schubert I, Michaelis A, Aurich O (1987b) Clastogenic adaptation of plant cells – reduction of the yield of clastogen-induced chromatid aberrations by various pretreatment procedures. Mutat Res 181: 157–171

Heindorff K, Schubert I, Rieger R, Michaelis A (1987c) Clastogenic adaptation of *Vicia faba* root tip meristem cells after consecutive treatments with S-phase dependent and S-phase independent agents. Biol Zentralbl 106: 439–448

Michaelis A, Ramshorn K, Rieger R (1959) Äthylalkohol – radiomimetisches Agens bei *Vicia faba* L. Naturwissenschaften 46: 381–382

Michaelis A, Takehisa S, Rieger R, Aurich O (1986) Ammonium chloride and zinc sulfate pretreatments reduce the yield of chromatid aberrations induced by TEM and maleic hydrazide in *Vicia faba*. Mutat Res 173: 187–191

Michaelis A, Rieger R, Nicoloff H (1988) Effects of ethidium bromide and nalidixic acid pretreatment on the induction of chromatid aberrations by TEM and maleic hydrazide in *Vicia faba* main root meristems. Mutat Res 208: 83–86

Neumann D, Nover L, Parthier B, Rieger R, Scharf KD, Wollgiehn U, Nieden U zur (1989) Heat shock and other stress response systems of plants. Biol Zentralbl 108: 1–156

Nicoloff H, Gecheff K, Rieger R, Michaelis A (1985) "Clastogenic adaptation" in barley: differential response of presoaked and dry seeds. Mutat Res 143: 83–85

Rieger R, Michaelis A (1965) Chemical induction of heat-reversible "potential lesions" in *Vicia faba* chromosomes. Nature 206: 741–742

Rieger R, Michaelis A (1988a) Heat shocks prior to mutagen treatment may result in protection against induction of chromatid structural changes in higher plants. Biol Zentralbl 107: 403–413

Rieger R, Michaelis A (1988b) Heat shock protection against induction of chromatid aberrations is dependent on the time span between heat shock and clastogen treatment of *Vicia faba* root tip meristem cells. Mutat Res 209: 141–144

Rieger R, Michaelis A, Nicoloff H (1982) Inducible repair processes in plant root meristems? "Below-additivity effects" of unequally fractionated clastogen concentrations. Biol Zentralbl 101: 125–138

Rieger R, Michaelis A, Nicoloff H (1984) "Clastogenic adaptation" of the *Vicia faba* root tip meristems as affected by various treatment parameters. Mutat Res 140: 99–102

Rieger R, Michaelis A, Schubert I (1985) Heat-shocks prior to treatment of *Vicia faba* root tip meristems with maleic hydrazide or TEM reduce the yield of chromatid aberrations. Mutat Res 143: 79–82

Rieger R, Michaelis A, Schubert I (1986) Reduction by heat shock of maleic hydrazide induced aberration yield is dependent on temperature and duration of heat pretreatment. Mutat Res 174: 199–204

Rieger R, Michaelis A, Aurich O (1987) Effects of benzamide pretreatment on clastogenic adaptation of *Vicia faba* root tip meristem cells to triethylenemelamine (TEM) and maleic hydrazide (MH). Mutat Res 191: 93–97

Samson L, Cairns J (1977) A new pathway for DNA repair in *E. coli.* Nature 267: 281–288

Scheller HV, Huang B, Hatch E, Goldsbrough PB (1987) Phytochelatin synthesis and glutathione levels in response to heavy metals in tomato cells. Plant Physiol 85: 1031–1035

Shall S (1984) ADP-ribose in DNA repair. Adv Radiat Biol 11: 1–69

Stepanyan NS, Krupnova GF, Zhestyanikov VD (1983) Reparation of cytogenetical damage in plant cells upon an unequal fractionation of injurious action. 1. Treatment by nitrogen mustard. Citologija 25: 958–964

Sybenga I, Kleijer G (1976) Below-additivity and "protective" effects of dose fractionation in *Crotalaria intermedia.* Mutat Res 34: 131–140

Teo J, Sedgewick B, Kilpatrick MW, McCarthy TV, Lindahl T (1986) The intracellular signal for induction of resistance to alkylating agents in *E. coli.* Cell 45: 315–324

Veleminsky J, Gichner T, Satava J (1983) Reduction in the frequency of N-methyl-N-nitrosurea-induced somatic mutations in *Tradescantia* by pretreatment with low doses of alkylating agents. Mutat Res 122: 229–234.

Yarosh DB (1985) The role of O^6-methylguanine-DNA-methyltransferase in cell survival, mutagenesis and carcinogenesis. Mutat Res 145: 1–16

Chromosome Aberrations in A-Bomb Survivors, Hiroshima and Nagasaki

A. A. AWA[1]

Contents

1 Introduction

In the course of chromosome study of atomic bomb survivors in Hiroshima and Nagasaki, we have obtained evidence relevant to the biological dose assessment in persons exposed to ionizing radiation (Awa 1974, 1975, 1983, 1984; Awa et al. 1971, 1978). The findings are summarized as follows:

1. Radiation-induced chromosome aberrations of the stable type were found to persist many years after exposure in cultured blood lymphocytes of A-bomb survivors in Hiroshima and Nagasaki.
2. The frequency of cells with chromosome aberrations increased with increasing radiation dose for individual survivors in the two cities, although there was a striking variability (or overdispersion) of the frequency of aberrant cells between survivors within a dose category.
3. The aberration frequency was consistently higher in Hiroshima than in Nagasaki in all dose ranges. The observed intercity difference was almost by a factor of 2. Furthermore, the shape of dose-response curves for chromosome aberrations seemed to be linear in Hiroshima, while it was dose-quadratic in Nagasaki. Therefore, the differences in chromosome aberration frequencies as well as in the shape of the dose-response relationship observed in the two cities were thought to be attributable to the difference in the proportion of a neutron component relative to the total kerma (kerma = kinetic energy released in material).
4. Stable chromosome aberrations, such as reciprocal translocations and inversions, were observed to predominate, and they contributed to a large extent to determining the dose-response relationship. In contrast, dicentrics and rings of the

[1] Radiation Effects Research Foundation, 5-2 Hijiyama Park, Minami-ku, Hiroshima 732, Japan

unstable type were found less pronounced in frequency in all dose ranges. This suggested that preferrential loss of cells with dicentrics and rings has occurred with lapse of time from the circulating lymphocyte population due to mitotic disturbances.

All of these observations were derived from the old A-bomb dosimetry system known as Tentative 1965 Dose (T65D). Since criticism was raised regarding the precision of the T65D computation system, it was agreed in 1981 to organize a US-Japan Joint Committee for the Reassessment of Atomic Bomb Radiation Dosimetry in Hiroshima and Nagasaki. After extensive collaborative work by the committee members, a new dosimetry system termed "Dosimetry System 1986 (DS86)" has become available for the majority of survivors in our major study samples at RERF, thus providing a data base for calculating free-in-air tissue kerma and organ doses (Roesch 1988).

The difference between the two dosimetry systems is rather small in Nagasaki, where gamma dose was the major component. In contrast, the difference is drastic in Hiroshima, which is characterized by a striking decrease in neutron dose and an increase in gamma dose as compared with the T65D. Thus, the neutron doses received by survivors of Hiroshima and Nagasaki were estimated to be considerably lower than estimated by T65D procedures. Another important feature of the DS86 is a considerable reduction of the transmission factor for Japanese-type houses by which a large number of survivors were shielded at the time of bombings.

It was desirable, using new DS86 estimates of individual exposure, to re-analyze various epidemiological and biological data obtained at RERF for the risk estimation of radiation exposure. By preliminary analysis of chromosome aberration data on Hiroshima and Nagasaki survivors, it was auticipated that the dose-response curves for aberration frequency would be steeper with the new dosimetry than with the T65D system (Awa 1988; Awa et al. 1984; Fujita et al. 1983; Sasaki 1983, 1988).

Preston and Pierce (1987) compared leukemia and non leukemic cancer mortality risk estimates under the old and new dosimetries, and reported that in terms of total kerma, the risk estimates for both types of cancer are 75–85% higher with the DS86.

Preston et al. (1988) and Awa et al. (1988) have confirmed the finding of Preston and Pierce by re-analyzing our cytogenetic data on 1245 A-bomb survivors collected between 1968 and 1980 to determine differences in the dose-response relationships for chromosome aberrations between the T65D and DS86 dose estimates and to assess differences between Hiroshima and Nagasaki. For a linear-dose response model, the average percentage of cells with chromosome aberrations increases more sharply with dose in Hiroshima than in Nagasaki. The magnitude of the intercity difference in the percentage of aberrant cells per gray is less for DS86 than for T65D, though the difference is statistically significant for both kerma and bone marrow dose with either dosimetry.

The main objective of this report is to confirm whether the observed intercity difference in the dose-response relationship for chromosome aberration frequencies is demonstrable in newly examined A-bomb survivors in Hiróshima and Nagasaki.

2 Materials and Methods

2.1 Samples

Subjects of the study were drawn from participants in the RERF Adult Health Study cohort, who visit the RERF clinic biennially for periodical health examinations. Blood specimens were collected between 1979 and 1980 from proximally exposed A-bomb survivors (112 in Hiroshima and 61 in Nagasaki). Twenty distally exposed survivors in Hiroshima with estimated kerma of 0 Gy served as control in this study.

Because of the lack of survivors with estimated doses of more than 3 Gy in Nagasaki, survivors with 3 Gy or more were excluded from the present analysis. We also excluded those who had received therapeutic radiation exposure or radioisotope treatment at any time in the past. In the present analysis, we used both DS86 kerma and bone marrow dose assigned to individual survivors. The mean bone marrow dose in the present study was about 80% of the mean DS86 kerma, with a further decrease in the proportion of neutron component.

2.2 Cytogenetic Procedures

Lymphocyte cultures from fresh blood specimens of the survivors were set up according to our standard protocol using Eagles MEM containing 20% fetal calf serum and phytohemagglutinin, and incubated for 52 h. Colchicine was added 2 h before the cultures were terminated. Microscopic slides were prepared according to the routine air-drying procedure and conventionally stained with Giemsa. Chromosome slides were coded and examined without knowledge of the survivor's exposure status. Cases were judged as eligible when we could analyze 100 metaphases. In order to compare more precisely the chromosome aberration data between Hiroshima and Nagasaki, all of the microscopic work was performed in Hiroshima by well-experienced examiners, and all chromosome aberrations observed were confirmed by a senior cytogeneticist.

2.3 Classification of Chromosome Aberrations

Chromosome-type aberrations scored in this study were classified into stable and unstable types; the former includes reciprocal translocations (interchanges), pericentric inversions (intrachanges), and deletions without an acentric segment, while the latter contains dicentrics (interchanges), rings with or without centromere (intrachanges), and acentric fragments of either a terminally or interstitially deleted type. Paracentric inversions are not detectable with the nonbanded conventional technique. Cells containing stable aberrations are referred to as Cs cells, while those with unstable ones as Cu cells.

3 Results

Figure 1 (top) shows the frequencies of cells with chromosome aberrations for individual survivors plotted against estimated DS86 kerma in Hiroshima. There are some cases whose aberration frequencies are grossly discrepant with regard to estimated dose, i.e., low-dose survivors with high aberration frequencies, and high-dose survivors with low aberration frequencies. They are termed "biological outliers". As noted in a previous report (Preston et al. 1988), there is appreciable overdispersion of aberration frequencies against estimated doses for individual survivors both in Hiroshima and Nagasaki. This means that cytogenetic outliers can be observed with

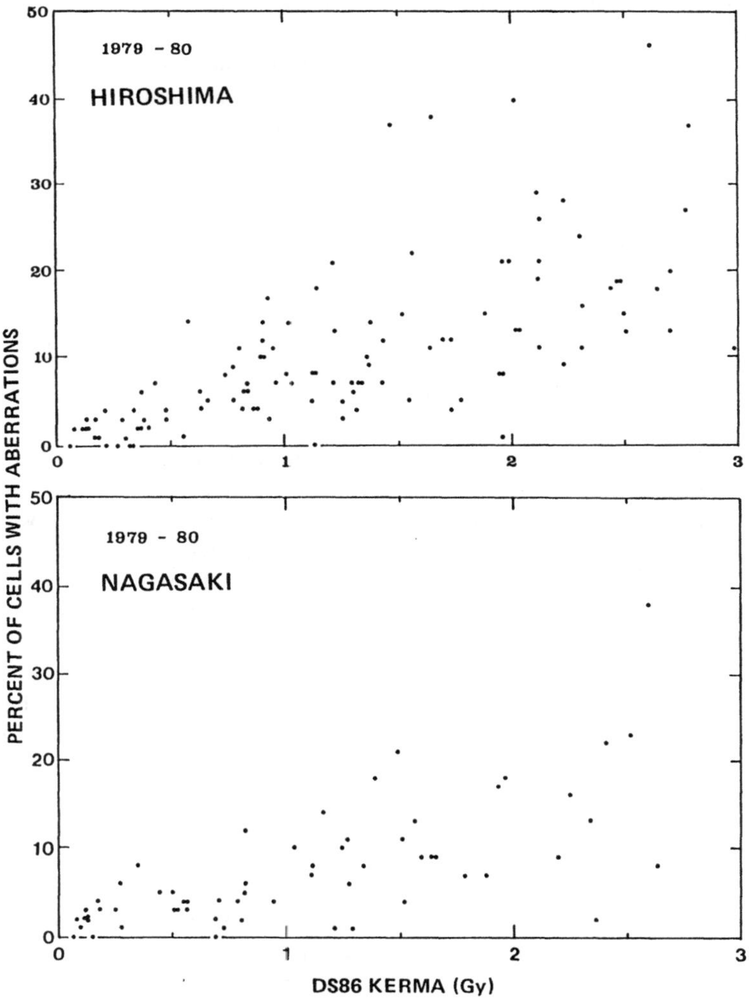

Fig. 1. Frequency of cells with chromosome aberrations (%) for individual survivors plotted against DS86 kerma in Hiroshima (*top*), and Nagasaki (*bottom*). Blood specimens were collected between 1979 and 1980

a consistent pattern and frequency on different samples. The majority of outliers are characterized by survivors with low aberration frequencies in the high dose range.

The pattern of the distribution in aberration frequencies of Nagasaki survivors is the same as Hiroshima (Fig. 1, bottom). Here again, survivors with low aberration values in the high dose range are seen more prevalently in Nagasaki.

Chromosome aberration data on Hiroshima and Nagasaki are shown in Table 1 a-b, with a further breakdown of proximally exposed survivors into several dose groups. Chromosome aberrations scored are (1) dicentrics and rings; (2) translocations and inversions; (3) Cs cells; and (4) total aberrant cells. Data are presented on the basis of kerma (Table 1a) and bone marrow dose (Table 1b) of DS86 estimates. The dose-response relationships for chromosome aberration frequencies seem to fit to a linear model in both cities, although in Nagasaki the quadratic term may not be ignored.

As shown in Table 1, chromosome aberration frequencies are consistently higher in Hiroshima than in Nagasaki, in terms of the percentage of both Cs cells and total aberrant cells. A preliminary analysis showed that, for a linear dose-response model, the slopes of the percentage of Cs cells in Hiroshima were 7.9/Gy for kerma and 10.5/Gy for bone marrow dose, respectively, and in Nagasaki they were 6.5/Gy for kerma and 8.1/Gy for bone marrow dose, respectively. Thus, the magnitude of the intercity difference ranged between 20 and 30% in both kerma and bone marrow dose, although the difference is not statistically significant, as was the case using the old dosimetry.

The above-mentioned intercity difference seems to be due in part to the difference in the amount of neutron dose between the two cities, since a proportion of neutron component relative to the total kerma is still larger by a factor of 3–4 in Hiroshima than in Nagasaki.

Linear dose-response regressions for frequencies of Cs cells were compared between the two cities, employing constant RBEs of 1, 5, and 10 for kerma, and 1, 10, and 15 for bone marrow dose, respectively. For the kerma, two regression lines from the two cities became overlapping at the RBE range of 5–10, while the two lines seemed to be closer at the RBE of 15 for the bone marrow dose. Exactly the same pattern was observed for the frequency of translocations and inversions per cell.

In terms of the frequency of aberrations per se, i.e., translocations and inversions, the patterns in the dose-response relationship for chromosome aberration frequencies, including the magnitude of intercity difference, are exactly the same as those derived from the aberrant cell frequencies as already described in the foregoing section.

It has been confirmed from the present observation that Cs cells predominated in all dose ranges, and thus they constituted more than 90% of the total aberrant cells both in Hiroshima and Nagasaki (Table 1). However, if we extend the number of metaphases to 500 per subject, a positive dose-effect relationship would be anticipated even for the frequency of dicentrics and rings. Our recent unpublished data on the frequency of dicentrics and rings, derived from 94 proximally exposed survivors in Hiroshima, have shown that a positive relationship is nevertheless observed between radiation doses and unstable chromosome aberrations.

Table 1. Frequency of chromosome aberrations in A-bomb survivors, Hiroshima and Nagasaki[a]

a. Dose group (mGy)	DS86 Kerma (mGy)			No. of cases	Total cells counted	dic+r	t+inv	Cs cells (%)	Total aberrant cells (%)
	Total	Neutron	Gamma						
Hiroshima									
0	0	–	–	20	2000	11	16	17 (0.9)	3 (1.7)
5– 499	263	4	259	25	2500	9	31	42 (1.7)	57 (2.3)
500– 999	801	26	775	23	2300	11	147	166 (7.2)	178 (7.7)
1000–1499	1241	42	1199	24	2400	15	218	221 (9.2)	237 (9.9)
1500–1999	1768	68	1700	15	1500	3	176	191 (12.7)	198 (13.2)
2000–2999	2387	110	2277	25	2500	19	476	491 (19.6)	516 (20.6)
Exposed total	1259	49	1210	112	11200	57	1048	1111 (9.9)	1186 (10.6)
Nagasaki									
5– 499	207	1	206	16	1600	11	27	29 (1.8)	47 (2.9)
500– 999	700	5	695	15	1500	3	45	50 (3.3)	57 (3.8)
1000–1499	1244	14	1230	12	1200	2	93	111 (9.3)	115 (9.6)
1500–1999	1703	18	1685	10	1000	1	86	101 (10.1)	104 (10.4)
2000–2999	2416	34	2381	8	800	7	119	122 (15.3)	131 (16.4)
Exposed total	1067	12	1055	61	6100	24	370	413 (6.8)	454 (7.4)

Table 1. (continued)

b. Dose group (mGy)	DS86 Bone marrow dose (mGy)			No. of cases	Total cells counted	dic+r	t+inv	Cs cells (%)	Total aberrant cells (%)
	Total	Neutron	Gamma						
Hiroshima									
0	0	–	–	20	2000	11	16	17 (0.9)	33 (1.7)
5– 499	230	2	228	28	2800	10	46	60 (2.1)	77 (2.8)
500– 999	739	10	729	34	3400	15	247	274 (8.1)	290 (8.5)
1000–1499	1254	20	1234	25	2500	13	269	276 (11.0)	291 (11.6)
1500–1999	1682	28	1654	20	2000	18	384	389 (19.4)	414 (20.7)
2000–2999	2225	41	2184	5	500	1	102	112 (22.4)	114 (22.8)
Exposed total	961	15	947	112	11200	57	1048	1111 (9.9)	1186 (10.6)
Nagasaki									
5– 499	224	0	224	21	2100	13	38	42 (2.0)	64 (3.0)
500– 999	734	3	731	18	1800	3	78	90 (5.0)	97 (5.4)
1000–1499	1267	6	1260	12	1200	1	113	135 (11.3)	138 (11.5)
1500–1999	1797	12	1786	10	1000	7	141	146 (14.6)	155 (15.5)
Exposed total	837	4	833	61	6100	24	370	413 (6.8)	454 (7.4)

[a] dic+r, Dicentrics and rings; t+inv, translocations and inversions; Cs, stable aberrations.

A somewhat elevated level of chromosome aberration frequencies, both stable and unstable types, was noted in 20 distally exposed survivors in Hiroshima; 5.5 dicentrics plus rings per 1000 metaphases in this study (Table 1), as opposed to the value of 1.3 dicentrics plus rings in the general human population (Bender et al. 1988). The reason for this increase remains unresolved; it will be discussed later.

4 Discussion

We noted in our original observations that the shape of the dose-response curve for aberration frequencies with the T65D system was more toward dose-squared in Nagasaki (Awa 1975; Awa et al. 1978). However, the quadratic term became less pronounced in the recent observation with the DS86, although the dose-squared term seemed to exist to a lesser degree. The reason for this discrepancy may be due in part to the exclusion of proximally exposed Nagasaki survivors from our study samples, because most of these survivors were exposed in the factories near the hypocenter at the time of the bombing, so that dose estimation with the new dosimetry for each of these survivors was virtually impossible due to their complex shielding situations. Their aberration frequencies were low relative to their high radiation dose estimates with the T65D.

There still is a striking variability of the frequency of cells with chromosome aberrations between survivors within a given dose category even in our recent samples. Such an overdispersion in aberration frequencies may either indicate differential radiosensitivity in A-bomb survivors, or may be due to random errors in the DS86 estimates assigned to individual survivors. Some biological end points recently studied at RERF point to the fact that no difference in radiation sensitivity is demonstrated among A-bomb survivors. On the contrary, (Sposto et al. 1990) have found that (1) the dose-response relationship for frequency of aberrant cells using the DS86 assigned dose is significantly steeper in the subsample of individuals who reported severe epilation after the bomb than in those who did not report severe epilation, and (2) there is substantially more variation between individuals at higher doses in the proportion of Cs cells than would be expected if there were no heterogeneity in the dose response between individuals. By performing statistical analyses of cytogenetic and epilation data over a range of dosimetry error models of the form developed by Pierce et al. (1989), it is concluded that random dosimetry errors in the range from 45 to 50% of true dose can explain the difference in dose response between severe and nonsevere epilation groups. This amount of dosimetry error also accounts for the overdispersion in aberration frequencies. This trend has also been observed between the occurrence of sever epilation and the subsequent risk of leukemia mortality in the A-bomb survivors (Neriishi et al. 1989).

As for differences in the dose-response relationships for chromosome aberrations between Hiroshima and Nagasaki, it is necessary to obtain better information on relative biological effectiveness (RBE) of neutron radiations using various biological data on A-bomb survivors. We are aware that it would not be appropriate to use the constant RBE value to adjust for the neutron component, because the dose-response curve for neutron radiations differ from that for gamma-rays; a linear dose-response for high-LET radiation, on the one hand, and a linear-quadratic dose-response for low-LET radiation, on the other. A more sophisticated analysis is required to make reasonable estimates of RBE of A-bomb neutron radiations.

It is worthwhile to make some comments on the increased level of chromosome aberration frequencies in distally exposed Hiroshima A-bomb survivors whose estimated kerma was virtually O. Such an increase in aberration frequencies in survivors with O-dose may be associated with cumulated radiation doses from frequent medical exposures to X-ray for diagnostic purpose. It is known that aged survivors have received medical X-ray exposure for diagnostic purposes, especially upper gastrointestinal fluoroscopy, more frequently than nonexposed people of the same age, because of the better medical care extended to A-bomb survivors. Thus, the cumulated radiation doses from medical exposure increase as the survivors become older. It is considered that the amount of cumulated medical X-ray doses per person are generally higher in Hiroshima than in Nagasaki.

As shown in Table 1, an elevated level of aberration frequency in our O-dose survivors was responsible for the increase not only of unstable aberrations but also of stable ones. Stable and unstable aberrations are thought to be induced with equal frequency, but the former are far less frequent because they are difficult to distinguish. The detection rate of stable aberrations by the conventional nonbanded method is influenced by the experience of the microscopist, and thus varies from observation to observation. Even so, the frequency of Cs cells in our control samples was higher than that of Cu cells, although our scoring efficiency for the detection of stable aberrations with the nonbanded staining technique has become greatly improved and thus consistent in our recent analysis (Ohtaki et al. 1982).

In this regard it is important to note here the recent work of Lucas et al. (1989), who reported the usefulness of in situ hybridization of repeat-sequence DNA probes, specific to the paracentromeric locus 1q12 and telomeric locus 1p36, to fluorescently stain regions of human chromosome 1p, for fast detection of structural aberrations involving chromosome 1p. They found that in their in vitro gamma-irradiation (Cs-137 and Co-60) experiments, background frequencies of translocations corresponded to the total human genome of 0.025–0.063, and those of dicentrics to 0.002–0.013, respectively, which were significantly higher than any expected from other studies. Although the reason for this discrepancy remained unknown, there is a striking similarity between the results of Lucas et al. (1989) and the present study, in that high background aberration frequencies are noted in nonirradiated persons.

The technique developed by Lucas et al. (1989) is based, in principle, on the use of fluorescence in situ hybridization with chromosome-specific DNA probes to increase the distinctiveness with which specific human chromosomes can be stained so that chromosome translocations and other structural aberrations involving the stained chromosomes can be rapidly detected with ease. Thus, this technique seems extremely advantageous (1) to precisely detect reciprocal translocations that have persisted predominantly in A-bomb survivors, and (2) to establish a better in vivo dose-response relationship for translocation frequencies in A-bomb survivors.

5 Summary

The present findings indicate that differences in chromosome aberration frequencies still exist between Hiroshima and Nagasaki both for DS86 kerma and bone marrow dose albeit, less than observed with the old dosimetry. It is likely that the contribution of neutron radiation plays a role in the observed intercity difference, although the reason for this still remains unresolved. There is a wide variability of the frequency of aberrant cells between survivors within a dose range. Random errors in the dose estimates assigned to individual survivors are very likely responsible, to a large extent, for the observed overdispersions in aberration frequencies.

Acknowledgements. I am very much indebted to Dr. Seymour Abrahamson, Chief of Research, Permanent Director of RERF, for his advice and encouragement, and for going through the manuscript. I am grateful to Mrs. S. Murata, Mr. M. Hiramoto, and Mr. T. Abe in Hiroshima Laboratory for their microscopic and technical help, to Mr. J. Takayama for his photographic work, to Dr. T. Honda, Mr. M. Itoh, Mrs. Y. Urakawa, and Mr. O. Kusumi in Nagasaki Laboratory for their technical assistance. Special appreciation is extended to Mrs. M. Utaka for her clerical assistance.

References

Awa AA (1974) Cytogenitic and oncogenetic effects of the ionizing radiations of the atomic bombs. In: Germa JL (ed) Chromosomes and cancer. Wiley, New York, p 637

Awa AA (1975) Review of thirty years study of Hiroshima and Nagasaki atomic bom survivors. II. Biological effects. G. Chromosome aberrations in somatic cells in somatic cells. J Radiat Res 16 (Suppl): 122–131

Awa AA (1983) Chromosome damage in atomic bomb survivors and their offspring – Hiroshima and Nagasaki. In: Ishihara T, Sasaki MS (eds) Radiation-induced chromosome damage in man. Liss, New York, p 433

Awa AA (1984) Radiations-induced chromosome aberrations in A-bomb survivors – a key to biological dosimetry. In: Prentice RL, Thompson DJ (eds) Atomic bomb survivor data: utilization and analysis. SIAM, Philadelphia, p 99

Awa AA (1988) Chromosomal aberrations in atomic bomb survivors. In: Takebe H, Utsunomiya J (eds) Genetics of human tumors in Japan. Gann monograph on cancer research No. 35. Japan Scientific Societies, Tokyo, p 175

Awa AA, Neriishi S, Honda T, Yoshida MC, Sofuni T, Matsui T (1971) Chromosome-aberration frequency in cultured blood-cells in relation to radiation dose of A-bomb survivors. Lancet 1971 ii: 903–905

Awa AA, Sofuni T, Honda T, Itoh M, Neriishi S, Otake M (1978) Relationship between the radiation dose and chromosome aberrations in atomic bomb survivors of Hiroshima and Nagasaki. J Radiat Res 19: 126–140

Awa AA, Sofuni T, Honda T, Hamilton HB, Fujita S (1984) Preliminary reanalysis of radiation-induced chromosome aberrations in relation to past and newly revised dose estimates for Hiroshima and Nagasaki A-bomb survivors. In: Eisert WG, Mendelsohn ML (eds) Biological dosimetry. Springer, Berlin Heidelberg New York, p 77

Awa AA, Ohtaki K, Ithoh M, Honda T, Preston DL, McConnex ME (1988) Chromosome aberration data for A-bomb dosimetry reassessment: Proceedings of the 23rd annual meeting of NCRP 8–9 April 1987, Washington DC, Proc No. 9 p, 185

Bender MA, Awa AA, Brooks AL, Evans HJ, Groer PG, Littlefield LG, Pereira C, Preston RJ, Wachholz BW (1988) Current status of cytogenetic procedures to detect and quantifiy previous exposures to radiation. Mutat Res 196: 103–159

Fujita S, Awa AA, Pierce DA, Kato H, Shimizu Y (1983) Re-evaluation of biological effects of atomic bomb radiation by changes of estimated dose. Proc IAEA Symp Biological effects of low-level radiation, IAEA, Vienna, p 55

Lucas LN, Tenjin T, Straume T, Pinkel D, Moore D II, Litt M, Gray JW (1989) Rapid human chromosome aberration analysis using fluorescence in situ hybridization. Int J Radiat Biol 56: 35–44

Neriishi K, Stram DO, Vaeth M, Mizuno S, Akiba S (1989) The observed relationship between the occurrence of acute radiation sickness and subsequent cancer mortality among A-bomb survivors in Hiroshima and Nagasaki. RERF Technical Report 18–89.

Ohtaki K, Shimba H, Awa AA, Sofuni T (1982) Comparison of type and frequency of chromosome aberrations by conventional and G-staining methods in Hiroshima atomic bomb survivors. J Radiat Res 23: 441–449

Pierce DA, Stram DO, Vaeth M (1989) Allowing for random errors in radiation exposure estimates for the atomic bomb survivor data. RERF Technical Report 2–89

Preston DL, Pierce DA (1987) The effect of changes in dosimetry on cancer mortality risk estimate in the atomic bomb survivors. RERF Technical Report 9–87

Preston DL, McConney ME, Awa AA, Ohtaki K, Itoh M, Honda T (1988) Comparison of the dose-response relationships for chromosome aberration frequencies between the T65D and DS86 dosimetry. RERF Technical Report 7–88

Roesch WC (ed) (1988) US-Japan Joint reassessment of atomic bomb radiation dosimetry in Hiroshima and Nagasaki. Final report, vol 1 and 2. Radiation Effects Research Foundation, Hiroshima

Sasaki MS (1983) Use of lymphocyte chromosome aberrations in biological dosimetry. In: Ishihara T, Sasaki MS (eds) Radiation-induced chromosome damage in man. Liss, New York, p 585

Sasaki MS (1988) Chromosomal approaches to the dose assessment in human exposures to ionizing radiation. Berzelius symposium XV, p 119

Sposto R, Stram DO, Awa AA (1990) An investigation of random errors in the DS86 dosimetry using data on chromosome aberrations and severe epilation. RERF Technical Report 7–90.

Biological Dosimetry of Absorbed Radiation Dose: Considerations of Low-Level Radiations

M. S. SASAKI, Y. EJIMA, and S. SAIGUSA[1]

Contents

1 Distribution of Dose Among Cells from Low-Level Radiations

Chromosome aberration analysis in peripheral blood lymphocytes provides a valuable means for dose assessment in persons exposed to ionizing radiation. To date, several methods have been proposed for the chromosome-based dose assessment (IAEA 1986). However, in spite of the growing importance in the risk assessment, the dose-yield kinetics of chromosome aberrations and their implications for dose assessment are not well established in exposures to low-level radiation.

The effectiveness of low-level radiation in producing biologically relevant effects cannot simply be extrapolated from the high dose effects. In the low dose range, the cellular event, resulting from a single pass of an ionizing particle, has a critical meaning for the biological consequences (Booz and Feinendegen 1988). A single ionizing particle traversing a sensitive target will provide a dose (in Gy) of

$$\overline{D} = \frac{1.602 \times 10^{-16}\,hL}{V}, \qquad (1)$$

where L is LET (keV/μm), h is an average path length in μm, and V is the sensitive volume in μm^3. The dose may be called the elementary dose. For the lymphocyte nucleus with 5 μm diameter, the elementary dose is about 2.3 mGy for ^{60}Co γ-rays and about 70 cGy for 5 MeV α-particle. Therefore, in the low dose range, the biological effects are practically the consequences of one radiation event in a critical target and the dose rate does not alter the dose to a cell but simply modifies the exposed fraction, which is expressed by

$$P_X = \frac{1}{1+e^{-D/\overline{D}}/\Sigma\, e^{-D/\overline{D}}(D/\overline{D})^r S^r/r!}, \qquad (2)$$

[1] Radiation Biology Center, Kyoto University, Yoshida-Konoecho, Sakyo-ku, Kyoto 606, Japan

where D is dose to organ as a whole, S is a probability to survive from a single hit by a charged particle and r is the number of hits per cell. Cells receiving hits by high LET radiations will have a significant reduction in survivability (interphase death), which facilitates the selective removal of the exposed cells and hence modifies the size of the exposed fraction (Fig. 1).

It may be noteworthy that even in the exposure to dose which is equivalent to the elementary dose, approximately 37% of the cells are not exposed at all, and moreover, in case of human exposure to low-level radiations, the daily dose is usually far below the elementary dose. Recently, experimental evidence has accumulated to show the cellular adaptive response to low-level radiations, which makes the cells less susceptible to the induction of chromosomal damage by the subsequent radiation dose (Wolff et al. 1988). However, in cases of practical low-dose exposures a lymphocyte might have a rare chance to be hit twice or more with in a limited time and, therefore, a modification of radiation effects by the intracellular adaptive response, if not the intercellular hormetic effects, is not likely.

2 Dose to Lymphocytes from Chronic Exposure to α-Particle

The α-particle is high LET radiation, and may provide useful information on the chromosomal damage associated with the elementary dose. For this study, we selected

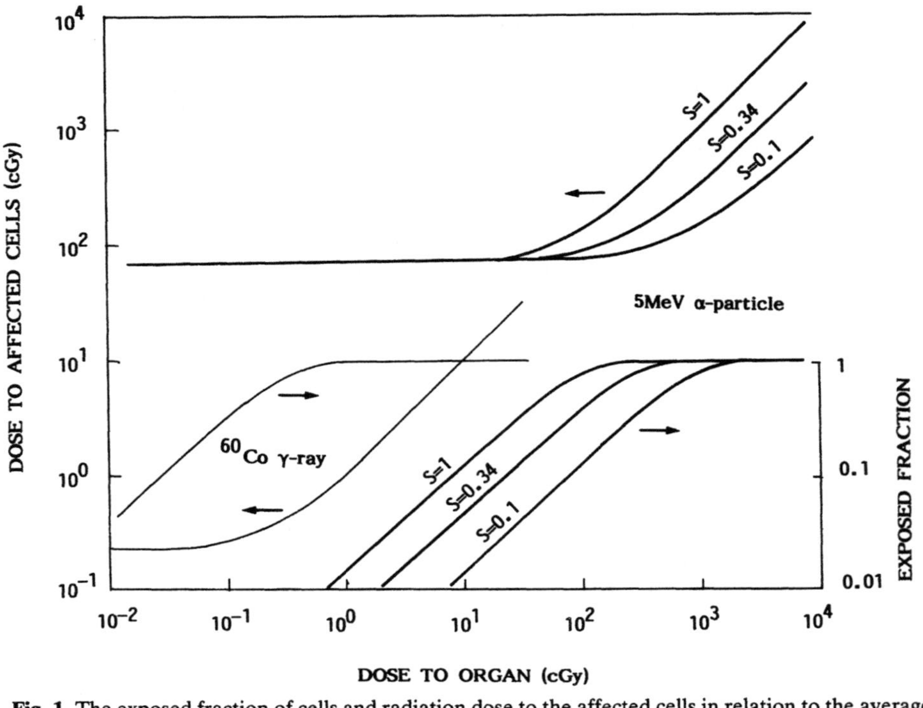

Fig. 1. The exposed fraction of cells and radiation dose to the affected cells in relation to the average total dose to organ. *Thin lines* for ^{60}Co γ-rays; *thick lines* for 5 MeV α-particles

persons who had been injected with Thorotrast. Thorotrast is a colloidal preparation of $^{232}ThO_2$ developed in the early 1930s as an X-ray contrast medium. Once injected, it is excreted at a negligible rate, taken up by reticuloendothelial cells, accumulated as granules or conglomerates mainly in such organs as spleen, liver, lymph nodes, and bone marrow, and continued to irradiate the neighboring cells with radiation from the radionuclides of the ^{232}Th decay series. The radiations involved are α-, β- und γ-rays, but about 95% of the dose is due to α-particles (Rundo 1955). For ^{232}Th and its daughters α-particle energy is 4.07–8.95 MeV, and the corresponding track lenght in water is 26.4–88.6 μm depending on the energy. The use of Thorotrast in Japan continued until about 1954; it was used mostly in 1937–1945. We studied chromosome aberrations in peripheral blood lymphocytes of 63 such persons; blood samples were obtained 30–40 years after injection. Details of the study have been published elsewhere (Sasaki et al. 1987).

Table 1 summarizes the results of the chromosome analyses. Unlike in vitro irradiation or total body exposures to relatively high doses of low LET radiation, the distribution of chromosome aberrations among cells does not fit Poisson distributions. The distribution is extremely overdispersed as evidenced by the very large values of relative variance (σ^2/m) and the dispersion index (u) of Papworth (Savage 1970) as compared with those expected from a Poisson distribution ($\sigma^2/m = 1$, $|u| = 0$). Such a distribution resembles that of partial body exposure. Therefore, obviously the aberration frequencies as expressed on a simple per-cell basis (m) cannot be used directly for dose assessment.

One may assume that the lymphocyte population is a mixture of unexposed and exposed cells to a single path of α-particle; the cells receiving two or more hits are likely to be killed. If this is the case, the distribution of chromosome aberrations should follow the 0-truncated Poisson distribution. However, goodness of fit to such a model is also poor particularly in groups of patients with high aberration frequencies (Table 1).

A more general model has been developed to estimate the fluence rate or dose rate to lymphocytes (Sasaki et al. 1987). The rationale of this method is that the lymphocytes are continuously replenished according to their life-span, and the age distribution of lymphocytes is approximatd by $F(t) = \lambda e^{-\lambda t}$, where $1/\lambda$ is the mean lifetime of lymphocytes (Norman et al. 1965; Buckton et al. 1967a). The longer-lived lymphocytes might have a higher chance to be hit by α-particles. Then, assuming Poisson statistics of the hits and chromosome aberrations, the observed distribution of aberrations among cells is expected by

$$P(r) = \sum_{k=o}^{\infty} \frac{a}{a+\mu}\left(\frac{\mu}{a+\mu}\right)^{k} \frac{(ka)^{r}}{r!} e^{-ka}, \quad (3)$$

where μ is the mean number of aberrations and expressed by

$$\mu = \frac{AaS}{A+\lambda-AS} \quad (4)$$

The parameters are as follows:

λ: rate of cells to be lost either by death or cell division.
μ: mean frequency of aberrations as expressed on a per-cell basis.

Table 1. Distribution of chromosome aberrations in blood lymphocytes from Thorotrast patients and statistical test for the goodness of fit to the models

μ[a]	No. of patients	No. of cells with indicated No. of dicentrics and rings						Mean (m)	σ^2/m	u[b]	Goodness of fit (p) to:		
		0	1	2	3	4	5 ≤				Poisson	O-truncated Poisson	Present model
0 –0.04	28	7356	111	31	8	1	0	0.027	1.580	35.65	<0.001	0.804	0.018
0.05–0.09	13	4013	114	45	25	4	0	0.070	1.907	41.61	<0.001	0.109	0.108
0.10–0.14	9	4125	210	92	39	13	4	0.130	2.007	47.72	<0.001	0.254	0.670
0.15–0.19	8	2890	203	88	33	14	5	0.174	2.011	40.67	<0.001	0.065	0.884
0.20–0.24	1	433	39	20	5	2	1	0.214	1.855	13.57	<0.001	0.071	0.923
0.25–0.29	2	640	62	27	12	5	4	0.259	2.219	23.64	<0.001	0.057	0.816
0.30–0.40	2	575	63	37	12	7	6	0.333	2.258	23.58	<0.001	0.033	0.756
Total	63	20032	802	340	134	46	20	0.102					

[a] μ Mean aberration frequency in each patient.
[b] u Dispersion index of Papworth.

A: mean number of α-particles passing through a cell nucleus per unit time.
k: number of paths of α-particle.
a: production coefficient for chromosome aberrations.
S: survival (interphase death) after a single path of α-particle.

Since μ is known for each person, production coefficient a can be estimated by the maximum likelihood method that satisfies Eq. (3) for the distribution of aberrations in all persons. Thus, a = 0.81 ± 0.03 was obtained, implying that approximately 0.8 dicentrics and rings are produced by a single path of α-particle. As seen in Table 1, this model gives the best fit to the observed distribution of chromosome aberrations.

Equation (4) implies that the mean number of aberrations, μ, is influenced by the lymphocyte survivability. When S is close to unity, as in case of low LET radiations, the mean aberration frequency is simply proportional to the fluence rate (dose rate) as expressed by aA/λ. However, when S is smaller than unity, the mean aberration frequency increases with the increase of fluence rate but soon reaches a plateau level where it remains constant against a further increase of fluence rate (Fig. 2). This may be a major reason for the lack of a direct dose-response relationship between chromosome aberration frequency in lymphocytes and body burden of Thorotrast (Buckton et al. 1967b; Kemmer et al. 1971, 1979; Sasaki et al. 1987).

Taking the production coefficient of about 0.8 dicentrics and rings as obtained here, 1500 days for the mean lifetime of small lymphocytes (Buckton et al. 1967a), and S = 0.34 for 4.9 MeV α-particles (Edwards et al. 1980), the fluence rate of an average A = 1.5 hits per lumphocyte is derived in the highest aberration group (μ = 0.33) for an average number of paths of α-particles per lymphocyte nucleus per year. Assuming 0.69 Gy per path through a lymphocyte nucleus with 5 μm diameter, this fluence rate corresponds to 1.04 Gy/year as an average dose rate to the lymphocytes.

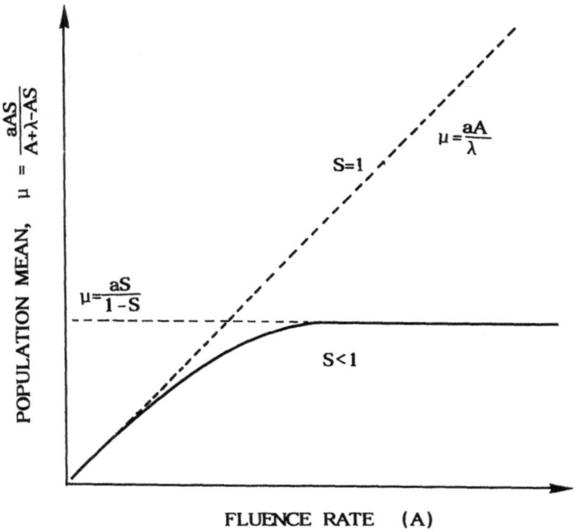

Fig. 2. The relationship between the mean aberration frequency (μ) and fluence rate to lymphocytes (A) in long-term exposure to radiation

3 Dose to Lymphocytes and Aberration Frequencies in Protracted X- and γ-Ray Exposures

For the protracted exposures to low LET radiations, such as X- and γ-rays, essentially the same model may be applied for the dose-yield kinetics of lymphocyte chromosome aberrations. However, for sparsely ionizing radiations, the modification by cell killing due to a single path of ionizing particle (or photon) may be ignored since S is close to unity. Chromosome aberrations may be formed at an increment of aA, but at the same time, if they are unstable aberrations, they may be eliminated according to the disappearance rate of lymphocytes as expressed in $Nu/Nu_o=e^{-\lambda t}$, thus giving a changing rate of $Nu/dt = aA - \lambda Nu$. Therefore, the frequency of cells with unstable aberrations (which is equivalent to the frequency of unstable aberrations at a low dose range) at time t from the commencement of exposure will be expressed by

$$Nu = \frac{aA}{\lambda}(1\text{-}e^{-\lambda t}), \text{ or } Nu= \frac{aA}{\lambda}(1\text{-}e^{-\lambda T/A}), \quad (5)$$

where T is the total accumulated dose. If t is long enough, the aberration frequency is expressed by $Nu=aA/\lambda$. This is a derivation of Eq. (4) at $S = 1$. On the other hand, stable aberrations are not lost when the cells undergo cell division. Therefore, the frequency of cells with only such aberrations, Ns, may not be influenced by the lymphocyte lifetime, and may be expected to be proportional to the accumulated dose as expressed by

$$N_s = bAt, \text{ or } N_s = bT \quad (6)$$

where the parameter b is a production coefficient for such cells.

Figure 3 shows the results of chromosome analysis in peripheral blood lymphocytes based on 230717 cells from 931 radiological technicians who were occupationally exposed to low-level X- and γ-rays. The frequency of unstable aberrations (dicentrics and rings) is not simply proportional to the total accumulated dose: It increases with the increase of the total dose in the first phase but soon reaches a plateau level where it stays constant against a further accumulation of dose (Fig. 3A). This is consistent with that expected from Eq. (5). Parametric fit gives a production coefficient of $a = 4.83 \times 10^{-2}$ per Sv at an average yearly dose of 15 mSv and lymphocyte mean lifetime of 1500 days. The production coefficient is comparable to that obtained in the in vitro experiments with X-rays (Lloyd and Edwards 1983). These observations may indicate that chromosome aberrations are induced proportionally to the radiation dose at a comparable efficiency as in the in vitro exposure and the plateau level can be readily explained by lymphocyte kinetics rather than the adaptive response of lymphocytes. The stable aberrations tend to increase with total dose, but a clear dose-response relationship was not obtained (Fig. 3B). This may be in part due to the intrinsic variability of the scoring criteria of this type of aberration, and also in part possibly because the accumulation of such cells is more closely related to the dose to lymphoproliferative organs. Therefore, the depth dose is more important.

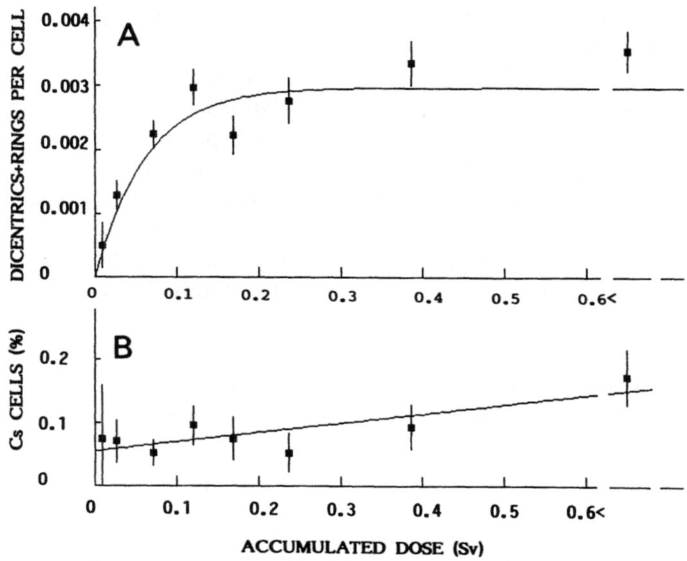

Fig. 3 A, B. Chromosome aberration frequencies in peripheral blood lymphocytes from occupationally irradiated persons (radiological technicians) as plotted against accumulated dose. **A** Dicentrics and rings per cell. **B** Percent of cells with stable aberrations (Cs cells)

4 Relative Biological Effectiveness (RBE) of Low-Level Neutrons and Its Implication to the Assessment of A-Bomb Radiation Dose

Epidemiological data of A-bomb survivors in Hiroshima have been the only source for risk evaluation of neutron exposure to human populations. However, recently, the dosimetric system (T65D), which has been used as a basis of risk estimation, has been questioned. As a result of the comprehensive reevaluation by the US-Japan Committee for the Reassessment of A-bomb Radiation Dosimetry, a new dosimetric system, DS86, was proposed (Roesch 1987). The difference between old and new dosimetric systems is minor in Nagasaki, where γ-rays are the major radiation component. The difference is dramatic in Hiroshima, where there is a considerable increase of γ-rays and a sharp decrease in neutron dose as compared with the old dosimetric system. According to the DS86, the contribution of neutrons to the organ-absorbed dose is nearly negligible in both cities. However, while the city difference is not statistically significant, the cancer mortality as well as chromosome aberration frequencies at the equi-dose level are consistently higher in Hiroshima than in Nagasaki (see Awa, this Vol.). The contribution of neutrons to the total bone marrow dose is less than 3% in Hiroshima and less than 1% in Nagasaki survivors (Shimizu et al. 1987). Can the city difference be explained by the RBE of neutrons? The difference in the neutron component between old and new dosimetric systems pertains not only to the quantity but also the energy spectrum is greatly shifted to the low energy range in the new dosimetric system. In order to discuss the city difference, the RBE of the low energy neutrons at low flux is particularly important. However, there is a serious lack of experimental data on low energy neutrons, particularly below 300 keV.

In view of the importance of RBE of low-energy neutrons, we conducted experiments on in vitro dose-response relationships of lymphocyte chromosome aberrations by neutrons with different energy spectra, i.e., thermal neutrons (E = 0.025 eV) obtained from the D_2O facility of the Kyoto University Research Reactor, fission spectrum neutrons ($\overline{E}$ = 1.98 MeV) obtained by the irradiating ^{235}U energy converter with thermal neutrons, and fission neutrons with a degraded broad energy spectrum ($\overline{E}$ = 0.10 MeV) obtained in the central cavity of the graphite reflector of the Kinki University Reactor. The contaminated external γ-rays at the sample position were 41,8, 50.7, and 46.7%, respectively.

Heparinized blood was irradiated with these neutrons, and chromosome aberrations were analyzed in the cultured lymphocytes. Assuming that the chromosome aberration yield is the sum of the aberrations by γ-rays and those due to neutrons, expressed by

$$Y = \alpha_\gamma D_\gamma + \beta_\gamma D_\gamma^2 + \alpha_n D_n + \beta_n D_n^2 + \delta D_\gamma D_n \tag{7}$$

and aberrations due to the interaction between damage caused by two types of radiation, $\delta D_\gamma D_n$, are negligible, the contribution of external γ-rays was subtracted to obtain dose-response relationships for neutrons. The yields of dicentrics and rings against neutron dose (D in Gy) thus obtained were $0.97D + 0.67D^2$, $1.01D + 1.00D^2$, and $0.69D + 1.14D^2$ for fission spectrum neutrons, degraded fission neutrons, and thermal neutrons, respectively. Figure 4 shows the RBE in terms of chromosomal effectiveness relative to that of ^{60}Co γ-rays. The relative chromosomal effectiveness

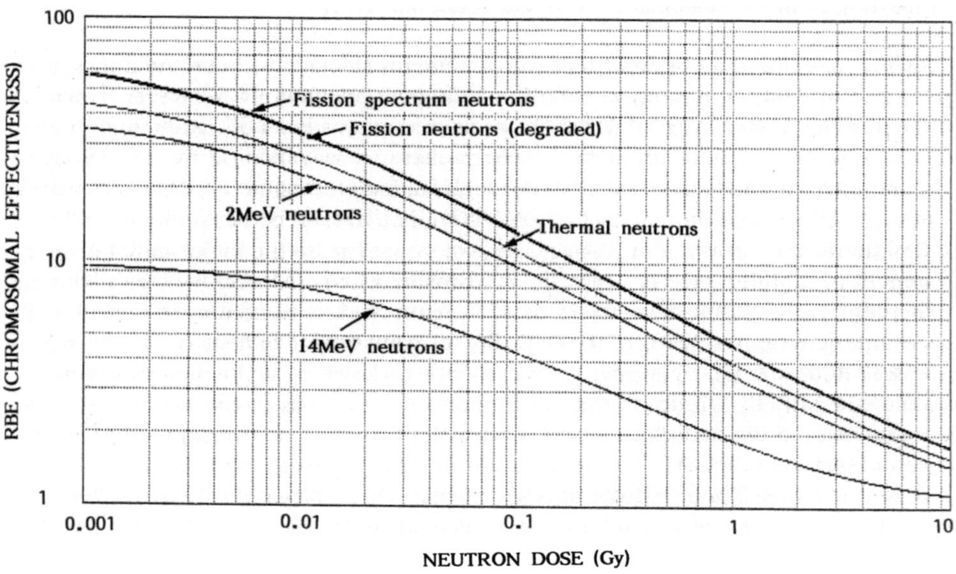

Fig. 4. RBE of neutrons as expressed by chromosomal effectiveness relative to ^{60}Co γ-rays. For calculation of RBE, dose-response relationships of dicentrics and rings were fitted to a model and rings per cell. **B** Percent of cells with stable aberrations (Cs cells)

of accelerator-produced monoenergetic 14 MeV and 2 MeV neutrons is also presented for comparison.

The fission neutrons showed the highest RBE values at all doses. It is interesting to note that in spite of the great difference in their energy spectrum fission neutrons showed identical RBEs whether they consisted mainly of the high energy flux or were greatly degraded toward the low energy range. At this stage of study the reason is not clear but the present results indicate that the RBE is not significantly influenced by the energy spectrum of the fission neutrons, and may provide a practical means for assessing neutron dose in Hiroshima without particular regard to the energy spectrum.

Recently, chromosome aberration data in Hiroshima and Nagasaki survivors were reanalyzed by Preston et al. (1988) according to the new dosimetric system. When aberration frequencies were plotted against the dose, the difference between cities decreased as compared with the one based on T65D. However, the aberration frequencies are still higher in Hiroshima than in Nagasaki. In Fig. 5A, data below 1.5 Gy are replotted and fitted to the linear-quadratic model as a function of bone marrow dose. When the fraction of cells with chromosome aberrations in Nagasaki dose in Nagasaki is negligible and the dose-effect relationship for γ-rays in Nagasaki $C_N = \alpha D + \beta D^2$ for the dose-response kinetics of this type of cell, $C_N = 1.10 \times 10^{-2}$, $\alpha = 3.08 \times 10^{-4}$, and $\beta = 1.86 \times 10^{-6}$ are obtained for the dose-independent, dose-linear, and dose-quadratic component, respectively. If we assume that the neutron dose in Nagasaki is negligible and the dose-effect relationship for γ-rays in Nagasaki

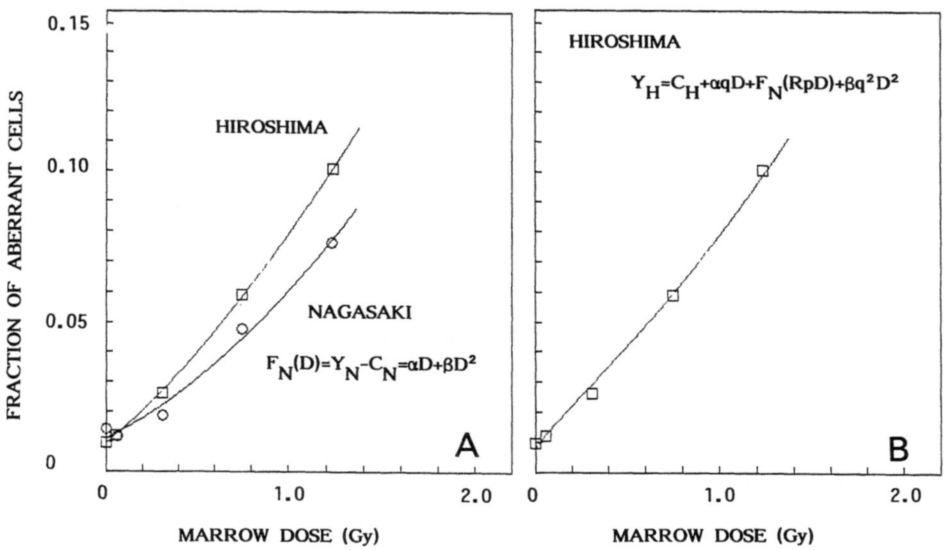

Fig. 5 A, B. The frequencies of lymphocytes with chromosome aberrations in A-bomb survivors replotted from the tables in Preston et al. (1988). **A** Comparison between survivors in Hiroshima and Nagasaki. **B** Estimation of the size of the neutron component in Hiroshima survivors based on the response kinetics for γ-rays obtained from Nagasaki survivors and RBE of fission neutrons. *Line* represents the dose-response relationship at the neutron component of 4,8%, which gives the best fit to the data

can also be applied for γ-rays in Hiroshima, the aberration frequencies due to mixed γ and neutron dose may be expressed by

$$Y_H = C_H + \alpha qD + F_N(RpD) + \beta q^2D^2 \quad (8)$$

where D is the total bone marrow dose in Gy. The third term is the contribution by neutrons in which p represents the fraction of neutron dose relative to the total dose, and pD and qD are each neutron and γ-ray dose in Gy at $p + q = 1$. R is the RBE of fission neutrons as a function of neutron dose, pD, and $F_N(RpD)$ is the yield of this type of cell at a γ-ray dose equivalent to RpD as expected from the dose-response relationship in Nagasaki. Since α and β are known from Nagasaki survivors and the RBE of the fission neutron may be used without any specific considerations to the energy spectrum, the fraction of neutrons, p, may be estimated by fitting Eq. (8) to the observed frequencies in Hiroshima by a maximum likelihood method (Fig. 5B). Thus, $p = 0.048$ was obtained. The value of ca. 4.8% for the neutron contribution appears to be slightly higher compared to the estimates based on the DS86 dosimetry system. However, the γ/neutron ratio may vary with the distance from the hypocenter, and may be different among survivors due to different shielding conditions. More detailed estimates of the neutron contribution may be possible by applying the dose information to the individual survivors.

References

Booz J, Feinendegen LE (1988) A microdosimetric understanding of low-dose radiation effects. Int J Radiat Biol 53: 13–21

Buckton KE, Court Brown WM, Smith PG (1967) Lymphocyte survival in man treated with X-rays for ankylosing spondylitis. Nature 214: 470–473

Buckton KE, Langlands AO, Woodcock GE (1967b) Cytogenetic damage following Thorotrast administration. Int J Radiat Biol 12: 565–577

Edwards AA, Purrott RJ, Prosser JS, Lloyd DC (1980) The induction of chromosome aberrations in human lymphocytes by alpha-radiation. Int J Radiat Biol 38: 83–91

International Atomic Energy Agency (IAEA) (1986) Biological dosimetry: chromosomal aberration analysis for dose assessment. IAEA Technical Report Series No. 260

Kemmer W, Muth H, Tranekjer F, Edelmann L (1971) Dose dependence of the chromosome aberration rate in Thorotrast patients. Biophysik 7: 342–351

Kemmer W, Steinsträsser A, Muth H (1979) Chromosome aberrations as a biological dosimeter in Thorotrast patients: dosimetric problems. Environ Res 18: 178–183

Lloyd DC, Edwards AA, (1983) Chromosome aberrations in human lymphocytes: effect of radiation quality, dose, and dose rate. In: Ishihara T, and Sasaki MS (eds) Radiation-induced chromosome damage in man. Liss, New York pp 23–49

Norman A, Sasaki MS, Ottoman RE, Fingerhut AG (1965) Lymphocyte lifetime in women. Science 147: 745

Preston DL, McConney MR, Awa AA, Ohtaki K, Itoh M, Honda T (1988) Comparison of the dose-response relationships for chromosome aberration frequencies between the T65D and DS86 dosimetries. Radiation Effects Research Foundation Technical Report, RERF TR pp 7–88

Roesch WC (ed) 1987 US-Japan Joint Reassessment of Atomic Bomb Radiation Dosimetry in Hiroshima and Nagasaki, vol 1. Radiation Effects Research Foundation, Hiroshima

Rundo J (1955) Considerations of the limits of radiation dosage from Thorotrast. Br J Radiol 28: 615–619

Sasaki MS, Takatsuji T, Ejima Y, Kodama S, Kido C (1987) Chromosome aberration frequency and radiation dose to lymphocytes by alpha-particles from internal deposit of Thorotrast. Radiat Environ Biophy 26: 227–238

Savage JRK (1970) Site of radiation induced chromosome exchanges. Curr Top Radiat Res 6: 130–194

Shimizu Y, Kato H, Schull WJ, Preston DL, Fujita S, Pierce DA (1987) Life span study report 11, part 1. Comparison of risk coefficients for site-specific cancer mortality based on the DS86 and T65D shielded kerma and organ doses. Radiation Effects Research Foundation Technical Report, RERF TR 12–87

Wolff S, Azal V, Wiencke JK, Olivieri G, Michaeli A (1988) Human lymphocytes exposed to low doses of ionizing radiations become refractory to high doses of radiation as well as to chemical mutagens that induce double-strand breaks in DNA. Int J Radiat Biol 53: 39–48

Use of Micronuclei in Biological Dosimetry of Absorbed Radiation Dose

M. Bauchinger and H. Braselmann[1]

Contents

1 Introduction

The induction of chromosome aberrations is one of several biological responses to ionizing radiation which have been investigated as a means of estimating an individual's average whole body dose. Chromosome analysis in peripheral lymphocytes dates back to 1966 when for the first time Bender and Gooch carried out dose estimations of three men exposed during the Recuplex accident. Since that time chromosome dosimetry has been steadily improved and is now the most reliable quantitative biological indicator system for real or suspected acute radiation exposures.

The main principles can be summarized briefly as follows. Among the different types of chromosomal aberrations, the dicentric chromosome is generally scored for routine investigations. The range in spontaneous dicentric frequencies observed in larger control populations is 0 to 2.8/1000 cells (Bender et al. 1988). In vitro dose-response curves for dicentrics, established with appropriate fitting procedures, are available for most radiation qualities relevant for radiation protection purposes. RBE-dependence (relative biological effectiveness) for the induction of dicentrics by high-LET (linear energy transfer) radiation can be clearly demonstrated. Several surveys of larger occupationally exposed populations have been carried out and individual overdose estimations exist for at least 1500 cases.

Provided that a sufficient number of cells – usually between 500 and 1000 – is analyzed, detection limits for acute X-ray and γ-ray exposure of 0.05 and 0.1 Gy, respectively, are possible. Chromosome dosimetry, however, is a slow procedure requiring highly skilled personnel. Only about 300 metaphases per day can be analyzed microscopically by an experienced observer. Though automated metaphase determination has made great progress, automatic aberration scoring is still a far from routine method.

[1] Institut für Strahlenbiologie, Gesellschaft für Strahlen- und Umweltforschung, 8042 München-Neuherberg, FRG

The easier, and far less time-consuming method of scoring micronuclei (MN) has been suggested as an alternative approach for quantifying chromosomal damage. In the following we discuss the extent to which the MN assay can be used with a similar precision to the chromosome analysis in the biological dosimetry of absorbed radiation dose.

2 The Micronucleus Assay

MN appear as separate, smaller nuclei in the cytoplasm in addition to the main nucleus of a cell. They originate from acentric chromatin elements or whole chromosomes that were not included into daughter nuclei during mitotic divisions.

Radiation cytogenetics has been concerned with MN induction since the early studies of Thoday (1951) and Evans et al. (1959) on Vicia faba root meristems. Fliedner et al. (1964) reported the occurrence of cytoplasmic chromatin clumps (karyomeres) in bone marrow cells from accidentally irradiated human beings. Countryman and Heddle (1976) and Countryman et al. (1977) published a method that permitted the demonstration of radiation-induced MN in human lymphocyte preparations.

Since MN are not expressed until the first post-irradiation interphase, the quantitative analysis is highly dependent on cell proliferation kinetics. For the construction of representative dose-response curves different sampling times were used in lymphocyte cultures to obtain the maximum response of irradiation in terms of induced MN (Countryman and Heddle 1976; Huber et al. 1983; Krepinski and Heddle 1983). Cells from different donors, however, respond differently to mitogenic stimulation of T-lymphocytes in vitro, so that it is not possible to establish a standardized culture time yielding maximum cell proliferation.

The first attempt to solve the kinetic problem of the MN assay was the development of the modified harlequin staining procedure or "BrdU method" by Pincu et al. (1984). Based on the principles of the FPG (fluorescence plus Giemsa) staining technique (Perry and Wolff 1974) cells are allowed to incorporate high BrdU concentrations (4×10^{-4} M) continuously during DNA synthesis. After staining, proliferating cells with blue nuclei can be distinguished from nonproliferating cells with red nuclei. However, detection, and thus scoring of MN, is not fully restricted to cells that have divided once in culture since cells in late S or G_2 of the first cycle are also included (Fenech and Morley 1985a).

With the technically less complicated cytokinesis-block (CB) method (Fenech and Morley 1985a, b), cells that have divided once only since irradiation can be precisely detected. By adding Cytochalasin B to the lymphocyte cultures cytokinesis is blocked without inhibiting nuclear division. Cytokinesis-blocked cells accumulate in the first division cycle and can be easily identified from their binucleate appearance. MN can thus be easily scored exlusively in CB cells. Thus, as in chromosome analysis, in which aberration scoring should be confined to first post-irradiation metaphases, quantitative MN analysis can be carried out under cell cycle controlled conditions. A more detailed description of the development of the human lymphocyte MN assay has been given elsewhere (Huber and Bauchinger 1990).

3 Dose-Effect Curves in Vitro

Background MN frequencies in lymphocytes obtained from different laboratories range from 0.002 to 0.036 per CB cell (Eastmond and Tucker 1989; Fenech and Morley 1986; Huber et al. 1983; Kormos and Köteles 1988; Littlefield et al. 1989; Mitchell and Norman 1987; Prosser et al. 1988; Ramalho et al. 1988; Thomson and Perry 1988), and from 0.008 to 0.025 per blue cell with the BrdU method (Pincu et al. 1984). As with a conventional MN assay (Fenech and Morley 1985c; Högstedt 1984; Huber et al. 1989; Norman et al. 1984, 1985), an increase in background frequency with donor age was observed with the CB method (Fenech and Morley 1986). Two further studies failed to show any significant age effect probably as a result of the smaller age range examined (Huber et al. 1989; Prosser et al. 1988).

In vitro exposure to low LET radiation has generally revealed a clear dose-dependent increase in induced MN incidences. The first dose response curve published for MN (induced by 150 kV X-rays and obtained from lymphocyte preparations with destroyed cytoplasm at 96 h culture time) was described by the traditional power law model $Y=kD^n$ with a dose exponent $n=1.2$ (Y, MN frequency; D, radiation dose and k, a constant) (Countryman and Heddle 1976). Later curves were presented for acute, sparsely ionizing radiations (X-rays with energies from about 100 kV upwards and gamma-rays) which fitted linear, $Y = c+\alpha D$ or linear-quadratic functions $Y = c+\alpha D+\beta D^2$, respectively, with c, α, β, denoting the estimated parameters for spontaneous and radiation-induced MN and D, the radiation dose in Gy. Huber et al. (1983) determined the intercellular distribution of MN induced by 220 kV X-rays applying a conventional preparation technique with preserved cytoplasm. Distributions of MN among cells were found to be overdispersed (dispersion index, variance/mean greater than 1), which is in contrast to dicentric data from low LET radiation which can be described by Poisson statistics. Thus, the actual distribution could be taken into account for the individual dose points in the derivation of variance estimates. A weighted least squares method was used for curve fitting. Data sets from a culture time of 48 h could be fitted to a linear model using the corresponding sample mean variance as a weight for each observation. Data sets from 84 and 96 h fitted a linear-quadratic model. At 60 and 72 h the data sets were statistically consistent with both models.

Tables 1–2 show published results of fitting MN data from acute low- and high-dose exposures to various types of low-LET radiation, derived with the BrdU or CB method. Corresponding dose-effect curves are shown in Figs. 1–2. Data were reanalyzed when curve fitting was not carried out with an appropriately weighted least squares method.

MN data assessed according to the CB method following high-dose exposures (up to 3.0 Gy and higher) of G_o lymphocytes to X-rays with energies of 200 kV (Kormos and Köteles 1988), 220 kV (Littlefield et al. 1989) and 250 kV (Prosser et al. 1988) fitted the linear-quadratic model. In the latter study, individual MN data on lymphocytes from two of the five donors still indicated, however, a sufficient fit to the linear model. A linear dose-response relationship was originally reported for 100 kV X-rays (Fenech and Morley 1985b) but recalculation of the data revealed a small but significant βD^2 term. A linear-quadratic curve has also been shown for Cs-137 exposure (Eastmond and Tucker 1989), though irradiation conditions are not

Table 1. Estimates of dose-effect coefficients ($Y=c+\alpha D+\beta D^2$) for MN induced in human G_o lymphocytes by acute low-LET radiation (high-dose exposures). MN data obtained with the CB method

Radiation quality	Dose range (Gy)	Dose rate ($Gy\ min^{-1}$)	Coefficients ± SE			α/β (Gy)	Ref.[a]
			$c \times 10^{-2}$	$\alpha \times 10^{-2} Gy^{-1}$	$\beta \times 10^{-2} Gy^{-2}$		
100 kV	0–4.0	4.0	0.6 ± 0.2	7.6 ± 0.7	0.7 ± 0.3	10.8	1
150 kV	0–2.0	1.0	3.4 ± 0.6	21.8 ± 1.0	–	–	7
150 kV [b]	0–3.0	1.0	2.2 ± 0.7 [c]	12.7 ± 0.7 [c]	–	–	8
200 kV	0–4.0	0.3	1.1 ± 0.6	13.9 ± 2.1	3.5 ± 0.7	4.0	2
220 kV	0–4.0	0.5	0.6 ± 0.2	9.4 ± 1.0	4.9 ± 0.4	1.9	3
250 kV	0–5.0	1.0	1.2 ± 0.2	11.7 ± 0.6	0.9 ± 0.2	13.4	4
Co-60	0–4.0	0.6	0.3 ± 0.4	9.5 ± 1.7	4.8 ± 0.6	2.0	6
Cs-137 [d]	0–4.0	0.3	1.0 ± 0.3	3.7 ± 1.2	4.8 ± 0.4	0.8	5

[a] Reference numbers correspond to curve numbers in Fig. 1.
[b] BrdU method, blue cells analyzed.
[c] Coefficients multiplied by a factor of 2 for comparison.
[d] Irradiation 24 h after culture initiation.

Table 2. Estimates of dose-effect coefficients ($Y=c+\alpha D$) for MN induced in human G_o lymphocytes by acute low-LET radiation (low-dose exposures). MN data obtained with the CB method

Radiation quality	Dose range (Gy)	Dose rate ($Gy\ min^{-1}$	Coefficients ± SE		Ref. [a]
			$c \times 10^{-2}$	$\alpha \times 10^{-2} Gy^{-1}$	
90 kV	0 –0.2	0.06	0.9 ± 0.3	9.5 ± 3.6	4
	0.2 –0.5		−2.7 ± 1.5	27.2 ± 4.5	
100 kV	0 –0.4	0.5	0.8 ± 0.1	8.0 ± 0.8	1
150 kV	0 –0.5	1.0	4.2 ± 0.5	13.8 ± 2.3	2
200 kV	0 –0.8	0.3	1.6 ± 0.5	14.3 ± 1.7	3
Co-60	0 –0.15	0.06	0.9 ± 0.3	2.8 ± 2.9	5
	0.15–0.5		−1.0 ± 0.9	14.7 ± 3.0	
Co-60 [b]	0 –0.8	0.06	1.9 ± 0.2 [c]	4.6 ± 0.7 [c]	6

[a] Reference numbers correspond to curve numbers in Fig. 2.
[b] BrdU method, blue cells analyzed.
[c] Coefficients multiplied by a factor of 2 for comparison.

directly comparable since radiation treatment was performed 24 h after culture initiation. Biphasic curves with linear slopes up to 3 and 2 Gy, respectively, were obtained with 150 kV X-rays for MN data sets derived with either the BrdU or CB method (Ramalho et al. 1988), whereby the latter was found to detect a higher frequency of MN. Mitchel and Norman (1987) expressed their findings in terms of cells with MN rather than MN per cell, and demonstrated a clear dose dependence with an ordinary linear regression analysis. A reanalysis of their data, using empirical variances as weights for the single dose points, shows a significant fit to the linear-quadratic model. Dose-response curves for induced MN scored in CB cells were linear for low-dose exposures (Fig. 2) of G_o lymphocytes to 100 kV X-rays with 0–

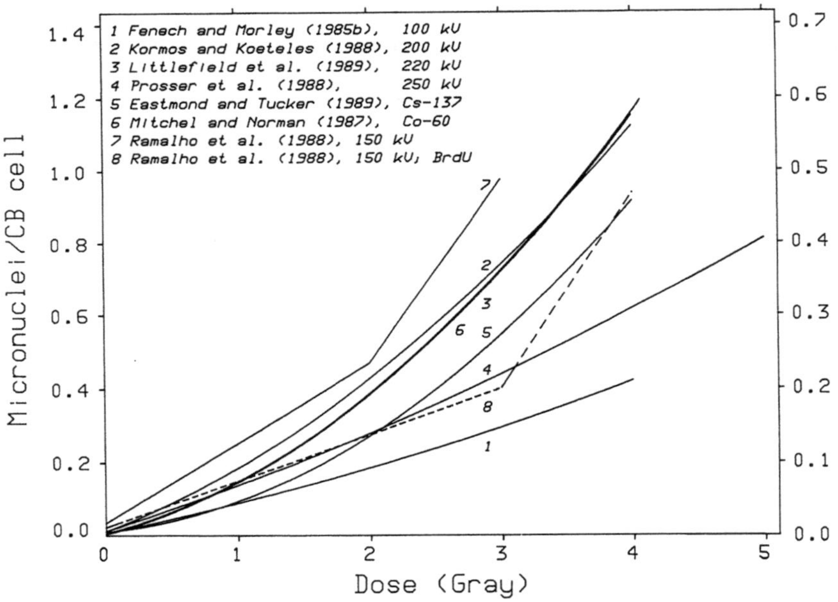

Fig. 1. In vitro dose-effect curves for MN induced in human G_o lymphocytes by acute low-LET radiation (high-dose exposures). *1–7* CB method ———; *8* BrdU method - - - -

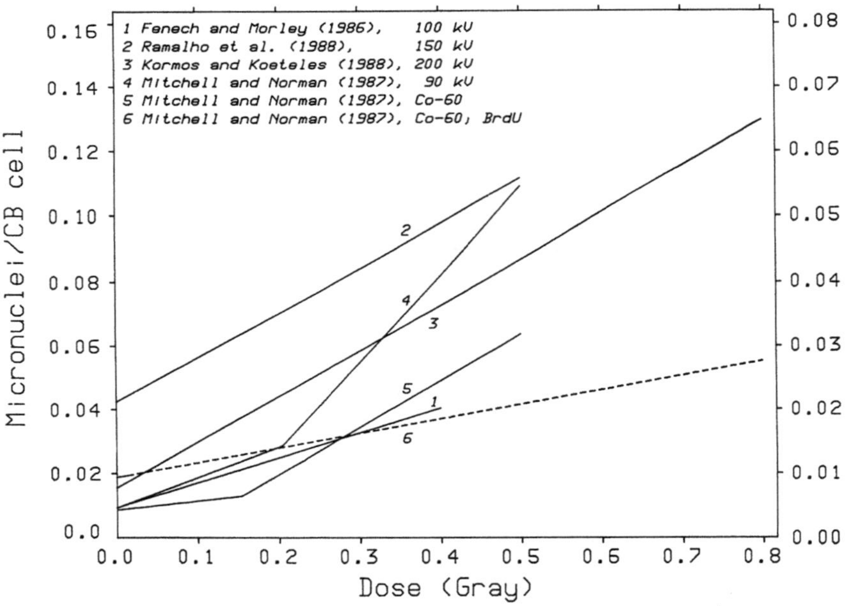

Fig. 2. In vitro dose-effect curves for MN induced in human G_o lymphocytes by acute low-LET radiation (low-dose exposures). *1–5* CB method ———; *6* BrdU method - - - -

0.4 Gy (Fenech and Morley 1986), to 150 kV X-rays with 0–0.5 Gy (Ramalho et al. 1988), and to 200 kV X-rays with 0–0.8 Gy (Kormos and Köteles 1988). 90 kV X-irradiation with 0–0.5 Gy and exposure to Co-60 gamma rays with 0–0.5 Gy (CB method) or 0–0.8 Gy (BrdU method) revealed biphasic curves represented by two straight lines with a significantly greater slope for the one in the range above 0.15 Gy (Mitchell and Norman 1987). A reanalysis of the latter curve, however, shows that the data can also be fitted linearly. There is no information on the intercellular distribution of induced MN in Fenech and Morley's data sets (Fenech and Morley 1985b; 1986). In all the other studies, except two (Eastmond and Tucker 1989; Ramalho et al. 1988), the distribution was found to be overdispersed as compared to Poisson. This is similar to the situation generally observed for acentrics produced by sparsely ionizing radiations, but different to that for induced dicentrics which conform to Poisson (Bauchinger et al. 1986; Braselmann et al. 1986; Edwards et al. 1979; Schmid and Bauchinger 1980).

No simple quantitative relationship between MN and acentric fragments scored at first post-irradiation metaphase can be invoked from this similarity. The genral observation of different shapes of dose-response curves for MN and acentrics per cell with lower slopes for the latter has been discussed in a theoretical approach by Savage (1988, 1989). According to his proposals, such discrepancies must be expected even in a sample of CB cells since the frequency of MN per cell will be a function of the probability of fragment inclusion within a nucleus. Further evidence for the lack of a numerical relationship between acentric and MN data comes from the comparison of yield-time curves for these cytogenetic end points. As suggested by Savage, in cases of acentric frequency, one deals with an "instantaneous" yield since acentrics are usually determined in metaphases accumulated during a short colcemid block. In contrast, the MN frequency is a "running average" generated from a cumulative MN score from all preceding divisions.

Recently, two studies have been published describing a comparison of estimates of coefficients from linear-quadratic dose-response curves established for X-ray-induced acentrics in first cycle metaphases and MN in CB cells. Prosser et al. (1988) found a two fold higher α-coefficient and a ten fold smaller β-coefficient for MN. Littlefield et al. (1989) observed no significant differences for the background or linear coefficients, but a two fold reduction of the quadratic term.The question of the mechanisms of MN formation and their contents has again been raised by these data.

Since with standard preparations one cannot discriminate between MN containing either acentrics or whole chromosomes, quantitative evaluation of MN data may be imprecise. Recent technical advances now provide a means for detecting centromeres by immunofluorescence staining in MN after antikinetochore antibody treatment (Moroi et al. 1980). With this technique it was shown that 34 to 61 % of spontaneously occurring MN contained kinetochores. In contrast, only 3 to 11% of radiation-induced MN revealed kinetochore fluorescence, indicating that the majority originated from acentrics (Eastmond and Tucker 1989; Fenech and Morley 1989; Thomson and Perry 1988). This conforms to the interpretations of curve fitting data for radiation-induced acentrics and MN given by Littlefield et al. (1989), suggesting that due to the common estimates of α virtually all acentrics are recovered as MN at low radiation doses. However, the significant differences of the dose-squared components indicate that the relative proportion of acentrics that will be expressed as independent MN decreases with increasing dose.

Assuming that with the CB method cell kinetics can be adequately taken into account in MN scoring, the considerable differences in the shapes of published dose-response curves for acute low-LET radiations are inexplicable (Figs. 1–2). They are also clearly evident from the 95% confidence ellipsoids for the linear-quadratic parameters of the corresponding curves (Fig. 3). Some of the dose-response relationships are also at variance with radiobiophysical expectations of the biological effectiveness of different radiation qualities. Such striking differences should not result within the narrow range of X-.ray energies.

Surprisingly, 100 kV X-rays are the less effective radiation quality in producing MN. Generally, increasing yields should be expected with decreasing energy of sparsely ionizing radiation. Furthermore, there is no explanation for the statistically significant different shapes of the 200 kV (Kormos and Köteles 1988) and 250 kV X-ray curves (Prosser et al. 1988). A comparison of filter conditions with 1 mm Cu and 1.2 mm Cu, respectively, shows that the variation cannot be attributed to different X-ray-bremsstrahlung spectra as could be demonstrated earlier for the induction of dicentrics and acentrics (Schmid et al. 1984). Neither can the variation be caused by the different dose rates, 0.29 Gy min^{-1} for 200 kV versus 1.0 Gy min^{-1} for 250 kV X-rays. At 4.0 Gy the exposure time for the 200 kV X-rays is still below 20 min which can usually be accepted as the limit for acute exposure. If one still expects to find a dose rate effect affecting the β-coefficient it should be reduced, rather than increased by a factor of 4 as observed for the 200 kV X-rays (Table 1).

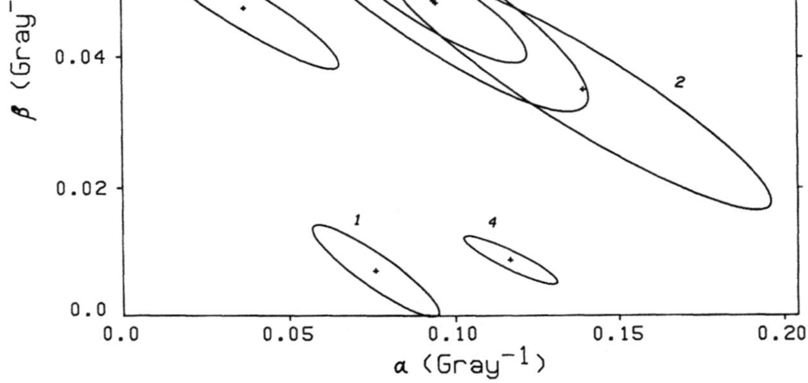

Fig. 3. 95% confidence regions for the dose-effect coefficients α and β ($Y=c+\alpha D+\beta D^2$) for MN induced in human G_0 lymphocytes by acute low-LET radiation (high-dose exposures)

A comparison of the curves for Cs-137 (Eastmond and Tucker 1989) and Co-60 gamma-rays (Mitchell and Normann 1987) also shows a distinct difference with an unexpected two fold higher α for the latter, although common values for β exist. The fairly concordant curves for Co-60 gamma-rays (Mitchell and Norman 1987) and 220 kV X-rays (Littlefield et al. 1989) are inconsistent with other biological dose-response data, e.g., for the induction of dicentrics where the linear coefficients for X-rays tends to be higher than that for gamma-rays. Similar discrepancies are also evident from low-dose data. 90 kV X-rays at doses up to 0.2 Gy (Mitchell and Norman 1987) and 100 kV X-rays (Fenech and Morley 1986) are less effective in MN induction than 150 kV (Ramalho et al. 1988) and 200 kV X-rays (Kormos and Köteles 1988). The corresponding α-values differ by a factor of about 1.5. Beyond 0.2 Gy the slope of the 90 kV curve sharply increases and at 0.5 Gy the MN frequency is similar to that induced by 150 kV X-rays (Ramalho et al. 1988).

4 Dose Estimation of in Vivo Radiation Exposure

Based on linear-quadratic in vitro calibration curves for Cs-137 gamma-rays (Krepinski and Heddle 1983) and 220 kV X-rays (Huber et al. 1983), conventional MN assays have been evaluated as insufficiently sensitive for detection of doses below 0.25 Gy. Krepinski and Heddle (1983) suggested that the MN test would be useful for biological dosimetry for acute doses ranging from 0.50 to 3.0 Gy. They also studied the efficiency of MN scoring and chromosome analysis in estimating chronic radiation exposure. In blood samples of three nuclear plant workers, chronically exposed to 400–600 mSv of mixed gamma- and tritium-radiation, the resolving power of the MN assay was less than that of direct metaphase analysis. When these results were published, however, cell cycle kinetics could not be adequately taken into account in the quantification of radiation-induced MN data. Curves established with an MN assay using the CB method indicate that it may well be possible to derive low dose estimates. Fenech and Morley (1986) claimed from their 100 kV X-irradiation experiments that the CB method is considerably more sensitive than chromosome analysis. They considered that 0.05 Gy could be unequivocally detected in vitro and thus in vivo. However, Prosser et al. (1988) considered from a reexamination of Fenech and Morley's data that this could be only achieved when an individual background MN incidence is available for the exposed subject. In reality probably doses above about 0.3 Gy might be detectable. Their own experiments with 250 kV X-rays gave a detection limit of 0.1 Gy, similar to that found by Kormos and Köteles (1988) for 200 kV X-rays. For Co-60 gamma rays (Mitchell and Norman 1987) significant differences between background and induced MN incidences could only be detected at doses above 0.15 Gy.

Whether individual dose estimates can be derived by MN scoring with a similar precision as by chromosome analysis cannot be evaluated at present since reliable data on in vivo exposures are completely lacking.

5 Conclusions

Provided that an MN assay takes cell proliferation into account, clear in vitro dose-response relationship can be obtained for radiation-induced MN. However, published dose-effect curves from different laboratories for acute low-LET radiations show considerable variations. Some of these curves are inconsistent with basic radiobiophysical expectations of the effectiveness of various radiation qualities. At present, this limits their application as calibration curves for biological dosimetry. Taking into account the considerable variation of background MN frequencies, it seems that it might be possible to estimate acute whole body exposures to low-LET radiation doses above 0.2 Gy. Since until now only X- and gamma-rays within a narrow energy range have been examined, there is still insufficient information on the effect of radiation quality. A similar lack of information exists for the effect of dose rate on MN frequency. It cannot yet be decided from existing data whether the MN assay can be established as an additional biological routine method for estimating an individual's radiation exposure. Further studies are required to analyze the influence of variables affecting the background and induced MN incidences. These include especially the influence of age and potential interindividual differences in radiosensitivity. The influence of confounding variables, such as smoking, exposure to medical and environmental clastogenic or spindle damaging agents (for review, see Huber et al. 1989), and virus infections on MN frequencies is of special importance in this approach.

References

Bauchinger M, Schmid E, Braselmann H (1986) Cell survival and radiation-induced chromosome aberrations. II. Experimental findings in human lymphocytes analysed in first and second post-irradiation metaphases. Radiat Environ Biophys 25: 253–260

Bender MA, Gooch PC (1966) Somatic chromosome aberrations induced by human whole-body irradiation: the "Recuplex" criticality accident. Radiat Res 29: 568–582

Bender MA, Awa AA, Brooks AL et al. (1988) Current status of cytogenetic procedures to detect and quantify previous exposure to radiation. Mutat Res 196: 103–159

Braselmann H, Bauchinger M, Schmid E (1986) Cell survival and radiation induced chromosome aberrations. I. Derivation of formulae for the determination of transmission and survival parameters of aberrations. Radiat Environ Biophys 25: 243–251

Countryman PI, Heddle JA (1976) The production of micronuclei from chromosome aberrations in irradiated cultures of human lymphocytes. Mutat Res 41: 321–332

Countryman PI, Heddle JA, Crawford E (1977) The repair of X-ray-induced chromosomal damage in trisomy 21 and normal diploid lymphocytes. Cancer Res 37: 52–58

Eastmond DA, Tucker JD (1989) Identification of auneuploidy-inducing agents using cytokinesis-blocked human lymphocytes and an antikinetochore antibody. Environ Mol Mutagen 13: 34–43

Edwards AA, Lloyd DC, Purrott RJ (1979) Radiation-induced chromosome aberrations and the Poisson distribution. Radiat Environ Biophys 16: 89–100

Evans HJ, Neary GJ, Williamson (1959) The relative biological efficiency of single doses of fast neutrons and gamma-rays on *Vicia faba* roots and the effect of oxygen. II. Chromosome damage: the production of micronuclei. Int J Radiat Biol 1: 216–229

Fenech M, Morley AA (1985a) Solutions to the kinetic problem in the micronucleus assay. Cytobios 43: 233–246

Fenech M, Morley AA (1985b) Measurement of micronuclei in lymphocytes. Mutat Res 147: 29–36

Fenech M, Morley AA (1985c) The effect of donor age on spontaneous and induced micronuclei. Mutat Res 148: 99–105

Fenech M, Morley AA (1986) Cytokinesis-block micronucleus method in human lymphocytes: effect of in vivo ageing and low dose X-irradiation. Mutat Res 161: 193–198

Fenech M, Morley AA (1989) Kinetochore detection in micronuclei: an alternative method for measuring chromosome loss. Mutagenesis 4: 98–104

Fliedner TM, Andrews GA, Cronkite EP, Bond VP (1964) Early and late cytogenetic effects of whole body irradiation on human marrow. Blood 23: 471–478

Högstedt B (1984) Micronuclei in lymphocytes with preserved cytoplasm. A method for assessment of cytogenetic damage in man. Mutat Res 130: 63–72

Huber R, Bauchinger M (1990) Development and perspectives of the human lymphocyte micronucleus assay. Adv Mutagen Res 1: 89–104

Huber R, Streng S, Bauchinger M (1983) The suitability of the human lymphocyte micronucleus assay system for biological dosimetry. Mutat Res 111: 185–193

Huber R, Braselmann H, Bauchinger M (1989) Screening for interindividual differences in radiosensitivity by means of the micronucleus assay in human lymphocytes. Radiat Environ Biophys 28: 113–120

Kormos C, Köteles GJ (1988) Micronuclei in X-irradiated human lymphocytes. Mutat Res 199: 31–35

Krepinski AB, Heddle JA (1983) Micronuclei as a rapid and inexpensive measure of radiation-induced chromosomal aberrations. In: Ishihara T, Sasaki MS (eds) Radiation-induced chromosome damage in man. Liss, New York, pp 93–109

Littlefield LG, Sayer AM, Frome EL (1989) Comparison of dose-response parameters for radiation-induced acentric fragments and micronuclei observed in cytokinesis-arrested lymphocytes. Mutagenesis 4: 265–270

Mitchell JC, Norman A (1987) The induction of micronuclei in human lymphocytes by low doses of radiation. Int J Radiat Biol 52: 527–535

Moroi Y, Peebles C, Fritzler MJ, Steigerwald J, Tan EM (1980) Autoantibody to centromere (kinetochore) in scleroderma sera. Proc Natl Acad Sci USA 77: 1627–1631

Norman A, Cochran S, Bass D, Roe D (1984) Effects of age, sex and diagnostic X-rays on chromosome damage. Int J Radiat Biol 46: 317–321

Norman A, Bass D, Roe D (1985) Screening human populations for chromosome aberrations. Mutat Res 143: 155–160

Perry P, Wolff S (1974) New Giemsa method for the differential staining of sister chromatids. Nature 251: 156–158

Pincu M, Bass D, Norman A (1984) An improved micronucleus assay in lymphocytes. Mutat Res 139: 61–65

Prosser JS, Moquet JE, Lloyd DC, Edwards AA (1988) Radiation induction of micronuclei in human lymphocytes. Mutat Res 199: 37–45

Ramalho A, Sunjevaric I, Natarajan AT (1988) Use of the frequencies of micronuclei as quantitative indicators of X-ray-induced chromosomal aberrations in human peripheral blood lymphocytes: comparison of two methods. Mutat Res 207: 141–146

Savage JRK (1988) A comment on the quantitative relationship between micronuclei and chromosomal aberrations. Mutat Res 207: 33–36

Savage JRK (1989) Acentric chromosomal fragments and micronuclei: the time displacement factor. Mutat Res 225: 171–173

Schmid E, Bauchinger M (1980) Analysis of primary processes in the formation of acentric fragments. Radiat Environ Biophys 17: 143–149

Schmid E, Bauchinger M, Streng S, Nahrstedt U (1984) The effect of 220 kVp X-rays with different spectra on dose response of chromosome aberrations in human lymphocytes. Radiat Environ Biophys 23: 305–309

Thoday JM (1951) The effect of ionizing radiations on the broad bean root. IX. Chromosome breakage and the lethality of ionizing radiations to the root meristem. Br J Radiol 24: 572–576, 622–628

Thomson EJ, Perry PE (1988) The identification of micronucleated chromosomes: a possible assay for aneuploidy. Mutagenesis 3: 415:41

Biological Dosimetry After Radiation Accidents

D. C. LLOYD and A. A. EDWARDS[1]

Contents

1 Introduction

For over 20 years the frequency of chromosomal, particularly dicentric, aberrations observed in cultured peripheral blood lymphocytes has been used as a biological dosemeter for assessing accidental overexposures to ionizing radiation. The technique has been developed into a routine procedure in radiological protection. It provides a supplement to physical methods of dosimetry and where these are not available the biological method may be the only means of quantifying dose. The technique has been amply described previously and a recent IAEA manual (1986) gives probably the most up-to-date and comprehensive account of the radiation-induced aberrations, the laboratory procedures and the way that doses are calculated from the aberration yields.

Experience gained from the study of over a 1000 cases worldwide of actual or suspected overexposure has demonstrated the usefulness of the method and also highlighted certain limitations. Some of these will be considered here. In recent years there have been some serious radiation accidents, notably at Chernobyl and Goiânia, that have involved large numbers of people. Many persons received significant, possibly life-threatening, overexposures and many more feared that they had. In this chapter some aspects of biological dosimetry in the aftermath of the large accident will also be discussed.

[1] National Radiological Protection Board, Chilton Didcot, Oxfordshire, OX11 ORQ, UK

2 Non-Uniform Exposure

The conversion of the chromosome aberration yield into dose by reference to an in vitro dose-response curve provides an estimate of the equivalent whole body dose. Accidental uniform irradiations are very rare events (Brewen et al. 1972). Experience has shown that acute irradiations from external sources almost always involves partial or inhomogeneous exposure. Particularly for highly localized doses, it would be desirable to refine the estimate of whole body dose to indicate the extent of the inhomogeneity. It is, however, not possible to detect an elevated yield of aberrations in the lymphocytes if the exposure is limited to a small volume of the body, e.g. a hand burn even if the local dose is so high as to require amputation. In such a case it is more appropriate to assess the severity by observation of skin responses such as erythema and to use imaging procedures such as microwave or infrared thermography (Daburon 1986) to delineate the extent of deeper tissue damage.

Analysis of chromosome aberrations in certain cells ensheathing the bases of hairs has been suggested as a local cytogenetic dosemeter (Wells et al. 1983). There are, however, technical problems concerning the numbers and quality of metaphase spreads to overcome before this can be used routinely. Other hair responses such as epilation or reduced shaft diameter (Sieber and Wells 1986) warrant further study.

If the exposure involves a larger volume of the body, sufficient lymphocytes may be irradiated for aberrations to be detected in the sample of metaphases scored. It is sometimes possible to demonstrate that non-uniform irradiation results in a distribution of aberrations that is overdispersed with respect to the expected Poisson distribution for that mean. Dolphin (1969) suggested that it may be possible to relate the magnitude of overdispersion to the heterogeneity of exposure. This idea has been refined and tested with simulated half-body irradiations (Lloyd et al. 1973). The procedure requires a correction factor to take account of in vitro and in vivo selection due to cell killing or mitotic delay that reduces the likelihood of the irradiated fraction of cells coming to metaphase by 48 h in culture. Work in progress (Natarajan and Lloyd) is investigating the limits of resolution of this approach. It may be possible to identify persons who have received high life-threatening exposure to much of the body with perhaps 5–10% of tissue being spared or only sublethally irradiated. This leads to the possibility of combining such information with other haematological parameters (Fliedner 1981) collected in the first few days after the exposure. Together these may indicate the likelihood that some areas of active bone marrow may have been spared. Given time and good nursing this may regenerate naturally, assisted perhaps by treatment with cytokines.

Interpretation of aberration distribution data in terms of partial body irradiation requires some simplifying assumptions regarding the in vivo distribution of PHA-responsive lymphocytes. Such information for humans is lacking and animal data (Liniecki et al. 1983; Savage and Breckon. 1985) indicate that a uniform distribution is most unlikely. It would be interesting if the aberration distribution data from the 13 highly irradiated patients at Chernobyl who received bone marrow transplantations (Baranov et al. 1989) were made available for reassessment. It may be possible to see evidence of overdispersion in those patients where regeneration of their own marrow led to immune reactions and in some instances may have contributed to their deaths. It should of course be stressed that distribution analysis of chromosome damage in lymphocytes cannot help with identification of the actual areas of the body that were exposed. These may be obvious if the doses exceed the erythema threshold.

3 Internally Incorporated Radionuclides

Cytogenetic dosimetry is best suited to estimating dose from penetrating radiation from external sources such as X-ray machines or industrial gamma radiography sets. Dose estimation is less certain if radionuclides have been taken into the body as these tend to localize in particular organs and tissues and so constitute another form of partial irradiation with the further complication of protraction of dose. Aberrations, sometimes overdispersed, can be observed in subjects who have incorporated radionuclides (Lloyd et al. 1976) and from these it may be possible to estimate dose to the pool of circulating lymphocytes. However, it is difficult to translate this into whole body dose and generally it is not possible to estimate doses of prime concern which are those to the organs or tissues where the material is predominantly located. There is scope for more research with chromosomal studies on patients given known quantities of a nuclide for which its distribution in the body is known by physical measurements and metabolic modelling.

Exceptions to this limitation are those chemical forms of the nuclides that tend to disperse fairly evenly throughout the body. Tritium, for example taken in as tritiated water, becomes distributed throughout the body water. Accidental intakes of tritiated water are rare but a recent industrial accident was described in which a person incorporated 35 GBq (~1 Ci) and was treated by forced diuresis (Lloyd et al. 1986). The cytogenetic estimate of dose (0.4 Gy) was in good agreement with effective dose equivalent of 0.45 Sv calculated from measurements of tritium in the urine.

Incorporated isotopes of caesium similarly result in a fairly uniform distribution of dose as they predominantly locate in muscle. At Goiânia Cs-137 was spread widely in the community leading to serious contamination of the environment (IAEA 1988). About 30 people received whole body doses which were determined by cytogenetics to be above 0.5 Gy (Ramalho et al. 1988a). However, gamma-ray body monitoring showed that in all but one patient (one of the four fatalities), internally incoporated material contributed relatively little to the total dose.

4 Dose Rate

Most researchers who have produced in vitro dose-response curves with X- or γ-radiation to use for accident dosimetry have irradiated blood acutely, i.e. in a few minutes. The resultant aberration yields generally fit well to the quadratic dose-effect model. Dose protraction or fractionation produces a lower yield of aberrations. Frequently, when attempts have been made to produce "chronic" curves, these have been done at a constant dose rate, thus each dose is delivered in a different time, and the resultant data fit poorly to the quadratic model (Lloyd and Edwards 1983). Dose delivery time is the important factor and if this is kept constant a much better fit to the model is obtained (Lloyd et al. 1984).

For estimating doses in a protracted accident in the absence of an in vitro curve produced with a comparable delivery time, one may utilize an acute curve with the dose squared term modified by Lea and Catcheside's (1942) time-dependent G-function. Use of this function combined with information on the likely time course of an accidental irradiation can provide a more realistic estimate of dose. In practice, where a continuous high dose is delivered over a long period of say 1 or 2 days, the G-function reduces the dose squared term to virtually zero so that the response is in effect linear. For brief intermittent exposure, where the interfraction interval exceeds about 6 h , radiation can be regarded as a sequence of isolated acute events in which induced aberration yields are additive. For shorter interfraction times an interaction term based on the G-function can be used.

A practical consideration is that patients may not be able to provide reliable information on their likely exposure regime. This was the problem encountered by Littlefield (1990) for the Juarez accident where she had to resort to quoting acute and chronic estimates of dose based respectively on the quadratic curve and on only its linear coefficient. These then provided the outer bounds to the most likely doses. At Goiânia many of the irradiations were protracted not only because of the incorporated caesium but also because about 10 days elapsed between the source being breached and commencement of remedial measures. Ramalho et al. (1988a) have published dose estimates for these patients using a gamma dose response curve $Y = 2.7 \times 10^{-2}D + 2.2 \times 10^{-2}D^2$. This was obtained at a constant rate of 0.12 Gy min^{-1}. The times required to give the higher doses may be partly responsible for the quadratic term being somewhat lower than about 6×10^{-2} which is generally found for shorter irradiations (Lloyd and Edwards 1983). This curve, however, is acute by comparison with the dose rates experienced by most of the patients and so possibly has led to some degree of underestimation of doses.

5 Speed in Obtaining a Dose Estimate

Chromosomal analysis is time-consuming although if several persons share the work by scoring replicate slides the required number of metaphases can be examined within a few hours. In the routine service that the NRPB provides we aim to produce a dose estimate on the third day after receipt of the blood specimen. Much of this time is taken up in culturing lymphocytes to metaphase. Generally this delay is acceptable and fits in with the clinical management of the patients. During this time the patient will be under surveillance and if a severe dose is suspected, for example from the observation of acute prodromal symptoms, then regular measurements such as for haematological changes will be under way. It is of course desirable to produce a dose estimate as soon as possible in order to reassure anxious subjects. Particularly if there is no other information, such as from physical dosimetry, persons involved in an accident with a radiation source often fear the worst.

If it were felt that obtaining a biological estimate of dose much sooner is essential, then the technique of prematurely condensed chromosomes could reduce the delay to just a few hours after receipt of the sample. The favoured method is to use polyethylene glycol to fuse human G_o lymphocytes to dividing cells (Pantelias and Maillie 1984). Currently, the technique is not used routinely for dosimetry; it needs fairly extensive calibration and in many laboratories reproducibility of the procedure is not always reliable.

6 Automation

There are a number of new developments in chromosomal analysis aimed at relieving tedium, reducing the reliance on very skilled technicians and perhaps reducing the number of expensive man hours involved. Automated metaphase finding has been successfully developed and several commercial systems are available. Experience in the NRPB since 1984 with one of the first Cytoscan instruments has shown that the system is very effective. It will automatically locate clusters of chromosomes on the slide and present them centered and in focus at high magnification to the operator who then completes the analysis by identifying the aberrations through the microscope eyepieces in the conventional way (Finnon et al. 1986). This has on average trebled the output of the technician.

The next step in automated analysis is to transfer a digitized image of the mataphase to the computer where it may be scored for abnormalities. The commercial systems have concentrated on perfecting semi-automated karyotyping for clinical cytogenetics. Development of software for handling non-constitutional aberrations ("dicentric hunting") has been pursued by a few academic research institutes. Some considerable success has been reported from Heidelberg (Lörch 1986) and Edinburgh (Piper et al. 1988). Both laboratories have produced experimental systems for field evaluation.

Because of the variability in the appearance of dicentrics no automated system can be guaranteed to find all of them. The false negative rate is considerably improved if from the outset those candidate aberrations are considered that conform to a certain strict definition encompassing only the more easily identified dicentrics. These criteria concern the overall size of the object, the minimum resolvable distance between the two centromeres and the closeness of one or both centromeres to the tips of the aberrant chromosome. A theoretical evaluation (Lloyd 1990) has indicated that approximately 65 % of dicentrics should be detected based on the design criteria of the Edinburgh system. At present 50 % efficiency at finding dicentrics is routinely achieved (J. Piper, pers. comm. 1989). Most of the analysis is done "hands off" thereby making full use of night-time and weekends. The operator is simply needed at the end of the procedure to confirm visually from a screen or, in a minority of cases, down the microscope, the validity of the candidate dicentric that the machine presents. Tentative estimates of performance allowing for the need to increase the number of cells examined to compensate for the machine's reduced efficiency in recognizing dicentrics give an improvement of output for operator time between 20-1 and 40-1.

7 In Situ Hybridization

In situ hybridization techniques are beginning to be used in experiments with radiation-induced aberrations. Meyne et al. (1989) have used fluorescence detection of an oligonucleotide label which is specific to repetitive DNA sequences at the centromeres of all human chromosomes. Highlighting centromeres simplifies the identification of aberrations such as the dicentric. Lucas et al. (1989) have also demonstrated the potential of using more specific, cloned DNA probes. They showed that the dose response for dicentrics involving just chromosome No. 1, detected with a centromeric and a 1p telomeric probe, agreed very closely with a standard conventional curve after making allowance for the relative differences in target size. Many probes are being developed and the ultimate would appear to be "chromosome painting" (Pinkel et al. 1988). These techniques open exciting prospects of much easier visualization of stable translocations for which coventional staining methods with banding are either too uncertain or laborious. The possibility exists to develop the in situ hybridization procedure for automated analysis. This may be with a microscope-based system or, if hybridization to chromosomes held in suspension can be perfected, with a flow system.

8 Micronuclei

A number of researchers (see Almassay et al. 1987 for review) have suggested counting micronuclei in blood lymphocytes as an alternative to counting aberrations. Because the images are simpler than metaphases, scoring micronuclei is faster, requires less skill and may be easy to automate. Fenech and Morley's (1985) method that uses cytochalasin B to ensure that by scoring binucleate cells analysis is restricted to those in their second in vitro interphase was a considerable advance. It overcame the major problem that micronucleus yields are dependent on the kinetics of cell cycling which in short-term lymphocyte cultures is notoriously variable and subject to many confounding factors. Pincu et al. (1984) proposed an alternative technique in which the cell cycle status of interphase nuclei could be detected by fluorescence plus Giemsa staining. For several reasons (Ramalho et al. 1988b) the cytokinesis blocked method is currently favoured.

The technique is still in the process of optimization. For example Littlefield et al. (1989) have recently shown that 3 μg ml^{-1} of cytochalasin B added to the cultures at 45 h, which was the time and concentration regarded as optimum by Fenech and Morley (1985), can fail to block cytokinesis in most of the cycling cells. By increasing the concentration to 6 μg ml^{-1} over 90% of the cells become binucleate. Dose-response curves have been published and there is considerable divergence of results between laboratories, for example with Littlefield et al. (1989); Kormos and Köteles (1988) and Ramalho et al. (1988b) showing yields at high radiation doses 1.5–2-fold higher than those obtained by Fenech and Morley (1985) and Prosser et al. (1988).

There is still uncertainty about just what objects become micronuclei. They could comprise whole chromosomes that have lagged behind at division. In addition, virtually the whole range of chromatid- or chromosome-type aberrations have the potential of becoming excluded as a micronucleus (Savage 1988). Thomson and Perry (1988) have used immunofluorescence staining with an anti-kinetochore antibody and concluded that X-ray-induced micronuclei in human lymphocytes are virtually all derived from acentric fragments. It is difficult to reconcile this with comparisons with the dose-response relationships between micronuclei and chromosomal aberrations (Prosser et al. 1988; Littlefield et al. 1989). There is general agreement that for higher doses of X-rays, when one would expect more than one aberration to be produced in a cell, the numbers of micronuclei observed is less than the number of acentric fragments. Thus the beta term obtained from fitting data to the linear quadratic model is significantly lower than that obtained for aberrations. A lowered linear coefficient is also found for alpha radiation (Bilbao et al. 1989) where even at low doses the passage of a single alpha particle across the cell is likely to produce multiple damage. This has led to the idea that two or more aberrations in a cell might "coalesce" to produce fewer micronuclei. If this were so one might expect the sizes of micronuclei induced by high radiation doses to be on average larger than at low doses. Littlefield et al. (1989) have shown that the size difference is only marginally significant so coalescence by itself is insufficient to explain fully the reduced numbers of micronuclei by comparison with acentric fragments. Another hypothesis (Savage 1988) is that with increasing radiation doses there is an increased likelihood that micronuclei could be included in one of the main daughter macronuclei.

A practical consequence of the lowered beta coefficient in the dose response for micronuclei is that to produce a dose estimate at high doses (>2 Gy) the number of binucleate cells that has to be scored needs to be larger, perhaps double, the number of metaphases scored for aberrations in order to obtain a dose estimate with the same statistical uncertainties.

Developers of automated systems for chromosomal analysis are well aware of the potential market for a system that can also handle the micronucleus assay. As yet, however, none of the commercial systems are able to provide this although Tates et al. (1988) have reported some significant advance along these lines. Image capture, digitization and analysis in the computer require very good preservation of cytoplasm and staining with wide spectral separation to distinguish nuclear from cytoplasmic material. These are somewhat more rigid requirements than are necessary for conventional scoring by a technician.

9 Other Biological Dosemeters

Alternative indicators of exposure have been proposed employing either biochemical or cytological end points. For recent reviews, see Kaul et al. (1986). It is highly desirable that other biological systems should be developed, particularly if they can make up for deficiencies in the chromosomal technique. Some examples of proposed assays are altered levels of amino acids and DNA metabolites in body fluids, changes in lectin binding to cell membranes, reduced blastic transformation of lymphocytes by lectins and changes in numbers of circulating CFU stem cells. In general these assays also exhibit a number of drawbacks such as poor sensitivity to low (<1.0 Gy) doses, fluctuating inter- or intradonor variability in response and in some cases a response to radiation that is too transient for practical application.

Currently, the non-cytogenetic methods that show most promise are assays for gene mutations. Of particular interest is the detection by flow cytometry of mutations at the glycophorin A locus on human chromosome number 4 (Langlois et al. 1987). However, at present it can be performed only on the 50% of the population who are heterozygous at this locus and a few months must elapse between exposure to a few hundred milligrays and assay. This latter restriction obviously prevents its use immediately after an accidental exposure when there may be a pressing need for a dose estimate to aid in clinical management of the patients. However, the technique may be useful as a longer term biological dosemeter for follow-up studies of irradiated populations. This has the potential of overcoming the problem with the dicentric that its yield in peripheral lymphocytes reduces with time after exposure (Buckton 1983). Analysis for "stable aberrations" is an alternative (Awa et al. 1978) but is too tedious and expensive to be considered for application to large numbers of subjects. The Chernobyl accident is a good example where such a method could be applied to a follow-up of many thousands of irradiated subjects who were not at risk of early injury, but where good estimates of personal radiation doses that were probably in excess of 0.5 Gy would be required, to compare with the rate of occurrence of cancer.

10 Dealing with a Large Accident

The faster speed with which cells may be examined for micronuclei is an attractive feature which could be especially useful for dosimetry after a large accident involving many people (Almassy et al. 1987). However, at Chernobyl and Goiânia the micronucleus assay was not established in the investigating laboratories so that dose estimation was done from chromosomal aberrations. The same approach was adopted after both of these incidents. Relatively small numbers of metaphases were analyzed at first, concentrating on those cases deemed to be most urgent. Supplementary scoring was done later to improve the accuracy of the initial, approximate dose estimates. At Chernobyl there was sufficient cytogenetics manpower available to score initially 50 metaphases from each of 154 persons. Supplementing the scoring caused not more than 5–10% changes in the first dose estimates (UNSCEAR 1988). At Goiânia generally 100–200 metaphases were eventually scored from 97 persons (Ramalho et al. 1988a).

It is worth considering the logistic aspects of biological dosimetry in the event of an accident involving many casualties. In such a situation physicians will have initiated triage within a few hours. This will be based on physical injuries, criteria like speed and onset of prodromal vomiting, and, if available, early measurement on dosemeters such as film badges. In particular people who have received large, possibly life-threatening doses need to be identified so that treatment can be concentrated where it is most needed. In the first few days after an accident, the role of biological dosimetry is to provide a quantitative confirmation of the sorting of patients. Highly accurate estimates of dose are not necessary and sufficient data can be obtained from the analysis of quite a small number of metaphases. An observation of 2 dicentrics in just 25 metaphases, for example, would indicate an equivalent whole body dose of about 1.0 Gy of gamma rays with lower and upper 95% confidence limits of respectively 0.27 and 2.0 Gy. An inevitably lethal dose would be apparent from examining even fewer cells. For example, 7.0 Gy acute determined from a yield of 30 dicentrics in just ten metaphases would carry confidence limits of 5.6 and 8.3 Gy which are quite adequate.

The choice of using chromosomal aberrations or micronuclei depends on the number of cases and the number of microscopists available. Preparations for chromosomal aberrations become available after only 48 h in culture whereas micronucleus preparations with the cytokinesis blocking technique currently require 72 h incubation. Thus in a laboratory with several trained personnel preliminary dose estimates based on 25 metaphases per case can be obtained from many casualties during the period 48–72 h before the micronucleus slides are ready. In Prosser et al. (1989), we have explored this more fully and concluded that the micronucleus method has no real advantage over conventional aberration analysis except in the most extreme case when numbers of casualties might be so large that they could overwhelm the resources of the emergency services.

After the first biological dose estimates have been made it will become apparent that priority should then be given to improving the data by scoring more cells for certain patients. For example there may be cases where there is an anomaly between the symptoms displayed and the early chromosomal data. Vomiting could be due to non-radiological causes such as hysteria or infection. Priority would also need to be given to those patients where the irradiation was high but inhomogeneous. Here decisions would soon need to be taken on managing possible bone marrow crisis. The cytogenetics objectives would be to increase the number of metaphases scored to permit aberration distributions to be analyzed by the deviation from the Poisson method described above.

In addition to treatment of injured persons, the physician has an important role to play in the counselling of patients. Indeed, Goiânia demonstrated that reassurance and psychological support was the treatment of choice for the vast majority of involved subjects. Having been reassured that they are not at risk of early injury, they will seek information of the likely longer term consequences of their exposure; in effect malignant disease. Clearly the demonstration of no aberrations in lymphocytes, indicating the dose was at most trivial is valuable. Because of the considerable statistical uncertainties at low levels of damage, it is necessary to score many more metaphases than the initial 25. This would only be considered after the more urgent cases described above have been completed. It is debatable whether the micronucleus technique is very useful here because of the somewhat variable background frequency. In the absence of an individual pre-radiation control, which will always be the case in an accident investigation, a population control of perhaps 10–20 micronuclei per 1000 binucleate cells (Fenech and Morley 1986) must be used. This imposes a lower limit to dose detection of approximately 0.3 Gy (Prosser et al. 1988).

For those persons where the possibility of induced malignancy is thought not to be trivial many would prefer a quantatative appraisal of their dose from which an estimate of risk can be inferred. This is best illustrated by an example. Suppose 5 dicentrics were seen in 100 cells, then using $Y = 1.57 \times 10^{-2}D + 5.0 \times 10^{-2}D^2$ (Lloyd and Edwards 1983) the estimated dose would be about 0.85 Gy. At this level of dose increases in cancer incidence and death have been observed in the Japanese A-bomb survivors. Stather et al. (1988) have reassessed the Japanese data bearing in mind the new dosimetry and the need to extrapolate for future cancers. They have deduced risks for acute high dose irradiation, averaged over the UK population, lying in the range 3.7% Gy^{-1} (absolute risk model) to 12.9% Gy^{-1} (relative risk model). This includes an extrapolation from the Japanese to the UK population in respect of age structure and spectrum of cancer deaths. These risk values can be further refined for the counselling of a particular individual by taking into account the person's age at the time of exposure. However, the above values translate into an excess risk that the person will die of cancer of 3–11%. One would perhaps present this calculation to the person by saying that their chance of dying of cancer is about 30% rather than the 25% that it was prior to the accident. For cancer incidence one would increase both of these values by approximately 50%.

If the accidental exposure was protracted, as for example at Goiânia, 5 dicentrics in 100 cells indicates a dose of 3.2 Gy (α term only) and converts to risk using low dose rate risk factors of 1.3 or 4.5% Gy^{-1}. These are derived from the acute factors assuming a dose rate reduction factor of about 3 (Stather et al. 1988). Calculating through leads to very similar risks of cancer death (4–14%) as for the acute example and so suggests that provided one's assumptions are consistent chromosomal aberration data can be used to obtain an estimate of risk without the need to know whether the irradiation was acute or chronic.

Clearly these types of calculation are cumbersome, involving two steps, namely; aberration yield to dose followed by dose to risk. Each carries considerable uncertainties of both statistical and subjective natures. It is a long-term objective of biological dosimetry to be able to take cells from an irradiated person and make a more direct measurement of stochastic risk; in effect to dispense with dose. In view of the rapidly expanding knowledge on the relationship between specific chromosomal lesions and certain cancers it seems probable that such an assay would employ cytogenetic analysis.

References

Almassy Z, Krepinsky AB, Bianco A, Köteles GJ (1987) the present state and perspectives of micronucleus assay in radiation protection. A review. Appl Radiat Isot 38: 241–249

Awa AA, Sofuni T, Honda T, Itoh M, Neriishi S, Otake M (1978) Relationship between the radiation dose and chromosome aberrations in atomic bomb survivors of Hiroshima and Nagasaki. J Radiat Res 19: 126–140

Baranov A, Gale PR, Guskova A et al. (1989) Bone marrow transplantation after the Chernobyl nuclear accident. New Eng J Med 321: 205–212

Bilbao A, Prosser JS, Edwards AA, Moody JC, Lloyd DC (1989) The induction of micronuclei in human lymphocytes by in vitro irradiation with alpha particles from plutonium-239. Int J Radiat Biol 56: 287–292

Brewen JG, Preston RJ, Littlefield LG (1972) Radiation-induced human chromosome aberration yields following an accidental whole-body exposure to ^{60}Co γ-rays. Radiat Res 49: 647–656

Buckton KE (1983) Chromosome aberrations in patients treated with X-irradiation for ankylosing spondylitis. In Ishihara T, Sasaki MS (eds) Radiation-induced chromosome damage in man. Alan R. Liss, New York, pp 491–511

Daburon F (1986) Biophysical methods for assessing the radiation dose causing lesions in the skin and subcutaneous tissues. Br J Radiol suppl no 19: 75–82

Dolphin GW (1969) Biological dosimetry with particular reference to chromosome aberration analysis. A review of methods. In: Handling of radiation accidents. International Atomic Energy Agency, Vienna, pp 215–224

Fenech M, Morley AA (1985) Measurement of micronuclei in lymphocytes. Mutat Res 147: 29–36

Fenech M, Morley AA (1986) Cytokinesis-block micronucleus method in human lymphocytes: effect of in vivo ageing and low dose X-irradiation. Mutat Res 161: 193–198

Finnon P, Lloyd DC, Edwards AA (1986) An assessment of the metaphase finding capability of the Cytoscan 110. Mutat Res 164: 101–108

Fliedner TM (1981) Strategien zur strahlenschutzmedizinischen ambulanten Versorgung von Betroffenen bei kerntechnischen Unfällen. Strahlenschutz Forsch Prax 22: 71–82

IAEA (19867 Biological dosimetry : chromosomal aberration analysis for dose assessment. Technical Report No 260. International Atomic Energy Agency, Vienna

IAEA (1988) The radiological accident at Goiânia. International Atomic Energy Agency, Vienna

Kaul A, Dehos A, Bögl W et al. (eds) (1986) Biological indicators for radiation dose assessment. MMV Medizin, München

Kormos C, Köteles GJ (1988) Micronuclei in X-irradiated human lymphocytes. Mutat Res 199: 31–35

Langlois RG, Bigbee WL, Kyoizumi S, Nakamura N, Bean MA, Akiyama M, Jensen RH (1987) Evidence for increased somatic cell mutations at the glycophorin A locus in atomic bomb survivors. Science 236: 445–448

Lea DE, Catcheside DG (1942) The mechanism of the induction by radiation of chromosome aberrations in Tradescantia. J Genet 44: 216–245

Liniecki J, Bajerska A, Wyszynska K (1983) Dose-response relationships for chromosome aberrations in peripheral blood lymphocytes after whole- and partial-body irradiations. I. Effects immediately after irradiation. Mutat Res 110: 83–101

Littlefield LG (1990) Cytogenetic problems in assessment of radiation exposure. In: The medical basis for radiation accident preparedness. Elsevier, Amsterdam (in press)

Littlefield LG, Sayer AM, Frome EL (1989) Comparisons of dose-response parameters for radiation-induced acentric fragments and micronuclei observed in cytokinesis-arrested lymphocytes. Mutagenesis 4: 265–270

Lloyd DC (1990) Advances in cytogenetic dosimetry. In: The medical basis for radiation accident preparedness. Elsevier, Amsterdam (in press)

Lloyd DC, Edwards AA (1983) Chromosome aberrations in human lymphocytes : effect of radiation quality, dose and dose rate. In: Ishihara T, Sasaki M (eds) Radiation-induced chromosome damage in man. Liss, New York, pp 23–49

Lloyd DC, Purrott RJ, Dolphin GW (1973) Chromosome aberration dosimetry using human lymphocytes in simulated partial body irradiation. Phys Med Biol 18: 421–431

Lloyd DC, Purrott RJ, Dolphin GW, Horton PW, Halnan KE, Scott JS, Mair G (1976) A comparison of physical and cytogenetic estimates of radiation dose in patients treated with iodine-131 for thryoid carcinoma. Int J Radiat Biol 30: 473–485

Lloyd DC, Edwards AA, Prosser JS, Corp MJ (1984) The dose response relationship obtained at constant irradiation times for the induction of chromosome aberrations in human lymphocytes by cobalt-60 gamma rays. Radiat Environ Biophys 23: 179–189

Lloyd DC, Edwards AA, Prosser JS et al. (1986) Accidental intake of tritiated water: a report of two cases. Radiat Prot Dosim 15: 191–196

Lörch T (1986) Ein System zur automatischen Detektion dizentrischer Chromosomen. Thesis, Heidelberg University

Lucas JN, Tenjin T, Straume T, Pinkel D, Moore D, Litt M, Gray JW (1989) Rapid human chromosome aberration analysis using fluorescence in situ hybridization. Int J Radiat Biol 56: 35–44

Meyne J, Littlefield LG, Moyzis RK (1989) Labeling of human centromeres using an alphoid DNA consensus sequence: application to the scoring of chromosome aberrations. Mutat Res 226: 75–79

Pantelias GE, Maillie HD (1984) The use of peripheral blood mononuclear cell prematurely condensed chromosomes for biological dosimetry. Radiat Res 99: 140–150

Pincu M, Bass D, Norman A (1984) An improved micronucleus assay in lymphocytes. Mutat Res 139: 61–65

Pinkel D, Landegent J, Collins C, Fuscoe J, Segraves R, Lucas J, Gray J (1988) Fluorescence in situ hybridization with human chromosome-specific libraries: detection of trisomy 21 and translocations of chromosome 4. Proc Natl Acad Sci USA 85: 9138–9142

Piper J, Tower S, Gordon J, Ireland J, McDougall D (1988) Hypothesis combination and context sensitive classification for chromosome aberration scoring. In: Gelsema ES, Kanal LN (eds) Pattern recognition and artificial intelligence. Elsevier, Amsterdam, pp 449–460

Prosser JS, Moquet JE, Lloyd DC, Edwards AA (1988) Radiation induction of micronuclei in human lymphocytes. Mutat Res 199: 37–45

Prosser JS, Lloyd DC, Edwards AA (1989) A comparison of chromosomal and micronuclear methods for radiation accident dosimetry. In: Proc 4th Int Symp SRP, 4–9 June 1989, Malvern, pp 133–136

Ramalho AT, Nascimento ACH, Natarajan AT (1988a) Dose assessments by cytogenetic analysis in the Goiânia (Brazil) radiation accident. Radiat Prot Dosim 25: 97–100

Ramalho AT, Sunjevaric I, Natarajan AT (1988b) Use of the frequencies of micronuclei as quantitative indicators of X-ray-induced chromosomal aberrations in human peripheral blood lymphocytes: comparison of two methods. Mutat Res 207: 141–146

Savage JRK (1988) A comment on the quantitative relationship between micronuclei and chromosomal aberrations. Mutat Res 207: 33–36

Savage JRK, Breckon G (1985) Chromosome aberration frequencies after partial-body irradiation of Syrian hamsters. Br J Radiol 58: 1105–1110

Sieber VK, Wells J (1986) The use of plucked hairs as a biological dosemeter. Br J Radiol Suppl No 19: 92–95

Stather JW, Muirhead CR, Edwards AA, Harrison JD, Lloyd DC, Wood NR (1988) Health effects models developed from the 1988 UNSCEAR report. Rep R226. National Radiological Protection Board, Chilton

Tates AD, Driel-Kulker AMJ van, Woldring VM, Engelse P den, Ploem JS (1988) Automated scoring of micronuclei in binucleated lymphocytes via computerised image analysis: feasibility studies. In: Micronuclei: origins, applications and methodologies workshop 23–25 May 1988, Manchester pp 54–55

Thomson EJ, Perry PE (1988) The identification of micronucleated chromosomes: a possible assay for aneuploidy. Mutagenesis 3: 415–418

UNSCEAR (1988) Sources, effects and risks of ionizing radiation. The United Nations Scientific Committee on the effects of atomic radiation. Report to the General Assembly. Annex G

Wells J, Charles MW, Warner TT (1983) Chromosomes from the epithelium of plucked telogen hairs. Hum Genet 63: 315–316

Dose Estimates and the Fate of Chromosomal Aberrations in Cesium-137 Exposed Individuals in the Goiania Radiation Accident

A. T. Ramalho, A.C.H. Nascimento, and P. Bellido[1]

Contents

1 Introduction

In September 1987, a radiation accident occurred in Goiania (Brazil) in which a radiotherapy ^{137}Cs source was removed from its housing and damaged, leading to external exposure and/or internal contamination of at least 100 persons (IAEA 1988). The activity of the ^{137}Cs source was 50.9 TBq (1375 Ci), with a dose rate of 4.6 Gy h^{-1} at 1 m. The description of the source, as well as the chronology of the accident, have been published earlier (IAEA 1988).

During the initial decision-making regarding the clinical management of the victims, it was essential to estimate the absorbed radiation dose of differently exposed individuals. In situations like this accident, in which physical measurements of dose were not feasible, dose estimation by a biological technique was the only method available.

The frequencies of induced chromosomal aberrations in peripheral blood lymphocytes of exposed individuals are currently being used as a validated biological dosimeter to estimate the absorbed dose of ionizing radiations. The dose is estimated by comparing the observed yields of unstable chromosome aberrations (i.e., dicentrics and centric rings) with standard dose response curves generated following in vitro irradiation of human lymphocytes (IAEA 1986). During the Goianian accident, we assessed the doses received by 110 individuals who could have been exposed to ^{137}Cs with doses higher than about 0.1 Gy, which has been suggested to be the lower limit of sensitivity of this technique (Lloyd and Purrott 1981). Of the 110 investigated cases, about 60 had some levels of internal contamination by ^{137}Cs. This was quantitatively estimated by colleagues in the Division of Internal Monitoring using radioactivity measurements in urine and feces or whole-body counting.

[1] Instituto de Radioprotecao e Dosimetria (IRD), Comissao Nacional de Energia Nuclear (CNEN), Av. das Americas Km 11.5, Barra da Tijuca, 22602 Rio de Janeiro, Brazil

The type of exposure in this accident was complicated since 2 weeks elapsed between breaking of the source and the discovery of the accident. Most of the affected individuals were exposed over a prolonged period, either fractioned and/or protracted at varying dose rates. Thus, the exposure was a chronic rather than an acute one and their total doses were, in many cases, due to exposure from external contamination plus internal incorporation of the nuclide. These factors make the estimation of the total absorbed dose very difficult.

2 Materials and Methods

The first 25 blood samples, comprising all the hospitalized patients, arrived within the first week following the discovery of the accident. Fifty samples were received during the next 4 weeks and the rest within the second month. From each individual, a 10-ml peripheral blood sample was collected in a heparinized vacuntainer. The erythrocytes were sedimented by centrifugation and 2 ml of the buffy coat containing leukocytes was collected and added to 10 ml of Hams's F-10 medium supplemented with fetal calf serum (25%) and phytohemagglutinin (Wellcome). The cultures were incubated for 48 h and colchicine was added 2 h before harvest. After hypotonic treatment with KCl (0.075 M) for 15 min, the lymphocytes were fixed in methanol-acetic acid (3:1) and dropped onto clean, cold, and wet slides. The slides were stained with aqueous Giemsa solution.

From each sample, depending on the frequency of aberrations, 100 to 500 cells were scored for the presence of unstable chromosome-type aberrations (i.e., dicentrics, centric rings, and acentric fragments). For individuals with a high frequency of aberrations (i.e., most exposed) an analysis of 100 cells was sufficient. However, in cases of low exposures, as evidenced by a low frequency of aberrations, 200 to 500 cells were scored.

The absorbed dose was estimated from the yield of dicentrics and centric rings (i.e., two-hit aberrations). Since no dose-response curve for a low dose rate ^{137}Cs was available, a calibration curve generated for Co-60 gamma rays (unpubli. results) at a dose rate of 0.12 Gy min^{-1} was used for the dose estimates. This Co-60 calibration curve fits the equation:

$$Y = (2.70 \pm 1.03) \times 10^{-2} D + (2.20 \pm 0.49) \times 10^{-2} D^2.$$

The alpha and beta coefficients agree quite well with other previously published curves for chronic gamma radiation (Lloyd and Edwards 1983).

3 Results and Discussion

The distribution of doses estimated for all 110 individuals analyzed so far is presented in Table 1. From these, 29 were considered to have received 0.5 Gy or more, and the details of their individual dose estimates, together with 95% confidence limits, calculated according to standard procedures (IAEA 1986), are presented in Table 2.

When the exposure to low LET radiation is uniformly distributed over the whole body, the induced dicentric chromosomes among the lymphocytes follow a Poisson distribution. Deviation from Poisson (i.e., overdispersion) may indicate a nonuniform

Table 1. Dose distribution and number of subjects exposed. The doses were estimated by cytogenetic dosimetry for the Goiania accident

Estimated dose (Gy)	Number of subjects
0 –0.49	81
0.50–0.99	8
1.00–1.99	8
2.00–2.99	4
3.00–3.99	1
4.00–4.99	3
5.00–5.99	2
6.00–6.99	2
7.00	1
Total	110

Table 2. Cytogenetic dose estimates for the subjects who received the highest doses in the Goiania accident

Subjects	Cells scored	F [a]	Estimated dose (Gy)	95% Confidence limits (Gy)	Poisson distribution	Internal dose (Gy)[b]
1-DAF	150	1.253	7.0	6.0 –8.4	Yes	0.17
2-RSA	100	1.030	6.2	5.2 –7.6	Yes	0.12
3-LNF[c]	26	0.962	6.0	5.1 –7.3	Yes	2.13
4-MGF[c]	170	0.824	5.5	4.6 –6.7	Yes	0.02
5-AAS[c]	100	0.760	5.3	4.4 –6.5	Yes	0.08
6-IBS[c]	200	0.570	4.5	3.5 –5.7	[d]	0.05
7-EDF	100	0.540	4.4	3.4 –5.6	Yes	0.03
8-MGA	150	0.533	4.3	3.4 –5.5	Yes	0.01
9-IAF	100	0.270	3.0	2.0 –4.1	Yes	0.06
10-GGS	130	0.262	2.9	2.0 –4.1	Yes	0.09
11-EBS	150	0.260	2.9	2.0 –4.1	Yes	0.04
12-WMP	250	0.228	2.7	2.0 –3.6	[d]	0.06
13-ERF	100	0.160	2.1	1.3 –3.2	Yes	<0.01
14-ODJ	100	0.100	1.6	0.7 –2.6	Yes	0.16
15-PRM	200	0.075	1.3	0.7 –2.1	Yes	<0.01
16-LUF	200	0.070	1.3	0.6 –2.0	[d]	0.17
17-DFJ	200	0.065	1.2	0.6 –2.0	Yes	<0.01
18-KSS	100	0.060	1.1	0.4 –2.0	Yes	0.04
19-RBG	250	0.060	1.1	0.6 –1.9	Yes	<0.01
20-OAF	100	0.050	1.0	0.2 –1.9	Yes	<0.01
21-LOS	170	0.050	1.0	0.1 –1.7	Yes	0.37
22-PNF	140	0.029	0.7	0.04–1.4	[d]	0.08
23-SPQ	200	0.025	0.6	0.1 –1.1	Yes	0.02
24-MPG	300	0.023	0.6	0.1 –1.1	[d]	0.01
25-AWJ	200	0.020	0.5	0.03–1.1	[d]	<0.01
26-DPS	200	0.020	0.5	0.03–1.1	Yes	<0.01
27-EAS	200	0.020	0.5	0.03–1.1	Yes	0.04
28-MNF	200	0.020	0.5	0.03–1.1	Yes	0.03
29-GAC	225	0.018	0.5	0.03–1.1	Yes	<0.01

[a] Frequency of dicentrics and rings per cell. [b] Internal dose delivered until blood collection.
[c] Deceased. [d] Overdispersed.

(i.e., partial body) exposure (Lloyd 1978). In the present accident, in spite of the fact that almost all the exposed individuals received an inhomogeneous dose, as evidenced by the presence of localized radiolesions in the skin, only in six cases did the distribution of exchange aberrations fail, by overdispersion, to conform to Poisson distribution. After an acute partial body exposure, it is sometimes possible to use a distribution analysis to estimate the dose and the proportion of the body exposed (IAEA 1986). In this accident the complicated (fractionated, protracted, and internal) nature of the exposure resulted, in most cases, in a distribution indistinguishable from Poisson. Thus, it was not possible to use such an approach to disentangle these different factors.

It is known that incorporated Cs distributes uniformly throughout the body, mainly in muscle (NCRP 1980; Rosoff et al. 1963). One can therefore expect that the incorporated Cs leads to a continuous and fairly even irradiation of the body, and thus of the lymphocytes in both vascular and extravascular pools. As a first approach, in the present accident, we can assume that the values estimated by cytogenetic dosimetry are basically those resulting from the exposure to ^{137}Cs gamma rays (external irradiation) with some possible contribution from the internally deposited material. The value of that contribution would be the integral of the dose rate emitted by the incorporated ^{137}Cs from the day of intake until the sampling of the blood. These values were calculated by the Division of Internal Monitoring based upon urine and feces analysis and are included in Table 2. The mathematical methodology used for these calculations has been published before (Bertelli and Lipsztein 1987). Except in four cases (Nos. 3, 14, 16, and 21) the doses due to internal contamination can be considered as negligible in relation to the external irradiation. Even in the highly internally contaminated individuals, there were no chromatid aberrations which are typical of irradiation during S- and G_2-phases. This indicates that the concentration of ^{137}Cs carried through into the cultures was too low to induce aberrations in vitro.

From the eight patients whose dose estimates exceeded 4 Gy, four died after about 40 days. For human exposure to low LET radiation, the $LD_{50/60}$, i.e., the lethal dose for 50% mortality occuring within 60 days, has been estimated to be approximately 3 to 4 Gy for an acute total body irradiation without medical treatment (Duncan and Nias 1977). However, when the exposure extends over days or weeks, a significantly higher radiation dose can be tolerated (Andrews 1980). Therefore, in the present accident, in which the exposure had been fractionated and protracted, it is not surprising that four medically supported individuals, who received estimated doses over 4 Gy, have survived so far.

4 Follow-Up Studies

Follow-up studies of the persistence time of lymphocytes containing unstable chromosome aberrations are in progress. It has been suggested that the unstable aberration yield in lymphocytes decays after exposure with a half-time of about 3 years (Lloyd et al. 1980). Other authors have reported different values of the mean lifetime (the mean time spent in G_0 or G_1 phase) of lymphocytes, in different surveys, namely, 530, 1000, or 1600 days (Sasaki 1983).

Due to the limited data in the literature, a follow-up study was started in order to investigate this disappearance rate. Blood samples from some heavily exposed Goianian patients, of different ages, were reexamined at different sampling times after the exposure. The results are presented in Table 3.

Table 3. Cytogenetic follow-up of some Goianian patients

Subject (sex/age)	Leukopenia	GM-CSF	Cells scored	Days after exposure	Freq.dic+ rings/cell
1 (M/37)	Severe	–	150	0[a]	1.253 ± 0.091
			100	30	0.980 ± 0.099
			190	100	0.457 ± 0.049
			150	220	0.389 ± 0.051
			270	340	0.265 ± 0.031
			300	470	0.123 ± 0.020
2 (M/22)	Severe	–	100	0[a]	1.030 ± 0.101
			100	30	0.940 ± 0.097
			50	100	0.460 ± 0.096
			240	470	0.066 ± 0.016
7 (M/42)	Moderate	yes	100	0[a]	0.540 ± 0.073
			240	340	0.037 ± 0.012
			100	470	0.029 ± 0.017
8 (F/57)	Severe	yes	150	0[a]	0.533 ± 0.059
			200	100	0.450 ± 0.047
			100	340	0.107 ± 0.033
10 (M/21)	Moderate	yes	130	0[a]	0.262 ± 0.045
			150	220	0.173 ± 0.034
			360	470	0.030 ± 0.009
11 (M/13)	Mild	–	150	0[a]	0.260 ± 0.042
			200	100	0.210 ± 0.032
			180	220	0.156 ± 0.029
			250	340	0.080 ± 0.018
			180	470	0.067 ± 0.019
12 (M/19)	Severe	yes	250	0[a]	0.228 ± 0.030
			50	220	0.180 ± 0.060
			360	340	0.036 ± 0.010
13 (M/46)	Mild	–	100	0[a]	0.160 ± 0.040
			200	340	0.015 ± 0.009
			220	470	0.005 ± 0.005
16[b] (M/14)	Mild	–	200	0[a]	0.070 ± 0.019
			200	30	0.100 ± 0.022
			250	100	0.112 ± 0.021
			100	220	0.042 ± 0.020
			250	470	0.008 ± 0.006
21[b] (F/28)	Mild	–	170	0[a]	0.050 ± 0.017
			100	30	0.070 ± 0.026
			80	110	0.100 ± 0.035
			160	340	0.012 ± 0.009
			250	470	0.016 ± 0.008

[a] First blood sample.
[b] High internal contamination.

Table 3 also includes information about the level of leukopenia for each patient (Brandao-Mello et al. 1989). In this work, leukopenia was considered to be mild for leukocyte counts lower than normal, but above 3000 mm^3; moderate between 2000 and 3000; and severe below 2000.

Leukopenia occurred during the first weeks following the accident. By the end of the first month, almost all the patients had already recovered from bone marrow aplasia or hypoplasia and presented normal blood cell counts. Patients Nos. 1 and 2 had extremely severe leukopenias (less than 100 leukocytes/mm^3). But their leukocyte counts were near to normal again when the second blood samples were collected (6000 leukocytes/mm^3).

In the course of the first month after the accident, some of the patients received a recombinant human granulocyte-macrophage colony-stimulating factor (rHuGM-CSF), a molecularly cloned hematopoietic growth factor that stimulates granulocyte progenitor cells (Brandao-Mello et al. 1989).

For the Goianian patients, the results so far show a half-time for the disappearance of unstable aberrations significantly shorter than values reported previously by others. The disappearance of lymphocytes containing unstable aberrations was fitted to a simple exponential function of time, using a "curfit" microcomputer program (Bevington 1969). Among eight individuals, the mean rate of disappearance was equal to 5.48×10^{-3} day^{-1}, with a standard deviation of the mean of 3.37×10^{-4}. The mean half-time was estimated to be around 130 days, with a range from 95 to 220 days. Although the simple exponential model was not rejected by normal statistical criteria, the temporal trend may be more complex.

For two patients who had high ^{137}Cs body burdens (case Nos. 16 and 21) the aberration yield first increased, in reasonable agreement with the estimated doses due to internal contamination, and, then, it began to fall, following the decorporation of ^{137}Cs.

5 Final Remarks

In assessing absorbed doses in radiation accidents by chromosome dosimetry, it is important to know the details of the exposure in terms of dose rate. This is necessary in order to choose a suitable calibration curve for estimating the dose. The choice of a calibration curve can modify the dose estimates by up to a factor of two. In the absence of any physical data, the biological dosimetry, coupled with other bioassay procedures, proved very valuable in the Goianian accident enabling the medical team to make initial decisions about the handling of the victims.

The cytogenetic follow-up performed so far in eight subjects highly exposed during the Goiania accident suggests that the disappearance half-time of lymphocytes containing unstable aberrations is significantly shorter than values reported previously by others. The results also suggest that the technique of cytogenetic dosimetry was able to detect increases in the frequency of chromosomal aberrations due to high ^{137}Cs body burdens. Further work is necessary to clarify these two assumptions.

Acknowledgments. We are grateful to Dr. C. E. Brandao Mello for collecting and sending the blood samples; Drs. Iris Ferrari, A. T. Natarajan (The Netherlands), and D. C. Lloyd (UK) for the contribution in scoring the slides; Dr. Luiz Bertelli for calculating the internal contamination values and Dr. G. Sigaud for the data analysis. We appreciate the help of Drs. A. T. Natarajan and D. C. Lloyd in the preparation of this manuscript. We are indebted to the "Leide das Neves Foundation" (Goiania), for their collaboration in our follow-up program. This work is part of an International Atomic Energy Agency (IAEA) coordinated Research Program.

References

Andrews GA (1980) Medical management of accidental total-body irradiation. In: Hubner KF, Fry SA (eds) The medical basis for radiation accident preparedness. Elsevier, Amsterdam, pp 297–310

Bertelli L, Lipsztein JL (1987) A mathematical simulation for the study of radionuclide kinetics in human body. Radiat Prot Dosim 18: 209–214

Bevington PR (1969) Data reduction and error analysis for the physical sciences. McGraw-Hill, New York, pp 204–246

Brandao-Mello CE, Oliveira AR, Valverde NJ, Farina R (1989) Medical management of the victims of Goiania Cs-137 accident. Radiation protection–theory and practice, session 6. 4th Int Symp Malvern, pp 121–124

Duncan W, Nias AHW (1977) Clinical radiobiology. Longman, London.

International Atomic Energy Agency (1986) Biological dosimetry chromosomal aberration analysis for dose assessment. Vienna, IAEA Technical Reports Series No. 260

International Atomic Energy Agency (1986) The radiological accident in Goiania. IAEA STI/PUB/815, Vienna

Lloyd DC (1978) The problems of interpreting aberration yields induced by "in vivo" irradiation of lymphocytes. In: Evans HJ, Lloyd DC (eds) Mutagen induced chromosome damage. Edinburgh University Press, Scotland, pp 77–88

Lloyd DC, Purrott RJ, Reeder EJ (1980) The incidence of unstable chromosome aberrations in peripheral blood lymphocytes from unirradiated and occupationally exposed people. Mutat Res 72: 523–532

Lloyd DC, Purrott RJ (1981) Chromosome aberration analysis in radiological protection dosimetry. Radiat Prot Dosim 1: 19–28

Lloyd DC, Edwards AA (1983) Chromosome aberrations in human lymphocytes: effect of radiation quality, dose and dose rate. In: Ishihara T, Sasaki MS (eds) Radiation-induced chromosome damage in man. Liss, New York, pp 23–49

National Council for Radiation Protection (1980) Management of persons contaminated with radionuclides. Washington DC, NCRP Rep No 65, pp 77–79

Rosoff B, Cohn SH, Spencer H (1963) Cesium-137 metabolism in man. Radiat Res 19: 643–654

Sasaki MS (1983) Use of lymphocyte chromosome aberrations in biological dosimetry: possibilities and limitations. In: Ishihara T, Sasaki MS (eds) Radiation-induced chromosome damage in man. Liss, New York, pp 585–604

Cytogenetic Studies in Male Germ Cells, Their Relevance for the Prediction of Heritable Effects and Their Role in Screening Protocols

I.-D. Adler[1]

Contents

1 Introduction

All chemicals that break chromosomes in germ cells also affect somatic cells. This statement was made by Holden (1982) when he surveyed the literature on 76 chemical compounds and found that 26 of these were active in both cell types. The same statement had already been made by Brusick (1980) and it seems to hold up to now (Adler and Ashby 1989). With this reassuring axiom one could be tempted to conclude that cytogenetic testing in germ cells may be a superfluous task, even when the goal is to label a chemical as a germ cell mutagen.

However, this conclusion would be based on the reversion of the above mentioned axiom, namely, that all somatic cell clastogens are also effective in germ cells. This is not true. With the literature information available at that time Holden identified 18 chemicals for which a positive response in somatic cells and a negative result in germ cells had been published. Had one used the somatic cell data to conclude that these chemicals are also germ cell mutagens more than 40 % of the compounds would have been mislabeled. This figure illustrates the need for cytogenetic germ cell data.

The data base on germ cell effects available to Holden was limited and at least one of the 18 compounds he stated to be germ cell negatives, procarbazine (Natulan),

[1] GSF-Institut für Säugetiergenetik, Ingolstädter Landstraße 1, 8042 Neuherberg, FRG

had been found to be a germ cell clastogen (Adler 1982) and had been known to induce gene mutations in male germ cells (Ehling and Neuhäuser 1979). Some of the other examples may prove likewise misjudged when the more recent literature is consulted or more accurate experiments are designed. These considerations further underline the necessity for cytogenetic germ cells studies. The tests require a thorough understanding of germ cell development and extensive cytogenetic training. They should be applied only when positive somatic cell data are available and a classification as germ cell mutagen is needed (MAK 1989). Furthermore, they should provide more specific answers than a plus or minus result.

2 Cytogenetic Germ Cell Tests

Germ cell cytogenetics is not one single test method but comprises an array of methods depending on the type of germ cells under study. The most commonly used species is the mouse, however, rats and Chinese hamsters have also been studied (Preston et al. 1981). Only limited infomation is available on germ cells of female mammals (Adler and Brewen 1982) due to the biology of female germ cell development. The accessibility of male germ cells is by far better and different stages of their development can be studied. The preparation techniques and criteria for microscopic analysis have been described in detail (Adler 1984) and are only briefly summarized below.

2.1 Differentiating Spermatogonia

Male germ cells develop from a population of slowly dividing stem cells. The timing of the various developmental stages for the mouse has been given by Oakberg (1960) and forms the basis for experimental design not only of cytogenetic studies but also for the assays in which genetic effects in progeny of male mice are evaluated, such as the dominant lethal test, the heritable translocation assay, and the specific locus test (Adler 1980; Ehling et al. 1978; Generoso et al. 1973; Russell 1951). The first stages of male germ cell maturation are differentiating spermatogonia. Their cell cycle is shorter than that of stem cells, i.e., 24–26 h. They remain in contact as chains of cells by plasma bridges (Schleiermacher and Schmidt 1973) which ensure synchrony of their further development. By successive steps of cell division they are termed $A_1 - A_4$, intermediate, and B-spermatogonia. The majority of spermatogonial mitoses are B-spermatogonia. Mitotic chromosomes of spermatogonia can be analyzed for chromosomal aberrations with almost the same ease as in bone marrow cells. Due to their cell cycle time a longer period of colchicine treatment is required for accumulation of sufficient numbers of metaphase cells. The adverse effect of colchicine on chromosome shape and chromatid alignment is minimal. Cell suspensions are most efficiently obtained by the use of 50% acetic acid. An example of a chromatid break in a spermatogonial metaphase cell is shown in Fig. 1A. The principle of analyzing first posttreatment mitoses applies to spermatogonia in the same way as to bone marrow cells. The majority of aberrations will be of the chromatid type since most chemical clastogens are S-dependent, i.e., a DNA-replication is required to translate the primary lesion into a chromosomal aberration (Bender et al. 1974).

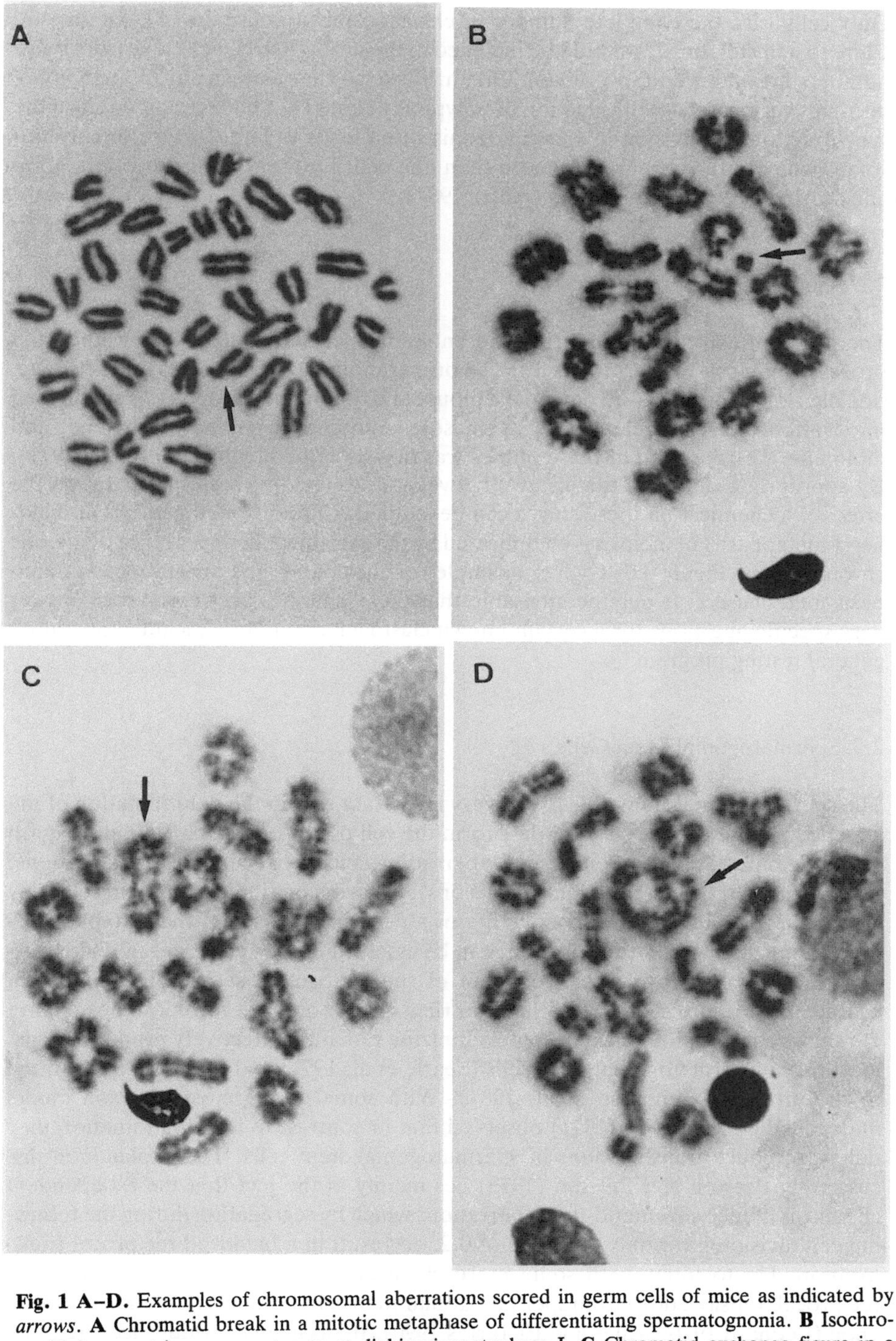

Fig. 1 A–D. Examples of chromosomal aberrations scored in germ cells of mice as indicated by *arrows*. **A** Chromatid break in a mitotic metaphase of differentiating spermatognonia. **B** Isochromatid fragment in a spermatocyte at diakinesis-metaphase I. **C** Chromatid exchange figure in a spermatocyte at diakinesis-metaphase I. **D** Ring of four chromosomes (translocation multivalent) in a spermatocyte at diakinesis-metaphase I

Only cells with the complete number of chromosomes (mouse 2n=40, rat 2n=42, Chinese hamster 2n=22) should be included in the evaluation. Deviations from these numbers are often due to preparational artifacts so that assessment of aneuploidy becomes very unreliable. Induction of polyploidy should not be ascertained since the synchrony of cell division in spermatogonia often leads to large groups of chromosomes which actually represent more than one cell. Further details for preparation and scoring have been published (Adler 1984).

2.2 Spermatocytes

The phase of cell proliferation in differentiating spermatogonia is followed by the process of meiosis in spermatocytes. The preparation of chromosomes for the reduction division requires 12–13 days in the mouse and this meiotic prophase is subdivided into various stages: preleptotene (S-phase), leptotene, zygotene, pachytene, and diplotene. The duration of these phases was first described by Monesi (1962, 1971). By appropriate timing of the interval between treatment and testes preparation the effect of a chemical on these stages can be studied. Chromosomal analysis at diakinesis-metaphase I of meiosis is complicated by the variability in appearance of meiotic chromosomes. Figure 1B, C gives examples of chromatid-type aberrations at diakinesis-metaphase I. It may be advisable that these analyses are reserved to answer very specific questions for a certain known clastogen and not be included in a more general testing program.

2.3 Spermatogonial Stem Cells

The spermatogonial stem cells play a very important role in the determination of the genetic risk of a chemical, since they form the cell pool from which all mature germ cells originate. Stable chromosomal aberrations such as reciprocal translocations, Robertsonian translocations (centric fusions), or inversions induced in stem cells may persist throughout the reproductive life of a male individual. Of these stable aberrations only reciprocal translocations can be assessed in spermatocytes at diakinesis-metaphase I of meiosis after treatment of spermatogonial stem cells. They form characteristic multivalents, i.e., chains or rings of four or more chromosomes. Figure 1D shows a ring of four chromosomes. Ionizing radiation effectively produces translocations in spermatogonial stem cells (Searle et al. 1971) and these are transmitted to the progeny (Generoso et al. 1984). With some of the most common model mutagens Léonard et al. (1971) observed that in contrast to ionizing radiation they failed to induce translocations in spermatogonial stem cells. The explanation discussed by Brewen and Preston (1973) lies mainly in the fact that the S-dependent chemicals induce chromatid-type aberrations which by segregation during the following cell division have only a chance of 0.25 to result in a balanced reciprocal translocation. Up to today with some more chemicals tested the lack of recovery of reciprocal translocations from spermatogonial stem cells remains a common phenomenon (Léonard 1976; Adler and El Tarras 1990). It has been so widely accepted that only very rarely chemicals are tested for translocation induction in spermatogonial stem cells. It may only be indicated for S-independent chemicals and polyfunctional alkylating agents to perform this analysis.

3 Quantitative Comparison of Clastogenic Effects

In order to compare clastogenic effects in somatic cells and germ cells it seems necessary to find a quantitative parameter in addition to the qualitative plus-minus statement. The quantitative comparison can only be applied if the experiments are performed under identical experimental conditions such as strains of animals, route of exposure, exposure regimen. The results described below fulfill these criteria with only a few exceptions which will be indicated specifically. The basis for the quantitative comparison is the doubling dose, that is the dose which induces as many events (aberrations, micronuclei, or heritable translocations) per cell generation as occur spontaneously. The doubling dose concept has been developed for radiation risk estimates (Lüning and Searle 1971) and is also applied to chemical risk assessment based on specific locus test data (Ehling 1980, 1984, 1989). Several conditions have to be fulfilled when doubling doses for different end points are calculated: first, the spontaneous and the induced events must be qualitatively the same; second, the dose-response relationship has to be linear and non-thresholded; and third, the germ cells of the experimental animal (mouse) and man must be equally sensitive. This third condition only applies if the doubling dose is used for risk estimates in relation to human exposure. In the present context the doubling dose is used as an indicator of effectiveness of a chemical in various assay systems within one species (mouse) and conditions 1 and 2 must be fulfilled. While for condition 1 the common observed end point is chromosomal aberrations, the justification of condition 2 has to be obtained from the experimental data. The studies, therefore, have to be designed to allow the calculation of a dose-response relationship.

The doubling dose (DD) is calculated from the linear equation $y=a+bD$ as the ratio of the spontaneous rate (a) and the linear regression coefficient (b): $DD=a/b$. For the following six chemicals data were available to perform this calculation. The comparison was made for chromosomal aberrations in bone marrow cells, the micronucleus assay, chromosomal aberrations in differentiating spermatogonia and spermatocytes. Additionally, where data were available the calculation of the doubling dose was also performed for the heritable translocation test. This test determines the rate of transmissible chromosomal damage, reciprocal translocations, in the next generation.

3.1 Acrylamide

It has been suggested that acrylamide may be a unique germ cell clastogen when only the limited data by Shiraishi (1978) were available and the strong dominant lethal and heritable translocation induction became known (Shelby et al. 1986, 1987). However, subsequently, it could be shown that acrylamide induces aberrations and micronuclei in bone marrow cells of mice (Adler et al. 1988). The doubling dose for micronuclei based on the linear dose-response relationship was calculated to be 23 mg/kg (Table 1). For chromosomal aberrations in bone marrow cells the data (Adler et al. 1988) were fitted to a linear dose-response relationship and a doubling dose of 10 mg/kg was calculated. Preliminary results in spermatocytes at preleptotene (Kliesch et al. 1989) led to the calculation of a doubling dose of 14 mg/kg based on a point estimate for 100 mg/kg acrylamide. Even though this calculation was hampered by the lack of complete dose-response data, the doubling doses for the somatic and germinal systems show that the chemical is not a unique germ cell clastogen and that the sensitivity of the aforementioned end points is very similar. Differentiating spermatogonia showed no aberration induction by acrylamide (Adler et al. 1988) so that no doubling dose can be calculated. The highest tested dose is given in parentheses in Table 1. Heritable translocations are induced in late spermatids which also showed the highest dominant lethal response (Shelby et al. 1986). The calculation of a doubling dose of 0.3 mg/kg for heritable translocations is based on a point estimate (Shelby et al. 1987). The experimental design was not a single treatment but a series of five daily doses given to late spermatids totaling 250 mg/kg. For the calculation of the doubling dose it was assumed that the effect of each dose of 50 mg/kg was additive, an assumption which has to be tested in future experiments. Therefore, at present this doubling dose is given in parentheses and needs revision when appropriate dose response data are available. Concluding, it can be said that differentiating spermatogonia are insensitive to acrylamide while bone marrow and meiotic as well as postmeiotic male germ cells are similarly sensitive.

3.2 Bleomycin

Bleomycin was included in the present comparison because it is an S-independent clastogen which breaks DNA in the linker region independently of the state of chromosome condensation (Tien Kuo and Hsu 1978a, b). Unfortunately, only limited data exist for aberration induction and the micronucleus test in mouse bone marrow with a single dose of 80 mg/kg (Kliesch, unpubl.). The doubling doses based on the point estimate of 11 and 13 mg/kg, respectively, are therefore given in parentheses in Table 1. A more extensive study of clastogenic effects with bleomycin was per-

Table 1. Doubling doses (mg/kg) for cytogenetic endpoints

Chemical	Somatic cells CA	MN	Germ Cells CA-SPG	CA-SPC	Progeny HT (SPTD)
Acrylamide	10	23	(>150)	14	(0.3)
Bleomycin	(11)	(13)	6	35	ND
Cisplatin	0.3	0.2	0.5	0.2	ND
Ethylnitrosourea	7	3	7	7	14
Mitomycin C	0.1	0.3	0.01	0.1	0.4
Procarbazine	2	2	50	ND	10

CA, chromosomal aberrations; HT, heritable translocations; MN, micronuclei; ND, no data; SPC, spermatocytes; SPG, spermatogonia; SPTD, spermatids. For references, see text. Figures in parentheses are based on point estimates.

formed in bone marrow cells of Chinese hamsters (Kürten and Obe 1975). Based on the linear dose response of these results, doubling doses of 2 and 0.5 mg/kg were calculated for chromosomal aberrations and micronuclei, respectively. Since these values were obtained with a different species they are not included in the table. The data for induction of chromosomal aberrations in differentiating spermatogonia of mice (Adler 1986) permit the calculation of 6 mg/kg as the doubling dose. Recently, an extensive study was performed on the clastogenic effect of bleomycin during meiotic prophase with the expectation that the S-independent chemical may induce aberrations during all prophase stages. It was found, however, that aberrations were only seen at diakinesis-metaphase I if preleptotene cells were treated (Poorman-Allen et al. 1990). The doubling dose derived from the linear dose response for preleptotene cells was 35 mg/kg (Table 1). Thus, one can conclude that spermatocytes during meiotic prophase are less sensitive to aberration induction by bleomycin than bone marrow cells and differentiating spermatogonia. Data on induction of heritable translocations in male germ cells which could provide a basis for inclusion in this comparison are not available.

3.3 Cisplatin

A series of experiments was performed with the bifunctional cytostatic agent cis-diaminedichloroplatinumII (cisplatin). Dose-response curves and doubling dose calculations for induction of chromosomal aberrations and micronuclei in bone marrow cells as well as chromosomal aberrations in differentiating spermatogonia and spermatocytes were published recently (Adler and El Tarras 1989, 1990). The doubling doses are all in the same order of magnitude and vary between 0.2 and 0.5 mg/kg cisplatin (Table 1). Data on the induction of heritable translocations are not available. It can be concluded with the present set of data that somatic and germinal cells are equally sensitive to the clastogenic effect of cisplatin. Cisplatin induced specific locus mutations in mouse spermatids and a doubling dose of 2 mg/kg has been published (Ehling 1989).

3.4 Ethylnitrosourea

One of the most thoroughly studied model mutagens is the monofunctional alkylating agent ethylnitrosourea (ENU) because it is the most potent mutagen in the mouse (Russell et al. 1979). The results for aberrant cells and for cells with micronuclei in mouse bone can be expressed by the equations $y=1.24+0.18D$ and $y=0.28+0.1D$, respectively, for the tested doses of 10, 30, and 50 mg/kg and the most sensitive sampling intervals of 6 and 18 h, respectively (Kliesch, unpubl.). These dose-response relationships permit the calculation of doubling doses of 7 mg/kg ENU for aberrations and 3 mg/kg for micronuclei in mouse bone marrow cells. A doubling dose of 2 mg/kg was calculated for aberration induction in differentiating spermatogonia (Adler 1986). For spermatocytes during preleptotene the response in aberrant meiotic cells can be expressed by the equation $y=0.6+0.9D$ (for the dose range of 32–125 mg/kg) which results in a doubling dose of 7 mg/kg ENU (Adler, unpubl.). In the heritable translocation assay early spermatids were the most sensitive germ cell stage (Adler et al. 1984). The results for 125 and 250 mg/kg can be expressed by the equation $y=0.001+8.1x10^{-5}D$ which gives a doubling dose of 14 mg/kg (Table 1). Since translocations are rare events, this estimate has large confidence margins. Thus, it emerges that the direct cytogenetic tests are more sensitive for the clastogenic effects of ENU than the tests with progeny. In addition, the doubling dose for specific locus mutations in spermatogonial stem cells was calculated to be 4 mg/kg (Ehling 1989) and is in the same order of magnitude as the ones for the direct cytogenetic assays in somatic and germinal cells.

3.5 Mitomycin C

The bifunctional antibiotic mitomycin C is one of the best-studied clastogens. The effects in bone marrow cells and germ cells are all compatible with a linear dose-response relationship and the resulting doubling doses vary between 0.01 mg/kg in differentiating spermatogonia (Adler 1974), 0.1 and 0.3 mg/kg in the bone marrow tests (Kliesch et al. 1981), and 0.1 mg/kg in spermatocytes during preleptotene (Adler 1976). The doubling dose for heritable translocation induction in spermatids (Adler 1980) was determined as 0.4 mg/kg based on the dose response of $y=3.6 \times 10^{-4}D$ for the doses of 2.5 and 5.0 mg/kg. It can be seen in Table 1 that differentiating spermatogonia are more sensitive to the clastogenic effect of mitomycin C than bone marrow cells or spermatocytes. Spermatids showed in the heritable translocation assay a similar sensitivity as the somatic cells and spermatocytes in the direct cytogenetic tests. It may be added that the doubling dose for specific locus mutations in stem cell spermatogonia is 1 mg/kg (Ehling 1989).

3.6 Procarbazine

The chemostatic drug procarbazine (Natulan) was an effective clastogen in the bone marrow tests (Kliesch et al. 1981) and less effective in differentiating spermatogonia (Adler 1982). This is reflected in the doubling doses of 2 mg/kg calculated for somatic cells and 50 mg/kg for spermatogonia (Table 1). No data exist on the clastogenic effect of procarbazine in meiotic prophase. From heritable translocation test data (Adler et al. 1984) a doubling dose of 10 mg/kg was calculated for spermatids. Procarbazine is an indirect mutagen and requires metabolic activation in the liver. The obviously higher sensitivity of bone marrow cells in comparison to germ cells may be due to better access of the chemical or its metabolite. For comparison it should be mentioned that specific locus mutations induced by procarbazine in stem cell spermatogonia resulted in a doubling dose of 110 mg/kg (Ehling 1989).

4 Conclusions

The task of a quantitative comparison of clastogenic effects in somatic and germinal cells described above has to be regarded as preliminary since not all values for doubling doses could be based on linear dose-response relationships but on point estimates and this uncertainty is indicated by values in parentheses in Table 1. The numbers given for doubling doses are estimates which, even if calculated from a linear dose-response relationship, have often very large confidence margins, therefore, for the present comparison the order of magnitude is important. Furthermore, one has to keep in mind that the resolving power of an assay depends largely on the experimental design which includes sample size, determination of the most sensitive stage of the cell cycle or stage of germ cell development, and the background frequency of the event under study. Even though the data used for the present calculations were all obtained in the same laboratory, they may not be as comparable as they appear at first glance. This criticism accounts for some of the obvious inconsistencies. Acrylamide was completely ineffective in differentiating spermatogonia, while it was clastogenic in all other systems. However, the data for spermatogonia were only gathered at one interval between treatment and sampling (24 h). It may well be that samplings at earlier or later times could reveal a positive response in this test.

With all possible criticism in mind, however, the laborious exercise seems warranted and will be continued. The large literature surveys during the past 5 years have clearly demonstrated that more quantitative approaches are necessary and simple plus-minus answers are no longer satisfying. In comparing test results on a quantitative basis one becomes aware of reasons for divergences and can study mechanisms in more detail to understand the action of the chemical under test.

The experiments required to determine doubling doses and to ascertain the effects in germ cells are more laborious and expensive than the base set of chemical testing for genotoxic effects. But they will be necessary in a validation process for chemicals already on the market or in our environment and for new chemicals which exceed a certain production margin. Acrylamide is possibly the best example how diverse the conclusions could be if only limited information on a small number of tests, and these partly incomplete, are considered. With the data of Shiraishi (1978) and Shelby et al. (1986, 1987) one could have concluded that the compound is a unique germ cell clastogen. On the other hand, given the data by Adler et al. (1988) as the only information, the reverse conclusion would have been drawn. Therefore, no individual germ cell test by itself will be sufficient to determine a germ cell mutagen.

The present comparison indicates that for four of the six chemicals the cytogenetic study of differentiating spermatogonia would have given an appropriate answer and allowed the conclusion that somatic and germinal cells were equally sensitive. The other two chemicals would have been declared less effective or ineffective in germ cells. The obvious question for the reason, aside from incomplete experimental data, would lead to one major factor, namely, the differential spermatogenic response which means that not all developmental germ cell stages respond in the same way to chemical mutagens. This phenomenon was first described for results of the dominant lethal assay (Ehling et al. 1972) and is also reflected in the doubling doses for specific locus mutations (Ehling 1989). The blood-testis barrier can prevent access of a chemical to spermatids and spermatozoa (Setchell and Main 1978). The biological processes of differentiation, DNA and RNA synthesis, replacement of chromosomal proteins, and repair capacities are characteristically different for the various germ cell stages. They contribute to the differential spermatogenic response as much as the primary action of the chemical mutagen itself. Therefore, by testing one single germ cell stage, such as differentiating spermatogonia, one cannot expect to obtain a general answer. The most feasable approach to identifying a germ cell mutagen is to perform a dominant lethal assay of postmeiotic and meiotic stages plus a cytogenetic assay of differentiating spermatogonia. This combination would cover the whole process of spermatogenesis. With the exception of ENU the doubling doses for specific locus mutations are one order of magnitude higher than those for the cytogenetic tests. The predictability of specific locus mutation yields from data of cytogenetic end points has to be determined by further quantitative studies.

References

Adler I-D (1974) Comparative cytogenetic study after treatment of mouse spermatogonia with mitomycin C. Mutat Res 23: 369–379

Adler I-D (1976) Aberration induction by mitomycin C in early primary spermatocytes of mice. Mutat Res 35: 247–256

Adler I-D (1980) New approaches to mutagenicity studies in animals for carcinogenic and mutagenic agents. I. Modification of the heritable translocation test. Teratogen Carcinogen Mutagen 1: 75–86

Adler I-D (1982) Male germ cell cytogenetics. In: Hsu TC (ed) Cytogenetic assay of environmental mutagens. Allanheld, Osmun & Co, New York, pp 249–276

Adler I-D (1984) Cytogenetic test in mammals. In: Venitt S, Parry JM (eds) Mutagenicity testing: a practical approach. IRL, Oxford, pp 275–306

Adler I-D (1986) Clastogenic potential in mouse spermatogonia of chemical mutagens related to their cell cycle specificities. In: Ramel C, Lambert B, Magnusson J (eds) Genetic toxicology of environmental chemicals, Part B. Genetic effects and applied mutagenesis. Liss, New York, pp 477–484

Adler I-D, Ashby J (1989) The present lack of evidence for unique rodent germ-cell mutagens. Mutat Res 212: 55–66

Adler I-D, Brewen JG (1982) Effects of chemicals on chromosome aberration production in male and female germ cells. In: Hollaender A, Serres FJ de (eds) Chemical mutagens. Principles and methods for their detection, vol. 7. Plenum, New York, pp 1–35

Adler I-D, El Tarras A (1989a) Clastogenic effects of cis-diamminedichloroplatinum. I. Induction of chromosomal aberrations in somatic and germinal cells of mice. Mutat Res 211: 131–137

Adler I-D, El Tarras A (1990) Clastogenic effects of cis-diamminedichloroplatinum. II. Induction of chromosomal aberrations in spermatocytes and spermatogonial stem cells. Mutat Res 243: 173–178

Adler I-D, Erbelding C, Lutzenberger B (1984) Procarbazine- and ethylnitrosourea (ENU)-induced heritable translocations in the mouse. Mutat Res 130: 175–176

Adler I-D, Ingwersen I, Kliesch U, El Tarras A (1988) Clastogenic effects of acrylamide in mouse bone marrow cells. Mutat Res 206: 379–385

Bender MA, Griggs HG, Bedford JS (1974) Mechanisms of chromosomal aberration production. III. Chemicals and ionizing radiation. Mutat Res 23: 197–212

Brewen JG, Preston RJ (1973) Chromosome aberrations as a measure of mutagenesis: comparisons in vivo and in vitro and in somatic and germ cells. Environ Health Perspect 6: 157–166

Brusick D (1980) Principles of genetic toxicology. Plenum, New York, p 130

Ehling UH (1980) Induction of gene mutations in germ cells of the mouse. Arch Toxicol 46: 123–138

Ehling UH (1984) Methods to estimate the genetic risk. In: Obe G (ed) Mutations in man. Springer, Berlin Heidelberg New York, pp 292–318

Ehling UH (1989) Quantifizierung des chemogenetischen Risikos. Naturwissenschaften 76: 194–199

Ehling UH, Neuhäuser A (1979) Procarbazine-induced specific locus mutations in male mice. Mutat Res 59: 245–256

Ehling UH, Doherty DG, Malling H (1972) Differential spermatogenic response of mice to the induction of dominant lethal mutations by n-propy methanesulfonate and isopropylmethanesulfonate. Mutat Res 15: 175–184

Ehling UH, Machemer L, Buselmaier W et al. (1978) Standard protocol for the dominant lethal test on male mice. Arch Toxicol 39: 173–185

Generoso WM, Russell WL, Gosselee DG (1973) A sequential procedure for the detection of translocation heterozygotes in male mice. Mutat Res 21: 220–221

Generoso WM, Cain KT, Cacheiro NLA, Cornett CV, Gosslee DG (1984) Response of mouse spermatogonial stem cells to X-ray induction of heritable reciprocal translocations. Mutat Res 126: 177–187

Holden HE (1982) Comparison of somatic and germ cell models for cytogenetic screening. J. Appl Toxicol 2: 196–200

Kliesch U, Danford N, Adler I-D (1981) Micronucleus test and bone-marrow chromosome analysis. A comparison of 2 methods in vivo for evaluating chemically induced chromosomal alterations. Mutat Res 80: 321–332

Kliesch U, El Tarras A, Ingwersen I, Adler I-D (1989) Clastogenicity of acrylamide in somatic and germinal cells of mice. Mutagenesis 4: 319–320

Kürten S, Obe G (1975) Premature chromosome condensation in the bone marrow of Chinese hamsters after application of bleomycin in vivo. Mutat Res 27: 285–294

Léonard A (1976) Heritable chromosome aberrations in mammals after exposure to chemicals. Rad Environ Biophys 13: 1–8

Léonard A, Deknudt GH, Linden G (1971) Failure to detect meiotic rearrangement in male mice given chemical mutagens. Mutat Res 13: 89–92

Lüning KG, Searle AG (1971) Estimates of the genetic risks from ionizing radiation. Mutat Res 12: 291–304

MAK (1989) Maximale Arbeitsplatzkonzentrationen und biologische Arbeitsstoffwerte. Mitteilungen der Senatskommission zur Prüfung gesundheitsschädlicher Arbeitsstoffe. VCH, Weinheim, pp. 14–15

Monesi V (1962) Autoradiographic study of DNA synthesis and the cell cycle in spermatogonia and spermatocytes of mouse testis using tritiated thymidine. J Cell Biol 14: 1–18

Monesi V (1971) Chromosome activities during meiosis and spermiogenesis. J Reprod Fert, Suppl 13: 1–14

Oakberg GF (1960) Irradiation damage to animals and its effect on their reproductive capacity. J Dairy Sci 43: 54–64

Poorman-Allen P, Backer LC, Adler I-D, Westbrook-Collins B, Moses MJ, Allen JW (1990) Bleomycin effects on mouse meiotic chromosomes. Mutagenesis (submitted)

Preston RJ, Au W, Bender MA et al. (1981) Mammalian in vivo and in vitro cytogenetic assays: a report of the US EPA's Gene-Tox Program. Mutat Res 87: 143–188

Russell WL (1951) X-ray induced mutations in mice. Cold Spring Harbour Symp Quant Biol. 16: 327–336

Russell WL, Kelly EM, Hunsicker PR, Bangham SC, Maddux SC, Phipps EL (1979) Specific-locus test shows ethylnitrosourea to be the most potent mutagen in the mouse. Proc Natl Acad Sci USA 76: 5818–5819

Schleiermacher E, Schmidt W (1973) The local control of mammalian spermatogenesis. Humangenetik 19: 75–98

Searle AG, Beechey CV, Evans EP, Ford CE, Papworth DG (1971) Studies on the induction of translocations in mouse spermatogonia. IV. Effects of acute gamma-irradiation. Mutat Res 12: 411–416

Setchell BP, Main SJ (1978) Drugs and the blood-testis barrier. Environ Health Perspect 24: 61–64

Shelby MD, Cain KT, Hughes LA, Braden PW, Generoso WM (1986) Dominant lethal effects of acrylamide in male mice. Mutat Res 173: 35–40

Shelby MD, Cain KT, Cornett CV, Generoso WM (1987) Acrylamide: induction of heritable translocations in male mice. Environ Mutagen 9: 362–368

Shiraishi Y (1978) Chromosome aberrations induced by monomeric acrylamide in bone marrow and germ cells of mice. Mutat Res 57: 313–324

Tien Kuo M, Hsu TC (1978a) bleomycin causes release of nucleosomes from chromatin and chromosomes. Nature 271: 83–84

Tien Kuo M, Hsu TC (1978b) Biochemical and cytological studies of bleomycin actions on chromatin and chromosomes. Chromosoma (Berl) 68: 229–240

Use of in Vivo Micronucleus Tests with Mammalian Cells for Clastogenicity and Carcinogenicity Studies

A. D. TATES[1], M. L. M. VAN DE POLL[2], M. VAN WELIE[1], and S. J. PLOEM[3]

Contents

1 Introduction

Almost two decades have elapsed since the introduction of the bone marrow micronucleus test (BMN test) by Boller and Schmid (1970). This test was primarily developed to simplify the rather difficult and time-consuming conventional procedure of analyzing structural and numerical chromosomal aberrations in metaphases. An added advantage of the technique was its usefulness for detection of cytogenetic damage in any species of mammal regardless of their karyotype. In the course of the past 19 years, much of the research efforts have been directed towards development, validation, standardization, and even automatization of procedures for evaluating clastogenic potential of chemical substances in polychromatic erythrocytes (PCEs) and normochromatic erythrocytes (NCEs) in bone marrow and/or peripheral blood of mice and other species of mammals including man (Schmid 1976; Jenssen and Ramel 1980; Heddle 1973; Heddle et al. 1983; OECD 1983; Ashby and Mohammed 1986; MacGregor et al. 1987; Collaborative Study for the Micronucleus Test 1986, 1988, 1989; Workshops 1988a, b; Wild 1988; Romagna and Staniforth 1989).

With the advent of an extensive data base on chemically induced micronuclei in cells from bone marrow and blood and increasing knowledge of organotrophic action of chemicals, it was realized that it would be worthwhile – and in fact mandatory – to study the induction of micronuclei in other types of cells. This has resulted in the development of micronucleus tests for male germ cells (Lähdetie and Parvinen 1981; Tates et al. 1983; Tates and de Boer 1984), hepatocytes (Tates et al. 1980; Cliet et al. 1989a, b), lymphocytes (Fenech and Morley 1985a, b), spleen cells (Shindo et al.

[1] Department of Radiation Genetics and Chemical Mutagenesis, State University of Leiden, Leiden, The Netherlands
[2] Division of Toxicology, Center for Bio-Pharmaceutical Sciences, State University of Leiden, Leiden, The Netherlands
[3] Department of Cytochemistry and Cytometry, State University of Leiden, Leiden, The Netherlands

1983), colon cells (Heddle et al. 1981; Heddle 1988/89), fetal liver cells (Cole et al. 1981), and exfoliated cells that can be readily obtained from numerous tissues, including buccal mucosa (scrapings of oral cells), bronchi (sputum), urinary bladder and ureter (centrifugation of urine), nasopharynx (sponge), esophagus (balloon or brush procedure), cervix (smears), etc. (Munoz et al. 1987; Stich and Rosin 1982; Stich and Anders 1989).

In the context of this chapter it is impossible to adequately review the enormous data base. Instead, we have selected some topics in the field of micronucleus research that are directly related to the work done in our laboratories.

2 Spermatid Micronucleus Tests (SMN) for Evaluating Clastogenicity of Chemicals/Radiation in Meiotic Cells of Mammals

In situations where human exposure to drugs, food additives, and industrial chemicals is likely to be high and where transmission of chromosomal damage to offspring can be foreseen, it is desirable to accumulate knowledge on the clastogenicity of such chemicals in mammalian germ cells. In addition, for purely scientific reasons, it is important to explore why some stages of gametogenesis are more sensitive for clastogens than others. From the literature it appears that chromosomal damage in germ cells can be measured more easily in males than in females.

Most test systems for measuring clastogenic effects in male germ cells are rather complex, time-consuming, and expensive. To overcome this problem Lähdetie and Parvinen (1981), Lähdetie (1988), Tates et al. (1983, 1989a) and Tates and de Boer (1984) developed micronucleus tests for quickly evaluating clastogenicity of chemical and physical agents in type-B spermatogonia and spermatocytes. With the SMN test developed by Lähdetie and Parvinen (dissection method; 1981), chromosomal damage induced in these stages of spermatogenesis in rats is detected as micronuclei in derived, early (= round) spermatids.

An essential step in this test involves the use of transillumination phasecontrast microscopy for selecting and dissecting specific regions of seminiferous tubules containing an association of cells with a relatively high proportion of the very early Golgi-phase spermatids. These are the cells in which micronuclei are to be detected.

Although this technique is very elegant for studies with rats, it has the drawback that it cannot be used for several other species of mammals, including the mouse. In contrast to the SMN test designed in our own laboratory (suspension method), early Golgi-phase spermatids are selected microscopically in slides carrying cells that have been isolated enzymatically from whole testes. An advantage of this test is that it can be used for any species of mammal (tests have been performed with germ cells from rats, mice, field voles, several species of monkeys, and man). Specific merits of the dissection and suspension methods have recently been compared by Lähdetie (1988), who concluded that both techniques provide a simple and rapid approach to the evaluation of clastogenic action of chemicals in male germ cells.

In order to optimize SMN tests for rapid identification of clastogenic chemicals, it was important to investigate whether one particular stage of spermatocyte development would generally be much more sensitive for induction of chromosomal damage than any other stage of spermatocyte differentiation. If this would indeed be the case, then clastogens could simply be picked up by analyzing micronuclei in spermatids sampled at a well-defined time interval after treatment with the chemical under investigation. From validation studies of Lähdetie (1983a, b; 1988) with Sprague-Dawley rats exposed to adriamycin (ADM), cyclophosphamide (CP), and 9,10-dimethyl-1,2-benzanthracene (DMBA) it could be concluded that the highest micronucleus frequencies are found in cells sampled approximately 20 days after treatment. From the literature on the rate of spermatogenesis in Sprague-Dawley rats it is known that such cells correspond with premeiotic G1-S phase which takes place during preleptotene and early leptotene stages of meiotic prophase.

In an experiment with ethylmethanesulfonate (EMS) Lähdetie (1983b) found that not preleptotene but diplotene-diakinesis (sampled 1 day after treatment) was the most sensitive stage for induction of chromosomal damage. Although this observation alone will be correct, we are of the opinion that it does not reflect the real sensitivity pattern. Because EMS was only tested at the high dose of 250 mg/kg, we believe that this dose resulted in partial selective elimination (or inhibition of further differentiation) of preleptotene spermatocytes (a stage known to be very sensitive to the toxic action of chemicals) so that the sensitivity of this stage could not be measured accurately.

In our experiments with Wistar rats (Tates et al. 1985a, b; 1986a) we determined the pattern of differential sensitivity of meiotic stages for ethylnitrosourea (ENU), mitomycin C (MTC), methylmethanesulfonate (MMS), and cyclophosphamide. In agreement with the forementioned observations of Lähdetie, we also interpreted our data (ENU, MMS, and MTC data are shown in Figs. 1–3) as indicating that spermatocytes in premeiotic G1 and S-phase are the most sensitive cells for induction of chromosomal damage. From the ENU data (Fig. 1) one may get the impression that cells corresponding with spermatocytes in diplotene-diakinesis are more sensitive than those in premeiotic G1 and S-phase. This is definitely not the case when stage sensitivities are compared at the dose of 35 mg/kg. Stage comparisons at higher doses are in fact impossible because these result in killing/inhibition of differentiation of G1 and S-phase cells.

We have no satisfactory explanation for the unexpected finding of high frequencies of micronuclei in cells sampled 1–3 days after treatment. More experiments are needed to clarify this observation. In particular, it would be of interest to explore whether similar results will be obtained for MMS when doses are tested that are so high that they result in killing of G1 and S-phase spermatocytes. Sufficient experimental data are not available to draw the conclusion that G1 and S-phase spermatocytes are also the most sensitive sparmatocyte stage for S-phase independent chemicals such as ADM and bleomycin (BLM). From our own studies with northern field voles exposed to X-rays or 1 MeV neutrons (Tates et al. 1989a) we know that spermatocytes in diakinesis, MI, and MII are the most sensitive stages, whereas spermatocytes in G1 and S-phase are somewhat less sensitive. It might well be true for other S-independent chemicals. Even then it is to be expected that a rather strong clastogenic effect of such chemicals will be found in G1 and S-phase spermatocytes. The general conclusion is that clastogenicity of chemicals in male germ cells can be tested reliably and most effectively by screening for micronuclei in cells that are in G1 and S-phase of spermatocyte differentiation at the time of treatment (sampling time of about 20 days after treatment).

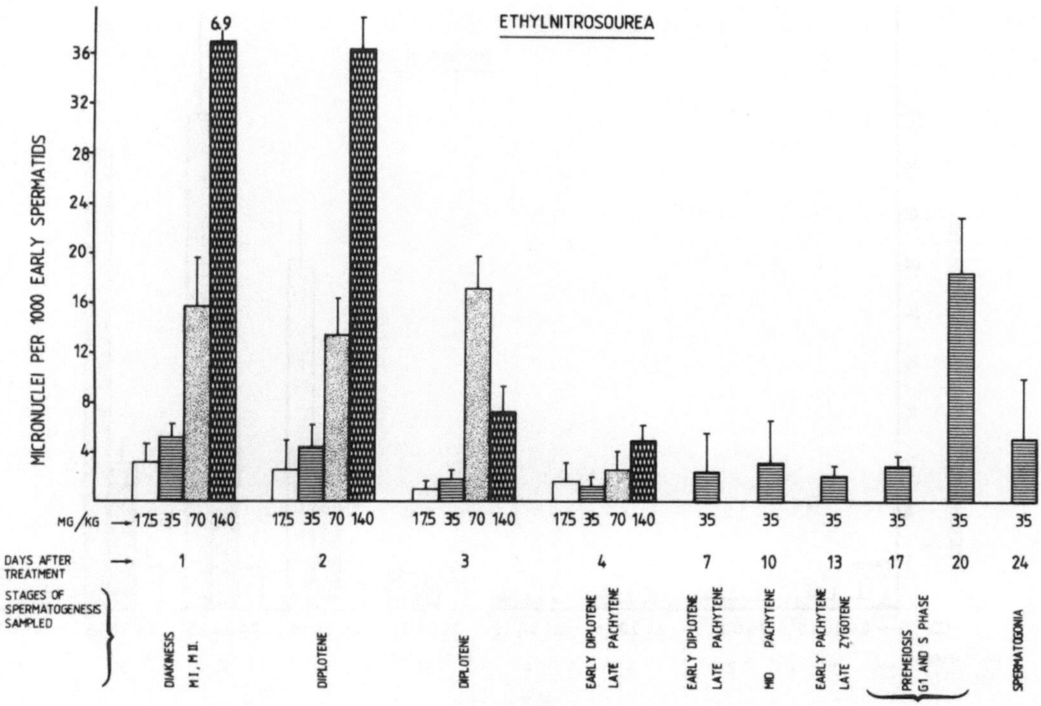

Fig. 1. Frequencies of micronuclei in Golgi-phase early spermatids sampled at different times after exposure to one or more doses of ethylnitrosourea (ENU). When spermatocyte stages were exposed to 35 mg/kg, G1- and S-phase spermatocytes were clearly the most sensitive spermatocyte stage for clastogenic effects of ENU. *Bars* indicate SD

For chemicals known or suspected to interfere with spindle function and/or formation it may be safer to include in the protocol an extra sampling time for cells that are in diakinesis/diplotene at the time of treatment (1 or 2 days after treatment). In view of the fact that G1 and S-phase spermatocytes are very sensitive to the toxic action of clastogens it is essential that micronucleus induction is studied at different dose levels which are not cytotoxic.

Recent results obtained by Régnier and Cordier in France (1989 pers. comm.) indicate that the forementioned conclusions are also valid for clastogenicity studies with mice albeit that other sampling times must be selected. These investigators modified our suspension method using different cell isolation and staining procedures. They proposed sampling times of 2, 9, and 16 days after a two-step (24-h interval) administration of the chemical under investigation. These sampling times correspond respectively with cells treated in diplotene/diakinesis, midpachytene, and preleptotene. The design of this protocol is based on testes with the following chemicals: ADM, BLM, diethylstilbestrol (DES), vinblastin (VLB), vincristin (VCR), MTC, cisplatin (CDDP), thio-TEPA (TTP), CP, busulfan (BSF), procarbazin (PCZ), ENU, 1,3-bis (2-chloroethyl)-1-nitrosourea (BCNU), and MMS. All chemicals were found to induce statistically significant increases in the occurrence of micronucleated sper-

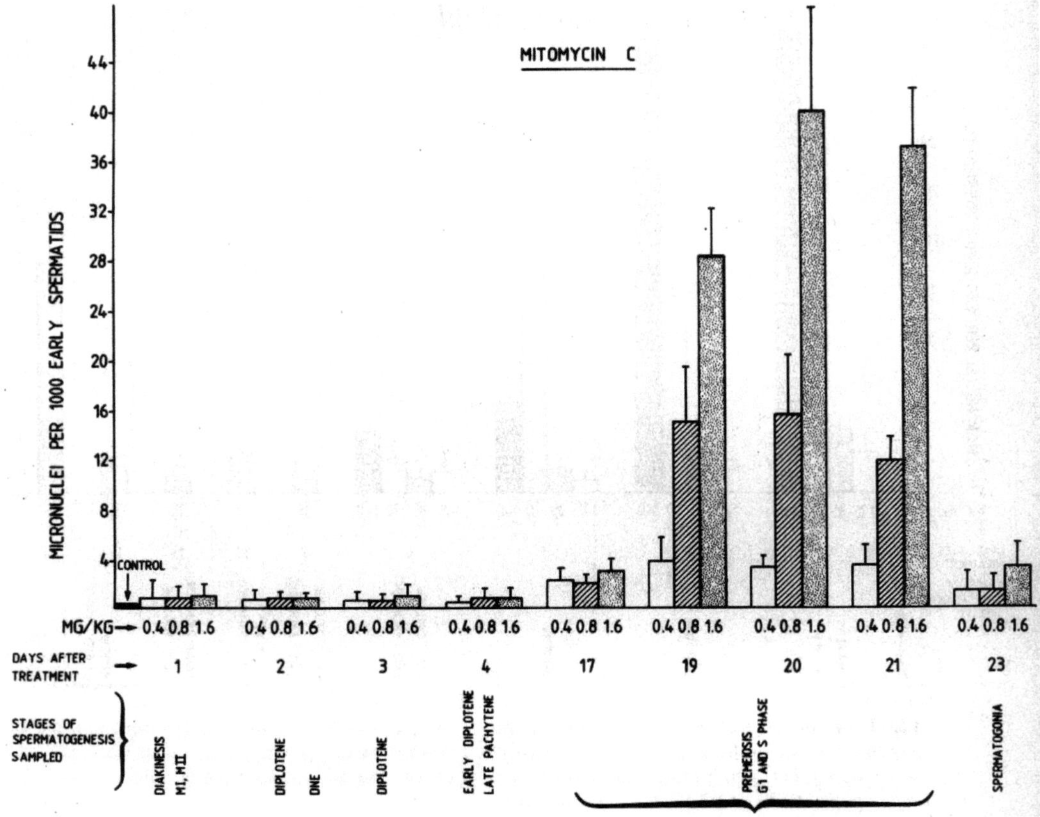

Fig. 2. Frequencies of micronuclei in early spermatids of rats exposed to different doses of mitomycin C. Note the high sensitivity of spermatocytes in G1- and S-phase. *Bars* indicate SD

matids upon treatment at one or more of the principal stages of meiosis (diplotene/diakinesis, midpachytene or preleptotene) (Table 1).

As has already been mentioned above, the suspension type of SMN test can very well be used to compare sensitivities of specific spermatogenic stages in different species of mammals. Due to the lack of sufficient data, such comparisons can hardly be made at present. Nevertheless, a comparison of our data for ENU and MMS in the rat, together with unpublished data of Régnier and Cordier for the same chemicals in the mouse, seem to indicate that G1 and S-phase spermatocytes of rats may at least be one order of magnitude more sensitive for induction of chromosomal damage than the corresponding cells in mice. This difference also seems to hold for diakinesis/diplotene cells.

It has often been argued that clastogenicity studies with bone marrow would have sufficient predictive value for effects to be expected in germ cells. Recently, however, data have been reported indicating that this statement certainly has no general validity.

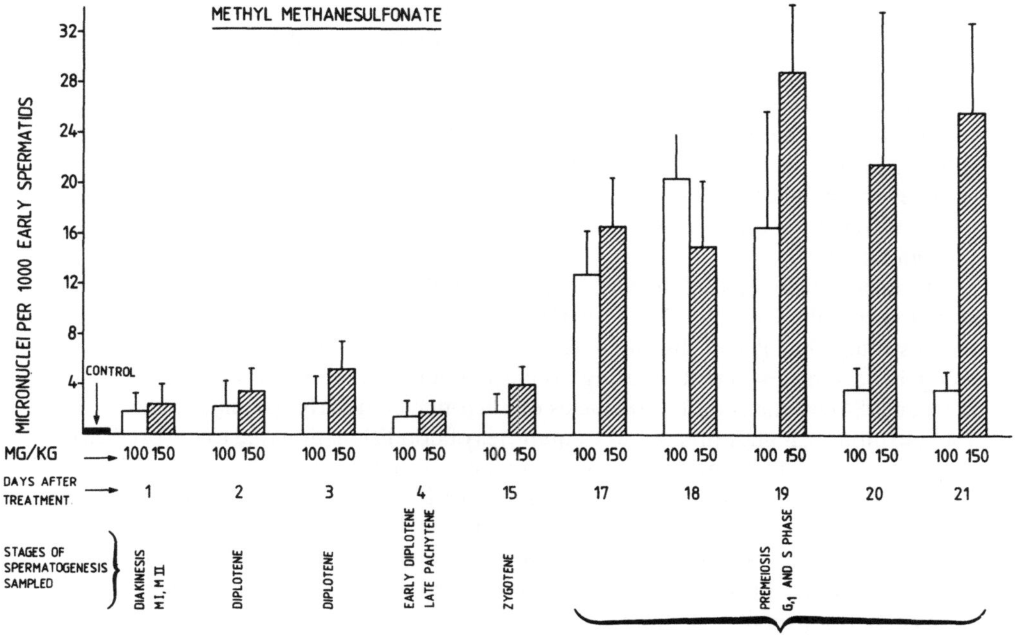

Fig. 3. Frequencies of micronucleated early spermatids at different time intervals after exposure of spermatogonia and differentiating spermatocytes to two doses of methylmethanesulfonate. G1- and S-phase spermatocytes were more sensitive than any other stage of spermatocyte development. *Bars* indicate SD

Table 1. Summary of results of the micronucleus test in spermatids and clastogenic potential (CP) of the positive compounds (J. F. Régnier and A. Cordier unpubl. data)

Days posttreatment Stage of meiosis diakinesis	3 Diplotene Result[a] CP[b]	9 Midpachytene Result CP	16 Preleptotene Result CP
TTP[c]	+ 0.0078	+ 0.0059	+ 0.0850
BSF	+ 0.0006	–	+ 0.0403
ENU	–	–	+ 0.0069
CDDP	–	+0.0330	+ 0.0504
BLM	+ 0.0130	–	–
DES	+ 0.0004	–	–
MMS	–	–	+ 0.0033
MTC	–	–	+ 2.0158
VCR	+ 6.6000	+ 2.0625	+ 2.4255
VLB	+ 6.9265	+ 0.5635	i[d]

[a] – Negative response; + positive response; i inconclusive response.
[b] CP expressed in micronucleated spermatids mmol kg^{-1} day^{-1}.
[c] Single treatment only, sacrifice 2, 9, or 16 days posttreatment.
[d] Very toxic at dose levels higher than 0.5 mg kg^{-1} day^{-1}.

Cliet et al. (1989b) compared the sensitivity of bone marrow, testis, and liver of mice for the following six carcinogens: 4-aminophenol (4-APOL), 4-aminobiphenyl (4-ABPYL), dimethylnitrosamine (DMN), diethylnitrosamine (DEN), 1-1-dimenthylhydrazin (1-1-DMH), and β-propiolactone (BPL). Only 4-APOL and 4-ABPYL were positive in the bone marrow micronucleus test (BMN), whereas all six carcinogens were clastogenic in testis and liver.

These data indicate clearly that one cannot simply rely on clastogenicity data for bone marrow cells to make predictions about effects in cells from other organs like testis and liver.

It can be concluded that the SMN test is a useful and rapid tool for demonstrating clastogenicity of chemicals. The SMN test would even be of greater use if it could be demonstrated that it can discriminate between chromosome breaking agents and aneugenic chemicals (agents causing nondisjunction). Several years ago (Tates and de Boer 1984) we measured frequencies of micronuclei in early spermatids of mouse stocks exhibiting high spontaneous frequencies of nondisjunction due to the presence of specific types of chromosome rearrangements. From these studies we concluded that the SMN test does not seem to be sensitive for demonstrating nondisjunction resulting from chromosome exchange formation. When we studied micronucleus formation in rats exposed to different doses of the spindle poison VCR (vincristine sulfate) we found little or no effect in cells treated at pachytene, diplotene, and diakinesis. Frequencies of micronuclei in bone marrow from the same rats were significantly enhanced and the size of the micronuclei was much larger than for typical clastogens. Similar findings have been reported by Lähdetie and Parvinen (1982). Only a few micronuclei in early spermatides were found. It seems that even low doses (0.2 mg/kg) completely inhibit the function of the spindle. These negative findings in the rat could not be confirmed in mice because Régnier and Cordier (1989a, b) observed a positive effect of vincristin and vinblastin in cells sampled 1–3 days after treatment.

Another approach to test the detection capacity of the SMN test for aneugens is to use CREST anti-kinetochore antibodies and immunoflueorescence to establish the presence of kinetochores in micronuclei from early spermatids (Degrassi and Tanzarella 1988; Hennig et al. 1988; Thomson and Perry 1988; Fenech and Morley 1989).

3 The Potential of the Micronucleus Test with Hepatocytes

Despite the general acceptance of the BMN test as an important tool for in vivo clastogenicity studies, several well-known liver carcinogens, such as DEN and DMN (see above), registered negative in this test although such compounds were clearly clastogenic in in vitro clastogenicity tests. This finding prompted the development of an in vivo micronucleus test with hepatocytes (LMN test), by means of which, the above-mentioned false negatives in the BMN test reacted positive provided the proper experimental design was used (Tates et al. 1980, 1989b).

An important feature of any type of micronucleus test is that micronuclei can only be detected in treated cells that have passed one or two mitoses depending on whether these cells were hit, respectively in, G1 and S-phase or in the G2 phase of the first cell cycle. Under normal circumstances mitotic activity in adult liver is very low. Therefore, proliferative activity must be artificially induced by either partial hepatectomy (PH) or treatment with a dose of 4-acetylaminofluorene (4-AAF; Braithwaite and Ashby 1988).

When using the LMN test, animals can be treated with the test chemical before or after the proliferation stimulus (Fig. 4). Treatment before PH has two advantages. Firstly, before PH the liver has a normal metabolic activation and detoxification capacity so that indirectly acting genotoxins can be converted into proximate or ultimate genotoxins/carcinogens. Secondly, by varying the time interval between treatment and PH (or 4-AAF), one can study persistence of preclastogenic damage in hepatocytes and thus the influence of intracellular repair processes on the realization of clastogenic damage.

When, in addtion, such experiments are combined with biochemical studies aiming at measurement of formation and persistence of specific DNA adducts, it becomes possible to obtain important information on the role of specific adducts in the formation of micronuclei. Following this procedure we were able to demonstrate that alkylating agents giving rise to substantial O-alkylation (e.g., DEN and ENU) induced long-lived preclastogenic lesions in rat liver cells (Tates et al. 1986b). We also found that damage induced by the ethylating agents DEN and ENU was more persistent than that caused by the methylating agent DMN (Fig. 5).

In the above-mentioned studies preclastogenic lesions were formed before PH and such lesions can only result in micronucleus formation when they persist until the peak of DNA synthesis which occurs 17 h after PH in Wistar rats. During DNA synthesis, preclastogenic lesions of most chemicals (= S-phase dependent chemicals) will be converted into chromosomal breakage events ultimately giving rise to micronuclei.

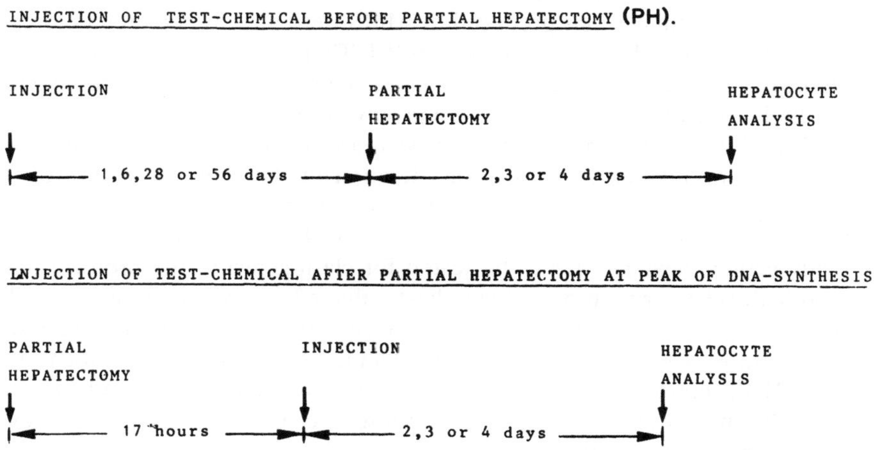

Fig. 4. Experimental design for cytogenetic studies with rat liver. Test chemicals were administered at different times before partial hepatectomy (above) or at the peak of DNA synthesis at 17 h after partial hepatectomy (below). Hepatocytes were sampled at different times after partial hepatectomy using a liver perfusion technique. Freshly isolated hepatocytes were fixed and stained and then analyzed for the presence of micronuclei in the cytoplasm

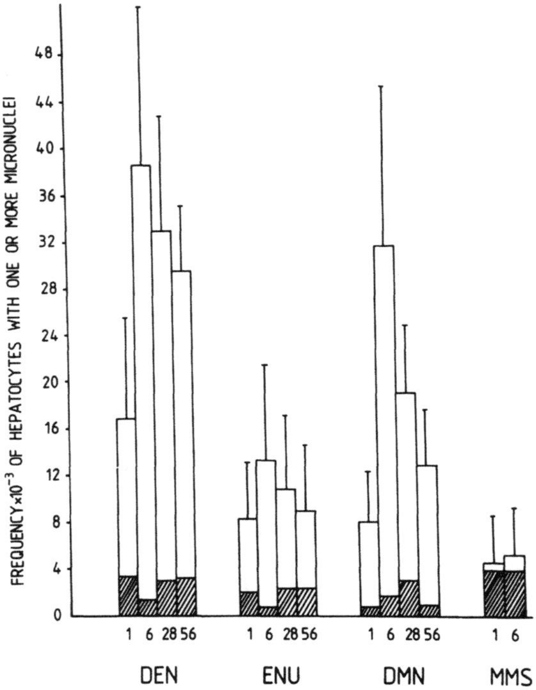

Fig. 5. Pooled data of experiments with *DEN*, *ENU*, *DMN* and *MMS*. Animals were exposed 1, 6, 28, or 56 days before partial hepatectomy (for MMS only at 1 and 6 days). Frequencies of micronuclei represent mean values for hepatocytes sampled 2–4 days after partial hepatectomy. *Dashed columns* relate to frequencies of micronuclei observed in untreated animals. *Bars* indicate SD (Tates et al. 1986b)

In view of the above considerations it is possible that a well-known clastogen such as MMS gives a negative response in the LMN test when administered shortly before PH (Tates and den Engelse 1989b). A simple explanation for such a result is that preclastogenic lesions produced by MMS were already repaired before treated hepatocytes entered DNA synthesis after PH. That this interpretation is correct could be proven in an experiment where the same chemical was given at the time of DNA synthesis after PH. In the latter case a positive response was found (Fig. 6).

Based on the findings briefly reviewed above, we propose that the LMN test must only be applied after due consideration of the chemical nature of the test chemical. In case of directly action chemicals the optimal procedure seems to be the administration of the chemical at the peak of DNA synthesis after PH. When indirectly acting chemicals are to be tested for clastogenicity, the protocol has to be modified. When prior information indicates that adducts will be short-lived, then the treatment before PH may give a negative result because adducts are repaired before DNA synthesis after PH. If, on the other hand, the chemical needs cytochrome P450 (Cyt P450) directed metabolic activation, then the administration of this chemical after PH may result in a negative response because Cyt P450 enzyme levels are severely reduced after PH.

However, also indirectly acting genotoxins which do not need Cyt P450 for their metabolic activation (e.g., N-OH-2-acetylaminofluorene = N-OH-AAF) can be evaluated after PH. Recently, we investigated the clastogenicity of N-OH-AAF and the role of its DNA adducts in the induction of micronuclei (van de Poll et al. 1989). In

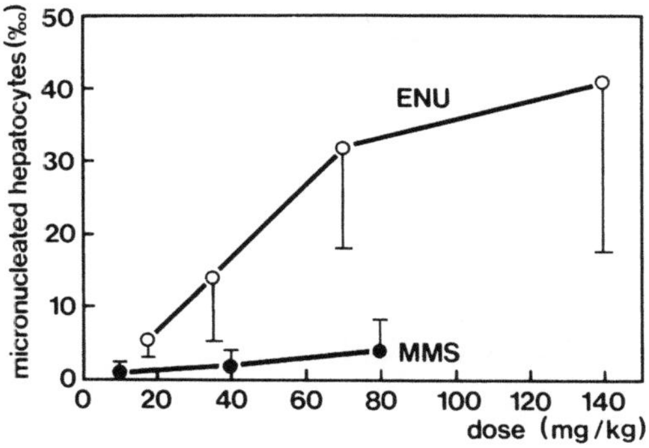

Fig. 6. Mean frequency of micronucleated hepatocytes (average obtained 1, 2, 3, or 4 days after injection) in partially hepatectomized rats as a function of *ENU* and *MMS* dose. The chemicals were administered 17 h after partial hepatectomy. *Bars* indicate SD (Tates and den Engelse 1989b)

this study N-OH-AAF was administered by i.p. injection to male Wistar rats 17 h after PH. Hepatocytes were then analyzed for the presence of micronuclei 7 h, 1, 2, 3 and 4 days after injection (Fig. 7). This treatment resulted in a dose-dependent increase of frequencies of micronucleated hepatocytes at days 3 and 4 (Fig. 8). Pretreatment of another group of rats with the sulfotransferase inhibitor pentachlorophenol (PCP) 45 min before injection of N-OH-AAF almost completely prevented the formation of micronuclei by N-OH-AAF. Parallel biochemical studies with tritium-labeled N-OH-AAF, administered 17 h after PH, showed that inhibition of sulfation of N-OH-AAF by PCP pretreatment prevented the formation of the N-acetylated DNA adducts N-deoxyguanosin-8-yl-AAF and 3-deoxyguanosin-N^2-yl-AAF by ~85%. Total adduct formation of DNA, however, was not lowered because of an increase in the formation of the deacetylated adduct, N-deoxyguanosin-8-yl-AF. On the basis of these cytogenetic and biochemical data, we conclude that micronucleus induction by N-OH-AAF is due to the formation of N-acetylated DNA adducts and not to the formation of deacetylated DNA adducts.

Recently, we accumulated evidence that the clastogenicity of three liver carcinogens, as indicated by results of LMN tests, is correlated with promotion activity but not with initiation efficacy (van de Poll et al. 1990). This idea is based on results from experimental studies with the structurally related liver carcinogens N-OH-AAF, N-OH-FAABP (N-OH-4-fluoro-4-acetylaminobiphenyl), and N-OH-AABP (4-acetylaminobiphenyl). These three compounds were compared for clastogenicity, initiation efficacy, and promotion activity in rat liver.

In the LMN test the chemicals were administered during the peak of DNA synthesis after PH and micronuclei were analyzed 2, 3, and 4 days after treatment (Fig. 7). The results shown in Fig. 8 indicate that N-OH-AAF was definitely a stronger clastogen than N-OH-FAABP, whereas very few micronuclei were induced by N-OH-AABP. Initiation capacity of the three chemicals was tested in a modified Solt and Farber protocol (Fig. 7). The compounds where administered 24 h after PH and the number of gamma-glutamyltranspeptidase positive (GGT+) foci that developed after promotion (aminofluorene-HCl in the drinking water in combination with a necrogenic dose of CCl4) was used as an index for initiating efficacy. At the dose tested the three carcinogens were equally effective (Table 2). However, when tested for promotion activity, after initiation with diethylnitrosamine, only N-OH-AAF was a very effective promotor (Fig. 7; Table 3). This was evidenced by a high number of

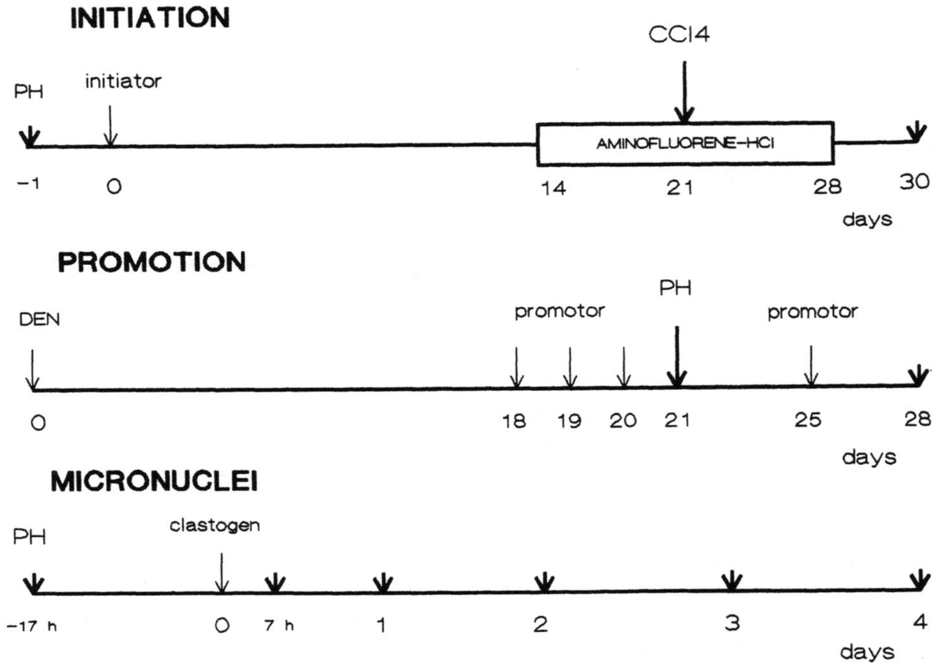

Fig. 7. Experimental design of studies on initiation/promotion capacity and clastogenicity of N-OH-AAF, N-OH-FAABP, and N-OH-AABP (see Fig. 8, Tables 2 and 3 and text for further details)

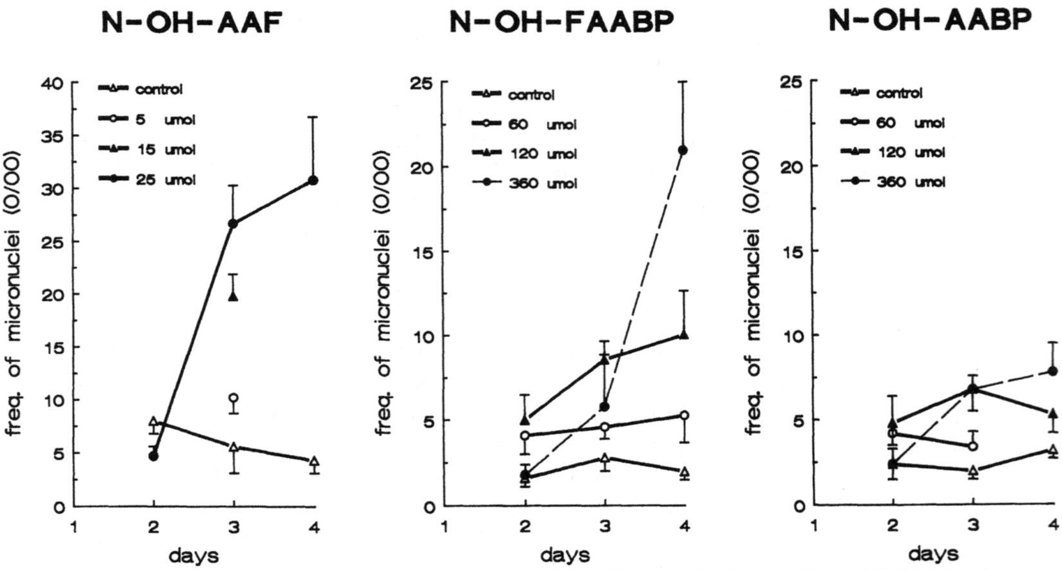

Fig. 8. Frequency of micronuclei induced in rat liver by different doses (μmol/kg) of N-OH-AAF, N-OH-FAABP, or N-OH-AABP injected 17 h after partial hepatectomy. Hepatocytes were isolated 2, 3, and 4 days after injection. *Bars* indicate SEM

large foci occupying a high percentage of liver volume. N-OH-FAABP proved to be a weak promotor, whereas N-OH-AABP was completely ineffective. In conclusion, analysis of the total data set indicates that clastogenicity of the three liver carcinogens does not predict initiating efficacy but correlates with their promotion activity.

Until very recently the liver MN test was exclusively used for studies with rats but Das and Roy (1988) as well as Cliet et al. (1989a) have shown that the test can also be applied successfully to studies with mice. Cliet et al. demonstrated that five procarcinogens (DMN, DEN, 4-APOL, 4-ABPYL, and 1-1-DMH) and the unstable, direct mutagen BPL gave a positive response in the mouse LMN test, whereas only 4-APOL and 4-ABPYL gave a positive result in the BMN test.

4 Possibilities for Automated Detection of Micronuclei Using Image Analysis Systems

Although manual screening for micronuclei in polychromatic erythrocytes (PCE) of bone marrow in laboratory mammals is simple and does not require special skill from

Table 2. Induction of GGT$^+$ foci in rat liver by N-OH-AAF, N-OH-FAABP, and N-OH-AABP (results are expressed as means ± SEM)

Initiator	No. of animals	Dose[a] (μmol/kg	Promotion[b]	No. of foci/cm^3	Volume (mm^3/cm^3
–	12	–	+	80 ± 27	1.0 ± 0.3
N-OH-AAF	13	30	+	148 ± 30[c]	8.4 ± 3.0[d]
N-OH-FAABP	10	120	+	197 ± 62[c]	8.7 ± 4.1[c]
N-OH-AABP	10	120	+	200 ± 37[d]	12.0 ± 4.1[d]

[a] Compounds were injected 24 h after a 2/3 partial hepatectomy.
[b] Promotion was by aminofluorene-HCl in the drinking water for 2 weeks in combination with a necrogenic dose of CCl_4 after 1 week.
[c] Significantly different from control at $p < 0.05$ (Wilcoxon's test).
[d] Significantly different from control at $p < 0.01$ (Wilcoxon's test).

Table 3. Promotion by N-OH-AAF, N-OH-FAABP, and N-OH-AABP in a modified Solt and Farber system (results are presented as means ± SEM)

Initiator	Promotor	No. of animals	Dose[a] (μmol kg^{-1} inj^{-1}	Foci cm^3	Volume (mm^3/cm^3)
DEN[b]	–	12	–	147 ± 20	0.3 ± 0.1
DEN	N-OH-AAF	11	40	464 ± 52[c]	22.2 ± 4.9[c]
DEN	N-OH-FAABP	6	120	392 ± 77[c]	3.7 ± 1.2[c]
DEN	N-OH-AABP	6	120	105 ± 18	0.3 ± 0.1
–	N-OH-AAF	6	40	0	0
–	N-OH-FAABP	5	120	6 ± 6	0.00 ± 0.0
–	N-OH-AABP	5	120	0	0

[a] The promotion regimen consisted of 3 times this dose at days 18, 19, and 20, a 2/3 partial hepatectomy (PH) at day 21, and finally a quarter of the dose at day 25.
[b] Diethylnitrosamine (DEN) was administered by i.p. injection at a dose of 200 μmol kg^{-1} at day 1.
[c] Significantly different from control (DEN/-) at $p < 0.005$ (Wilcoxson's test).

the analyst, it cannot be denied that scoring of micronuclei on a routine daily basis is not one of the most pleasant jobs. Furthermore, when such work involves evaluation of clastogenicity of newly developed chemicals, it may be necessary – for statistical reasons – to analyze substantially more cells than the usual number of 1000–2000 PCEs per animal. Therefore, it is of great practical interest to develop methods for automated scoring of micronuclei.

Presently, most automatization efforts have been directed towards the detection of micronuclei in PCE from bone marrow. Thus far two approaches have been used, namely, image analysis and flow cytometry. Progress in this field was recently discussed during the Automated Micronucleus Scoring Workshop in Milpitas (USA) in November 1988 (Workshop 1988b).

It then became clear that the work of most investigators was still in an exploratory phase and as of this writing the situation has not substantially changed, so that no firm statements can be made concerning methods that will eventually be the most useful and practical ones. Nevertheless, the prospects seem to be promising.

Most progress was reported by Romagna and Staniforth (1989). They tested a number of image analysis systems for their capacity to automatically analyze micronuclei. An important breakthrough in their procedure for slide preparation has been the development of a simple technique for separation of PCEs plus NCEs (normochromatic erythrocytes) from nucleated cells. This can be accomplished by passing the cell suspension through a cellulose column. The resulting eluate consists almost exclusively of PCEs and NCEs which is of great help for successful application of image analysis procedures (and flow cytometry techniques as well). Enrichment of the PCE fraction of anucleated cells by up to 70–80% and removal of residual debris can be accomplished by using a Percoll step gradient. The latter step is very useful for studies with rats because it results in the removal of mast cell granules which closely resemble micronuclei. A final important step for image analysis involves flattening of PCEs and NCEs by means of cytocentrifugation on polylysine-coated slides. Romagna and Staniforth use the May-Gruenwald/Giemsa stain for automated analysis of slides. It is known, however, that this staining procedure does not result in a stoichiometric reaction with DNA and batches of stains do not always have constant qualities. Therefore, it seems necessary to improve the staining technique.

The procedure for scoring micronuclei in PCE has also potential for detection of micronuclei in peripheral blood from rats and man. Studies with rats have the advantage that blood of the same treated rats can be sampled at different times after treatment. In man such studies can only be performed with blood from splenectomized donors because micronucleated cells from normal subjects are captured by the spleen (Schlegel et al. 1986).

Romagna and Staniforth (1989) tested image analyzers from Kontron Zeiss (IBAS 2000) and Leitz (TAS Plus and MIAMED). These are all instruments for evaluation of morphological and densitometric structures on the basis of gray level information. All systems employ autofocus and a black and white camera. Since hard- and software are still being improved, it is premature to indicate one of the instruments as being superior to the other. In any case preliminary validation tests with IBAS 2000 and TAS Plus showed that both systems are suitable for micronucleus analysis. The MIAMED image analyzer seems to have very good prospects for the purpose of large-scale screening. The software of this instrument can distinguish between PCE and NCE populations for each slide individually, allowing fully automated sequential analysis of 16 slides on one microscope stage.

Several groups are exploring the use of flow cytometry techniques for scoring of micronucleated PCEs. Leary and Tometsko (1986) use a FACStar PLus machine. Malaria parasites in red blood cells are being used as a source of DNA and RNA for model studies with fluorochromes like propidium iodide. In pilot studies with MMS-treated mice, reasonably good agreement was found between frequencies of micronucleated PCEs counted manually and automatically. When micronucleated PCEs were sorted by the instrument 87% of sorted cells did carry a micronucleus. Norppa and Hayashi (pers. by Hayashi et al. 1989) carried out collaborative studies with the EPICS V (Coulter) flow cytometer, using DAPI staining preceded by RNase treatment, to reveal micronuclei in PCEs. Percentages of micronucleated erythrocytes in mutagenized mice were determined manually and with the flow technique and results seem to correlate very well. Results obtained are very promising because clastogens can be selected more easily, rapidly, and accurately than manually.

In the future, better quantitative data can be obtained when micronuclei will be scored in preparations devoid of nucleated cells. In summary, it looks as if automated measurement of micronuclei in PCEs form bone marrow and peripheral blood will soon be possible with either image analysis procedures or flow cytometry techniques.

It seems quite clear that instrumentation per se will not be the limiting factor for wide application of such techniques. Other aspects such as instrument costs, stability of preparation techniques, and artifact rejection precedures seem to play a vital role in the eventual wider acceptance of such techniques.

With the present state of the art, the automated scoring of micronuclei in lymphoxytes seems to be considerably more difficult than the forementioned problem of automatically detecting micronuclei in PCEs and NCEs.

Firstly, there is the problem of automated selection of the population of lymphocytes in which micronuclei must be identified. In peripheral blood almost all lymphocytes are in the Go phase of the cell cycle. For the demonstration of chromosome breakage events in such cells it is necessary that they are stimulated to divide in vitro. Breakage events resulting in micronucleus formation will thus be demonstrable only in cells that have passed the first metaphase. Therefore, scoring of micronuclei must be restricted preferably to cells between first and second metaphase. Selection of such cells can be accomplished by two methods. The first, simplest and most reliable method is the cytochalasin-B method (CB) developed by Fenech and Morley (1985a, b). It involves administration of cytochalasin-B to the cell culture, utlimately resulting in the formation of binucleated cells which represent the required cell population for micronucleus analysis. It has been argued that cytochalasin-B is rather cytotoxic and somewhat clastogenic. Therefore, some investigators favor the bromdesoxy-uridine (BrdU) method as an alternative method for identifying cells between first and second metaphase.

Secondly, from the technological point of view, automated identification of micronuclei in a selected population of nucleated mono- and/or binucleated lymphocytes is certainly much more complicated than registering micronuclei in anucleated PCEs or NCEs. At present we have the impression that this difficult task can only be accomplished successfully with image analysis techniques.

First results of preliminary studies on scoring micronuclei in lymphocytes by computerized image analysis were recently reported by Fenech et al. (1988). Investigators compared possibilities of the BrdU and CB method. With their microcomputer video-image analysis system, designed for densitometric measurements, recognition of binucleated cells proved to be impossible. The BrdU method, however, was found to be amenable to automation although a considerable amount of exploratory work still has to be performed.

In the Department of Cytochemistry and Cytometry at Leiden we are carrying out pilot studies directed towards automated detection of micronuclei in binucleated lymphocytes using the CB method and the Leytas-MIAC image analysis system (Leyden Texture Analysis System using the Leitz Modular Image Analysis Computer and a multichannel Leitz autoplan microscope). Because preparation of slides with reproducible qualtity is essential for a successful image analysis procedure, we are standardizing slide preparation as much as possible. At the moment cytochalasin-B treated cells are first given a brief hypotonic shock. They are then fixed in methanol-acetic acid (3:1) plus a trace of formaldehyde. The next step involves resuspension in Carbowax. Cells are then counted and transferred to poly-L-lysine coated slides by means of a centrifugation bucket (Hettich, W.Germany). Nuclei and micronuclei in air-dried slides are stained with Gallocyanin [1% in $KCr(SO_4)_2 \cdot 12H_2O$] after prior hydrolysis of RNA with 5N HCl. After rinsing the slides in water, the cytoplasm is stained with Naphtol Yellow-S (0.5% in 1% acetic acid). The two types of stain used have nonoverlapping absorption spectra which is very useful for successive image analysis of nuclei plus micronuclei and cytoplasm.

Image analysis takes place at a total magnification of 60 x, using an objective of 63 x with a numerical aperture of 1.40. All cells in a field of 512 x 512 μm are analyzed simultaneously; pixel separation is 0.375 μm. Because binucleated cells are big and micronuclei are small, analysis at three focus levels is necessary. Identification and registration of micronucleated, binucleated lymphocytes occurs via the following steps: (1) storage of TV images at three focus levels of stained DNA (nucleus and micronuclei) of each cell; (2) composition of one gray value image by summation of three TV images from the same cell; (3) conversion of the gray value image into a binary image; (4) storage of the cytoplasmic image taken at the absorption maximum of the protein stain (Naphtol Yellow-S); (5) skeletonization steps of the binary image to exclude mononucleated cells and artifacts; (6) verification of the location of selected nuclei and micronuclei within a common cytoplasm; (7) storage of automatically selected cells in the computer for eventual visual inspection by the investigator. Presently, progress has been such that exceptional cells can be selected in an automated fashion. The whole system, however, needs more extensive validation and software development before it can be used routinely.

A drawback of the forementioned system is its realtively high cost. As an alternative, we are now also exploring the potential of a personal computer equipped with an image processing board and a processing board for microscope control (focus plus stepping motors for movement of the stage). Advantages of this configuration are lower costs for equipment, maintenance, and software development as well as easier operation procedures. At present, the system operates in such a way that one slide is analyzed at a time and field by field. A TV camera transmits the microscope image to the image analyzing board. The image is digitized and transformed in the following steps: (1) image enhancement; (2) segmentation (separation of objects of interest form the background); (3) detection of micronuclei and nuclei in binucleated lymphocytes; and (4) verification of the location of micronuclei and nuclei within a common cytoplasm.

Acknowledgments. We thank Dr. P. den Engelse from Microscan B.V., Leiden, for his help in developing an automated image cytometry test for detecting micronuclei in binucleated lymphocytes. We would also like to thank Drs. Régnier and Cordier for permitting us to use part of their data on micronucleus induction in male germ cells. We are also very grateful for the technical assistance of A. Bos, D.A.M. van der Hulst, N. de Vogel, A.M. Schröder-Rotteveel, I. Neuteboom, and V.M. Woldring. These investigations were financially supported by grants from the Dutch Cancer Foundation (Koningin Wilhelmina Fonds), the European Community Environmental Research Program. J.A. Cohen Institute of Radiopathology and Radiation Protection, and Shell International.

References

Ashby J, Mohammed R (1986) Slide preparation and sampling as a major source of variability in the mouse. Mutat Res 164: 217–235

Boller K, Schmid W (1970) Chemische Mutagenese beim Säuger, das Knochenmark des chinesischen Hamsters als in vivo-Testsystem, Hämatologische Befunde nach Behandlung mit Trenimon. Humangenetik 11: 34–54

Braithwaite I, Ashby J (1988) A non-invasive micronucleus assay in the rat liver. Mutat Res 203: 23–32

Cliet I, Fournier E, Melcion C, Cordier A (1989a) In vivo micronucleus test using mouse hepatocytes. Mutat Res 216: 321–326

Cliet I, Fournier E, Melcion C, Cordier A (1989b) In vivo micronucleus tests in mouse. Comparison of the sensitivity of three target organs (bone marrow, liver and testis) to six carcinogens. Environ Mol Mutagen 14: Suppl 15:39 (abstract No 108)

Cole RJ, Taylor J, Cole J, Arlett CF (1981) Short-term tests for transplacentally active carcinogens. 1. Micronucleus formation in fetal and maternal mouse erythroblasts. Mutat Res 122: 347–353

Collaborative Study Group for the Micronucleus Test (1986) Sex difference in the micronucleus test. Mutat Res 172: 151–163

Collaborative Study Group for the Micronucleus Test (1988) Strain difference in the micronucleus test. Mutat Res 204: 307–316

Collaborative Study Group for the Micronucleus Test (1989) Difference between intraperitoneal and oral gavage application in the micronucleus test. Mutat Res 223: 329–344

Das RK, Roy B (1988) A simplified method for micronucleus preparation from hepatic cells. Stain Technol 63: 71–74

Degrassi F, Tanzarella C (1988) Immunofluorescent staining of kinetochores in micronuclei: a new assay for the detection of aneuploidy. Mutat Res 203: 339–345

Fenech M, Morley AA (1985a) Measurement of micronuclei in human lymphocytes. Mutat Res 147: 29–36

Fenech M, Morley AA (1985b) Solutions to the kinetic problem in the micronucleus assay. Cytobios 43: 223–246

Fenech M, Jarvis LR, Morley AA (1988) Preliminary studies on scoring micronuclei by computerized image analysis. Mutat Res 203: 33–38

Fenech M, Morley AA (1989) Kinetochore detection in micronuclei: an alternative method for measuring chromosome loss. Mutagenesis 4: 98–104

Hayashi M, Yoshimura I, Sofuni T, Ishidate M Jr (1989) A procedure for data analysis of the rodent micronucleus test involving a historical control. Environ Mol Mutagen 13: 347–356

Heddle JA (1973) A rapid in vivo test for chromosomal damage. Mutat Res 18: 187–190

Heddle JA, Blakey DH, Duncan AMV, Goldberg MT, Newmark H, Wartgovich MJ (1981) Micronuclei and related nuclear anomalies as a short-term assay for colon carcinogens. In: Gastrointestinal Cancer: endogenous factors. Banbury Report No. 7

Heddle JA, Hite M, Kirkhart B, Mouvranin K, MacGregor JT, Newall GW, Salamone MR (1983) The induction of micronuclei as a measure of genotoxicity. A report of the US. EPA Gene-Tox program. Mutat Res 123: 61–118

Heddle A (1988/89) Nuclear anomalies in colonic epithelium. Workshop on micronuclei, Manchester 1988

Hennig UGG, Rudd NL, Hoar DI (1988) Kinetochore immunofluorescence in micronuclei: a rapid method for the in situ detection of aneuploidy and chromosome breakage in human fibroblasts. Mutation Res 203: 405–414

Jenssen D, Ramel C (1980) The micronucleus test as a part of a short-term mutagenicity test program for the prediction of carcinogenicity evaluated by 143 agents tested. Mutat Res 75: 191–202

Lähdetie J, Parvinen M (1981) Meiotic micronuclei induced by X-rays in early spermatids of the rat. Mutat Res 81: 103–115

Lähdetie J, Parvinen M (1982) Meiotic micronuclei as indicators of mutagenesis. Prevention of occupational cancer. Intern Symp, Occupational Safety and Health Series 46, Intern Labour Officce, Geneva, pp 483–488

Lähdetie J (1983a) Meiotic micronuclei induced by adriamycin in male rats. Mutat Res 119: 79–82

Lähdetie J (1983b) Micronuclei induced during meiosis by ethylmethanesulfonate, cyclophosphamide and dimethylbenzanthracene in male rats. Mutat Res 120: 257–260

Lähdetie J (1988) Induction and survival of micronuclei in rat spermatids. Comparison of two meiotic micronucleus techniques using cyclophosphamide. Mutat Res 203: 47–53

Leary JF, Tometsko AM (1986) An automated mouse micronucleus assay for detection of clastogenic chemicals in the environment. 17th Ann Meeting Environ Mutagen Soc, Baltimore MD, 9–13 April, Environ Mutagen 8, 5

MacGregor JT, Heddle JA, Hite M et al. (1987) Guidelines for the conduct of micronucleus assays in mammalian bone marrow erythrocytes. Mutat Res 189: 103–112

Munoz N, Hayashi M, Bang LJ, Wahrendorf J, Crespi M, Bosch FX (1987) Effect of riboflavin retinol and zinc on micronuclei of buccal mucosa and of oesophagus: (a randomized double-blind intervention study in China). J Natl. Cancer Inst 79: 687–691

OECD (1983) Genetic toxicology: micronucleus test. In: OECD guidelines for testing of chemicals. Document 474, pl-6

Poll MLM van de, Hulst DAM van de, Tates AD, Mulder GJ, Meerman JHN (1989) The role of specific DNA adducts in the induction of micronuclei by N-hydroxy-2-acetylaminofluorene in rat liver in vivo. Carcinogenesis 10: 717–722

Poll MLM van de, Hulst DAM van der, Tates AD, Meerman JHN (1990) Correlation between clastogenicity and promotion activity in liver carcinogenesis by N-hydroxy-2-acetylaminofluorene, N-hydroxy-4-fluoro-4-acetylaminobiphenyl and N-hydroxy-4-acetylaminobiphenyl. Carcinogenesis 11: 333–339

Régnier JF, Cordier A (1989a) Micronucleus test on mouse spermatids. I. Description of an experimental protocol. (submitted)

Régnier JF, Cordier A (1989b) Micronucleus test on mouse spermatids. II. Validation of an experimental protocol. (submitted)

Romagna F, Staniforth CD (1989) The automated bone marrow micronucleus test. Mutat Res 213: 91–104

Schlegel R, MacGregor JT, Everson RB (1986) Assessment of cytogenetic damage by quantitation of micronuclei in human peripheral blood erythrocytes. Cancer Res 46: 3717–3721

Schmid W (1976) The micronucleus test for cytogenetic analysis. In: Hollaender A (ed) Chemical mutagens, principles and methods for their of detection, vol 4, Plenum Press, New York London, pp 31–53

Shindo Y, Hirano F, Maeda H, Takeda U (1983) The micronucleus test with mouse spleen cells. Mutat Res 121: 53–57

Stich HF, Anders F (1989) The involvement of reactive oxygen species in oral cancers of betel quid/ tobacco chewers. Mutat Res 214: 47–61

Stich HF, Rosin MP (1982) Micronuclei in exfoliated human cells as an internal dosimeter for exposures to carcinogens. In: Stich HF (ed) Carcinogens and mutagens in the environment, vol 2. CRC, Boca Raton, pp 17–25

Tates AD, Boer P de (1984) Further evaluation of a micronucleus method for detection of meiotic micronuclei in male germ cells of mammals Mutat Res 140: 187–191

Tates AD, Engelse L den (1989b) The role of short-lived lesions in the induction of micronuclei in rat liver by ethylnitrosourea and methyl methanesulphonate: the importance of experimental design. Mutat Res 210: 271–279

Tates AD, Neuteboom I, Hofker M, Engelse L den (1980) A micronucleus technique for detecting clastogenic effects of mutagens/carcinogens (DEN, DMN) in hepatocytes of rat liver in vivo. Mutat Res 74: 11–20

Tates AD, Dietrich AJJ, Vogel N de, Neuteboom I, Bos A (1983) A micronucleus method for detection of meiotic micronuclei in male germ cells of mammals. Mutat Res 121: 131–138

Tates AD, Bos A, Neuteboom I, Vogel N de (1985a) Micronuclei in hepatocytes and early spermatids at different time intervals after exposure of rats to a single dose of mitomycin C. Mutat Res 147: 323–324

Tates AD, Menkveld GJ, Engelse L den (1985b) Time dependent induction of chromosomal damage in rat hepatocytes and spermatocytes in relation to alkylation damage of DNA. Abstract Fourth Intern Conf Environ Mutagens Stockholm Sweden 1985

Tates AD, Vogel N de, Rotteveel AHM (1986a) Clastogenic effects of genotoxins in rat spermatocytes. XVIth Ann Meeting Eur Environ Mutagen Soc Brussels Belgium 25–30 August, Abstract

Tates AD, Neuteboom I, Rotteveel AHM, Vogel N de, Menkveld GJ, Engelse L den (1986b) Persistence of preclastogenic damage in hepatocytes of rats exposed to ethylnitrosourea, diethylnitrosamine, dimethylnitrosamine and methylmethanesulphonate. Correlation with DNA O-alkylation. Carcinogenesis 7: 1053–1058

Tates AD, Vogel N de, Rotteveel AHM, Leupe F, Davids JAG (1989a) The response of spermatogonia and spermatocytes of the Northern vole Microtus oeconomus to the induction of sex-chromosome nondisjunction, diploidy and chromosome breakage by X-rays and fast fission neutrons. Mutat Res 210: 173–189

Thomson EJ, Perry PE (1988) The identification of micronucleated chromosomes: a possible assay for aneuploidy. Mutagenesis 3: 415–418

Wild D (1988) Summary report on the performance of the in vivo micronucleus assays. In: Ashby J, Serres FJ de, Shelby MD, Margolin BH, Ishidata M Jr, Backing GC (eds) Evaluation of short-term test for carcinogens. Report of the International Program on Chemical Safety, Collaborative Study on in Vivo Assays, vol 1. Cambridge University Press, Cambridge pp 1.145–1.148

Workshop on Micronuclei: Origins, Applications and Methodologies (1988a) Sponsored by the Health and Safety Executive, held at ICI, Alderley Park, UK, 23–25 May 1988

Workshop on Automated Micronucleus Scoring (1988b) held at Embassy Suites Hotel, Milpitas, California, November 17–18 (no abstracts)

In Vitro Chromosomal Aberration Test – Current Status

M. ISHIDATE, Jr.[1]

Contents

1 Introduction

The majority of chemical mutagens and/or carcinogens interact with cellular DNA and are extremely potent in inducing chromosome damage in mammalian cells. A recent review of chromosomal aberration tests in vitro, on 951 chemical substances, from over 240 published papers indicated that about 93% (91/98) of well-known carcinogens were positive (clastogenic) in the system with or without metabolic activation, using either primary cultured cells or established cell line cells (Ishidate et al. 1988). Of the substances which were positive in the gene mutation assays in bacteria (Ames test), about 97% (176/181) were also positive in the chromosome test, while of those negative in the Ames test, about 90% (122/135) were positive in the chromosome test (Table 1).

These results support the previous findings of an international collaborative study on the evaluation of short-term tests for carcinogens organized by IPCS/WHO (Ashby et al. 1985). This study indicated that a large number of genotoxic chemicals would be considered nontoxic if tested only by the Ames test, and that the chromosomal aberration assay is a valid supplementary assay for the assessment of genotoxic potential.

In the OECD guidelines for genetic toxicology for chemicals, it was emphasized that the test systems used should cover two genetic end points, gene mutation, and

---

[1] Division of Genetics and Mutagenesis, National Institute of Hygienic Sciences, 1-18-1 Kamiyoga, Setagaya-ku, Tokyo 158, Japan

Table 1. Comparison of results from the in vitro clastogenicity test and other biological assays (Ishidate et al. 1988)

Biological assay		In vitro chromosome test		
		+	–[a]	Total
Ames test	+	176	5	181
	–	122	13	135
	Total	298	18	316
Micronucleus test	+	43	2	45
	–	52	8	60
	Total	95	10	105
Carcinogenicity test	+	91	7	98
	±	56	3	59
	–	18	2	20
	Total	165	12	177

[a] Only those substances which were negative both with and without metabolic activation were used.

chromosomal aberration (OECD 1988). During the past decade, this concept has been commonly accepted. An in vitro cytogenetic test is adopted in the genetic toxicology guidelines of various countries for the registration not only of new drugs or pesticides, but also of other new industrial chemicals (Berry and Litchfield 1985).

The present chapter introduces the guidelines recommended at present, and discusses some technical problems and quantitative evaluations on the data which have been accumulated.

2 Recommended Guidelines for the in Vitro Chromosomal Aberration Test

In Japan, the in vitro chromosomal aberration test is required as a complementary test to the Ames test. These two in vitro tests and one in vivo test (e.g., the micronucleus test in mice) are recommended as a minimum package of mutagenicity tests for new drugs (Ministry of Health and Welfare 1984; revised, 1989). A similar strategy has been adopted in other guidelines for new industrial chemicals (Ministry of Health and Welfare 1986), or new pesticides (Ministry of Agriculture, Forestry, and Fisheries 1985). The Ministry of Labour (1979) requires that the in vitro chromosome test only be used when the chemical showed a high mutagenic potency in the Ames test (e.g., more than 1000 revertant colonies/mg).

The recommended guidelines for the in vitro chromosomal aberration test on new drugs are as follows:

1. *Cells used.* Primary cells or established cell lines can be used.
2. *Dose levels.* At least three successive doses should be used (Note 1).
3. *Controls.* Solvent-treated cells are used as a negative control. Well-known clas-

togens should be employed in positive controls (both systems with and without metabolic activation).

4. *Metabolic activation.* An appropriate metabolic activation method (S9 mix, etc.) should be used concurrently.
5. *Experimental procedure.* Chromosome preparations are made at appropriate times after treatment (Note 2).
   At least two plates per dose are used. One hundred metaphase cells per plate should be observed and analyzed for structural aberrations and polyploid cells (Note 3).
6. *Presentation of Results.* The incidence of metaphase cells with aberrations (aberrant cells), or the frequency of aberrations per cell should be presented in tables.

Note 1. The highest concentration should be the dose in which cell proliferation (or cell division) is inhibited by more than 50%. When no cytotoxic effect is observed, the highest concentration should be limited to about 10 mM.

Note 2. Two sampling times are required, since a possible delay of the cell cycle by the treatment is expected.

Note 3. For structural aberrations, cells with different types of chromatid as well as chromosome aberrations should be classified.

3 Critical Problems in the Procedure

3.1 Cell Type to Be Used

Either established cell lines or primary cultured cells such as human peripheral blood lymphocytes are acceptable (Evans et al. 1980). Chinese hamster fibroblast cell lines, however, are most commonly used for routine work. There are several advantages in using these cell lines: (1) a relatively small number of large chromosomes, rapid cell cycle and proliferation in a simple culture medium; (2) the same cell lines available for different experiments; (3) the cells are relatively restistant to S9 mix; (4) the cells are sensitive to a variety of chemicals.

A Chinese hamster ovary cell line (CHO) has been used for cytogenetic studies under the NTP program in the USA (Galloway et al. 1987) and its standard experimental protocol was introduced by Galloway et al. (1985). However, due to different kinds of CHO cells (e.g., CHO, CHO-WBL, CHO-K1), sometimes investigators produce different results even when the same agent is treated. In Japan, a Chinese hamster lung fibroblast cell line (CHL; Koyama et al. 1970) is widely used. The cells have 25 chromosomes (more than 80% of cells) and their doubling time was estimated to be 15–17 h in Eagle's MEM medium supplemented with 10% calf serum (without antibiotics).

There are some differences in the standard protocol between the CHO and CHL cell systems. The main differences are shown in Table 2. In the CHO system, cells are cultured in McCoy's 5A medium supplemented with 10% fetal calf serum. The cells are treated with a test agent for 8–12 h (sometimes 12–26 h) in the system

Table 2. Differences in the protocol

	CHO system	CHL system
Culture	McCoy's 5A 10% FCS	Eagle's MEM 10% CS
Cytotoxicity	Hemocytometer 50% Inhibition	Monocellater 50% Inhibition
Treatment (−S9) (+S9)	8–12 h (12–26 h) 2 h	24 h and 48 h 6 h
Concentration	3 doses (log 10)	3 doses (ratio 2)

without S9 mix, but only for 2 h with S9 mix for metabolic activation. The concentration of S9 in the medium is about 2%, which is about one–third of that used in the CHL system. On the other hand, in the CHL system, cells are cultured in Eagle's MEM supplemented with 10% calf serum (inactivated). The cells are treated with a test agent for 24 and 48 h without S9 mix (the direct method), but for 6 h with S9 mix for metabolic activation (Ishidate 1988)

Recently, a comparative study between the CHO and CHL systems was conducted on 25 different chemicals, as part of the USA-Japan collaborative research program (Sofuni et al. 1989). Results indicated that the CHL system seemed to be more sensitive than the CHO system to a certain number of chemicals tested, as summarized in Table 3.

3.2 Treatment with Test Agents and Sampling Time

The maximum concentration should be selected by a preliminary cytotoxicity test, since chromosomal aberrations are usually observed at the dose very close to the cytotoxicity. The cytotoxicity can be measured from the cell density on a plate 24 or 48 h after treatment, if an apparatus called "Monocellater" (Olympus) is used. This instrument was designed at our laboratory to measure the color absorption of the cell monolayer stained by 0.1% crystal violet.

For the chromosome test, at least three doses (at ratio 2 as a rule) including the 50% cell growth inhibition dose should be used. At least two plates per dose are recommended. If the test agent shows no cytotoxicity, a maximum of 2–3 mg/ml can be used. For soluble test material (e.g., salt solution), the maximum dose should be limited to 10 mM, since the osmolality in the medium usually starts to increase from this dose level (Fig. 1; Ishidate et al. 1984).

Possible changes in physiological conditions, such as pH (Cifone et al. 1987) or osmolality of the culture medium, may not directly affect chromosome damage but may stimulate indirectly the incidence of chromosomal aberrations. For insoluble material, a suspension in CMC sodium may be recommended, since if the size of the particle is small enough, the cell can interualize it by phagocytosis, although according to the OECD guidelines, "Relatively insoluble substances should be tested up to the limit of solubility." In the system with S9 mix, ethanol should not be used as a solvent,

Table 3. Overall results of chromosomal aberration tests in the CHL and CHO test systems (Sofuni et al. 1989)

Cell system	Test results[a]			
	+	$+^w$	±	–
CHL	14 (56%)	6 (24%)	2 (8%)	3 (12%)
CHO	11 (44%)	1 (4%)	0	13 (52%)

[a]Number of chemicals (% of chemicals).
+ Positive results; $+^w$ weakly positive ± questionable results; – negative results.

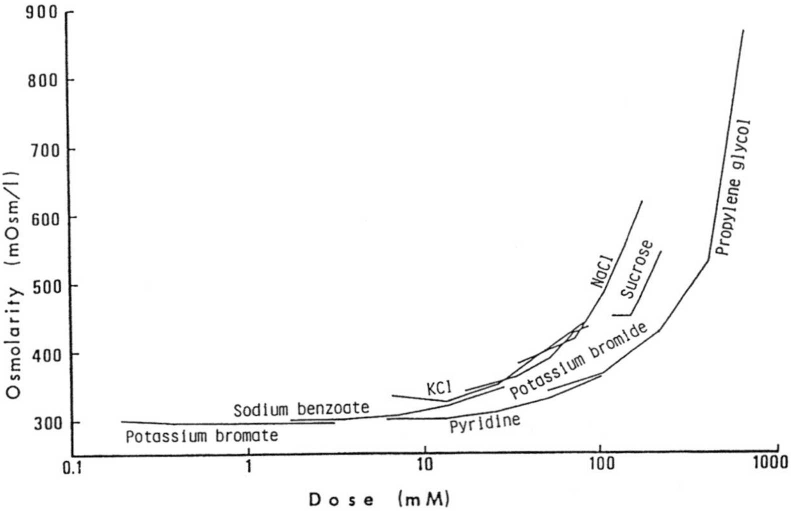

Fig. 1. Osmotic pressure versus concentration for various chemicals in culture medium, determined by a freezing-point technique

since a possible false positive effect can be expected due to the formation of acetaldehyde in the medium.

Treatment time is a critical problem in the protocol. In the system without S9 mix, treatment may be given continiously up to the time of harvesting and cells can be exposed to the test agent throughout the whole cell cycle. If the cells cannot proliferate and no metaphases are observed, then the cells can be washed and cultured in fresh medium. In the system with S9 mix, however, cells cannot be exposed to the test agent for a long period because of the cytotoxic effect of S9 itself. In the cell suspension method introduced by Natarajan et al. (1976), the cells in suspension can be incubated with S9 mix for 1–3 h only. On the other hand, in the plate method, where cells are exposed to the S9 mix in a monolayer, the cells can be treated for a longer time (3–6 h), and more aberrations can be expected at lower dose levels of certain chemicals, as shown in Table 4.

Table 4. Chromosomal aberration test in vitro combined with S9 mix

Compound	Suspension method [a]		Monolayer method [b]	
	Dose (mg/ml)	Incidence (%)	Dose (mg/ml)	Incidence (%)
Benzo(a)pyrene	0.5	22.0	0.04	47.6
3-Methylcholanthrene	2.0	3.0	0.08	12.3
Dimethylnitrosamine	4.0	77.3	1.0	73.3
7,12-Dimethylbenz(a)anthracene	0.5	25.0	0.1	60.6
3'-methyl-4-dimethylaminoazobenzene	8.0	10.0	0.05	25.6
Phenacetin	3.2	51.0	0.8	23.3

[a] Treated for 3 h in shaker.
[b] Treated for 6 h on plate.

The best sampling time after treatment may depend greatly on the chemical tested and the cell type used. In principle, chromosomal aberrations are observed at the first posttreatment metaphase (M_1). However, the cell cycle may be prolonged by the treatment and sometimes more aberrations are found at a much later stage. There were many chemicals which induced more aberrations after exposure for 48 h rather than for 24 h in the CHL system; e.g., 6-mercaptopurine was negative at 24 h, but positive at 48 h and induced 42% aberrant cells at a dose of 0.001 mg/ml, in the system without S9 mix. Compounds such as maleic anhydride, sodium dehydroacetate, 4-nitro-o-phenylenediamine, or phenylbutazone showed a similar response. It should be noted that about 21.8% (56/257) of clastogens, which were positive in the system without S9 mix, induced more aberrations 48 h after treatment than 24 h (Ishidate 1988). These observations suggest that there may be chemicals which induce very extensive mitotic delay at clastogenic doses or may be clastogenic only when cells have passed through more than one cell cycle since the beginning of treatment (Thust et al. 1980).

According to the UK-EMS guidelines (1989), "if negative or equivocal results were obtained with a single harvesting time at 1.5 normal cycle time, the repeat test should include an additional sample at approximately 24 h later." The Japanese guidelines, however, require two sampling times from the beginning and propose to repeat an experiment only when the results are marginal or show no dose responses.

Some chemicals induce polyploid cells alone or together with structural aberrations and these effects can be detected only at the second posttreatment metaphase (M_2). Diethylstilbestrol (DES) is a good example. It induces no structural changes but induces 5 and 85% polyploid cells at 24 and 48 h, respectively, after treatment with 0.01 mg/ml in the system without S9 mix (Sawada and Ishidate 1978).

The chemicals which induced not only structural aberrations but also polyploid cells are listed in Table 5, a few carcinogens are also included.

In the UK guidelines, for example, scoring such polyploid cells is not required, since the significance of these is unknown and there are many tissues in man in which there is naturally occurring polyploidy without any obvious deleterious effects (Scott, pers. comm., 1989). There is a possibility, however, that polyploid cells which are

Table 5. Chemicals which induced mainly polyploid cells in the chromosome test without S9 mix

Compound	Dose (mg/ml)	Polyploid (% at 48 h)	Structural changes	Ames test
Bendroflumethazide	0.2	41	+	–
Diethylstilbestrol	0.015	81	–	–
Ethenzamide	0.5	15	+	–
Ethylvanillin	0.25	43	–	–
Hexane	0.5	10	–	–
Hydrochlorothiazide	0.5	12	+	–
Metformin	2.0	27	+	–
Nitrobenzene	0.5	12	–	–
o-Nitrotoluene	0.25	14	–	–
m-Nitrotoluene	0.25	23	–	–
p-Nitrotoluene	0.25	53	–	+

chemically induced may be different from those occurring naturally, in addition agents inducing polyploid could also induce aneuploid cells. As shown in Fig. 2, DES could induce aneuploid cells, when CHL cells were treated with 7.5 μg/ml for 48 h, and the modal number of chromosomes shifted from the diploid to the tetraploid range (Ishidate and Sofuni 1985).

3.3 Negative and Positive Controls

Solvent-treated cells are used as a negative control. Untreated cells should also be checked periodically for the spontaneous incidence of chromosomal aberrations in the cell line used.

In the system with S9 mix, a control group in which cells are treated with the test agent but not with S9 mix should be included, since some chemicals will be positive in this control but negative with S9 mix at a similar dose, indicating that this agent may be detoxicated or inactivated by the metabolic enzymes.

For the positive control, mitomycin C, N-methyl-N'-nitro-N-nitrosoguanidine (MNNG), 4-nitroquinoline-N-oxide (4-NQO), or methyl methane sulfonate (MMS) can be used for the system without S9 mix, while cyclophosphamide (CP), dimethyl-nitrosamine (DMN), benzo(a)pyrene (B(a)P), or 7,12-dimethylbenzo(a)-anthracene (DMBA) can be used for the system with S9 mix.

3.4 Presentation of Results

Classification of chromosomal aberrations should be made according to internationally accepted criteria (Evans and O'Riordan 1975; ISCN 1985). The criteria, especially for chromatid- or chromosome-type gaps, are difficult and vary among different investigators. It is recommended, therefore, that the investigator notes his or her own criteria in the final report. The incidence of cells with aberrations (% aberrant cells) seems to be easier to count and more practical in routine work than the exact

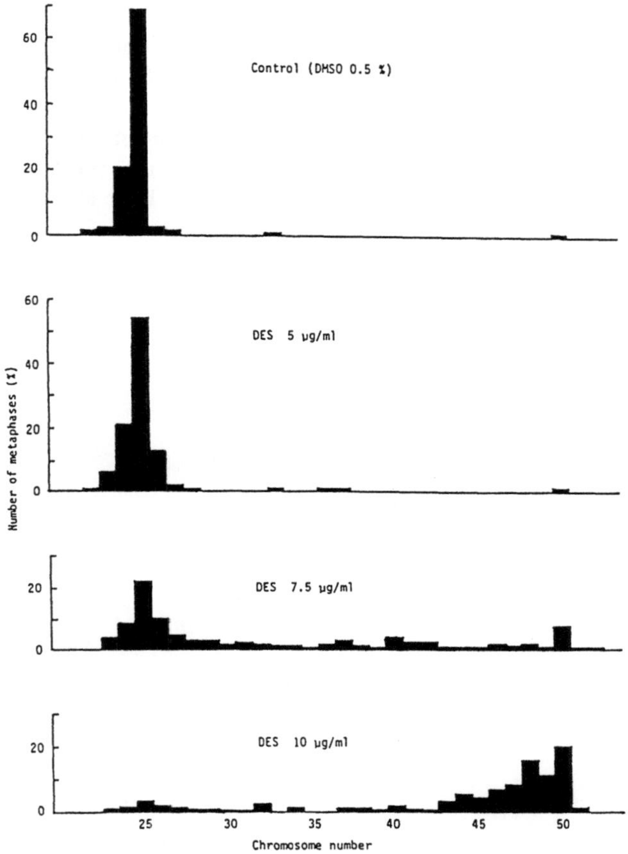

Fig. 2. Chromosome number distribution in CHL cells after treatment with different doses of diethylstilbestrol (DES) for 48 h

count of chromosomal aberrations per cell. For example, an aberrant cell with three structural aberrations, e.g., one chromatid gap and two chromatid breaks, can be counted as one aberrant cell with two types of chromatid aberrations, gap, and break. The incidence of gaps is commonly excluded from the total incidence of aberrations. However, it should be noted that there are some clastogens which induce mainly gaps and breaks but few exchanges. A comparative study indicated that well-known carcinogens may induce more exchange-type rather than break-type aberrations at relatively low doses and usually at later sampling times. The calculation of such cells with exchange-type aberrations may be important when the clastogenic potency is compared among different environmental chemicals.

4 Quantitative Evaluation of Clastogenic Potency

A carcinogen, 4-NQO, induces 20% aberrant cells even at a very low dose, 0.0003 mg/ml. On the other hand, a food additive, propylene glycol induces 20% aberrant cells only at a high dose, 20.0 mg/ml. Both are positive but there is about an 80000-

fold difference in their clastogenic activities. If the D_{20} value, which is the dose (mg/ml) at which aberrations are detected in 20% of metaphase cells, was calculated and compared among different clastogens, a 10^6 times difference exists as shown in Fig. 3.

Many investigators do not take note of the mutagenic potency of the test chemical but are only interested in qualitative results, i.e., positive or negative. Unfortunately, the data from Gene-Tox or the NTP, USA, have also been evaluated mainly in such

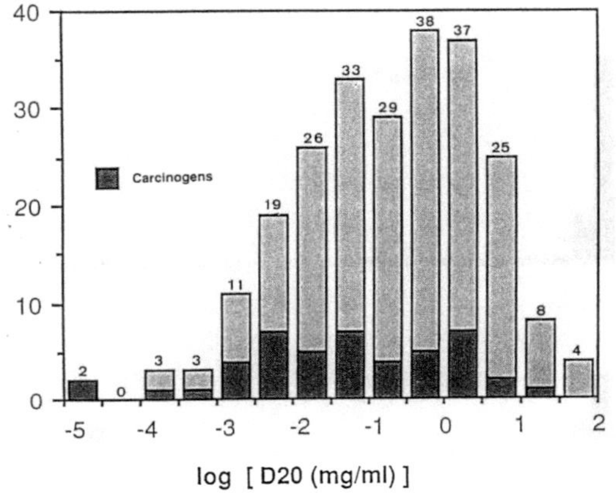

Fig. 3. Distribution of test substances which were clastogenically arranged according to their D_{20} values (*dark areas* indicate number of carcinogenic substances)

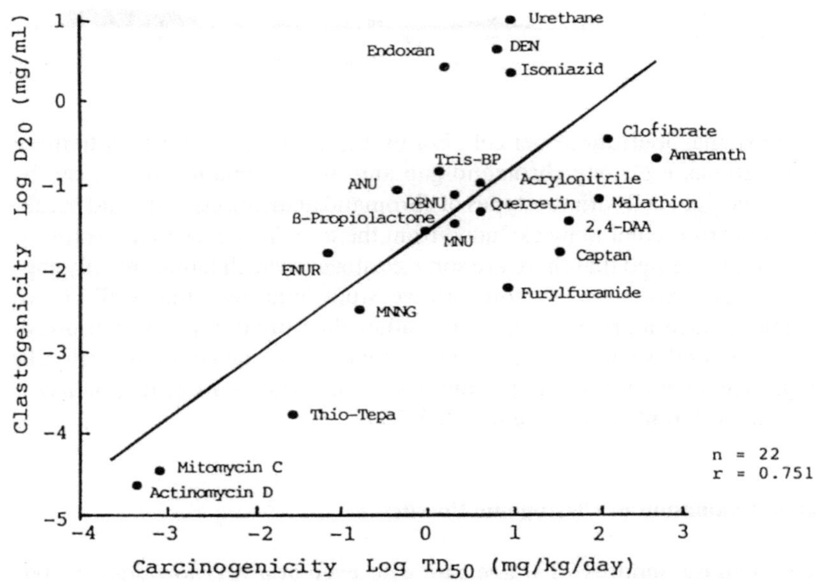

Fig. 4. Quantitative relationship between clastogenic activity (D_{20} value) in the chromosomal aberration test with CHL cells and carcinogenic potential (TD_{50} value) in rodents

Table 6. Specific clastogenic activity in vitro (D_{20} value) and the micronucleus test in mice

Compound tested	Chromosome D_{20} (mg/ml)	test	Micronucleus Dose (mg/kg)	test i.p.)
Mitomycin C	0.00001	+	3	+
4-NQO	0.00032	+	80	+
5-FU	0.00038	+	100	+
MNNG	0.0028	+	50	+
ENU	0.063	+	50	+
Potassium bromate	0.071	+	100	+
Sodium nitrite	0.32	+	200	–
Fast Green FCF (crude)	2.0	+	2000	–
Potassium bromide	3.7	+	500	–
Acid Red	4.7	+	1600	–
Propylene glycol	20.0	+	15000	–

a qualitative way and discussed in terms of the validity of short-term tests from a viewpoint of sensitivity, specificity, predictability, or accuracy in relation to their carcinogenicity. If the clastogenic potencies of chemicals were calculated as their D_{20} values and compared with their carcinogenic potency, estimated as TD_{50} values, indicating the doses (mg/kg/day) at which tumors were induced in 50% of animals some positive correlation is found as shown in Fig. 4, in which carcinogens were ramdomly selected.

The significance of using appropriate statistical treatment on the in vitro data is still questionable. It may not be so important if the data are treated in a quantitative rather than qualitative way. In the Japanese guidelines, only clearly positive results with a dose response are taken into consideration; a result is positive if the total incidence, including gaps, is more than three times greater than the control. Compounds, which induce aberrations at doses higher than 10 mM (or 5 mg/ml), may be categorized as weak clastogens or false positives. There is evidence indicating that substances which are positive only at very high dose levels in vitro may induce no positive effects in the micronucleus test in rodents (Table 6).

5 Further Problems

5.1 Correlation Between Chromosomal Aberrations and Stable Chromosomal Mutations Induced by Chemicals

The majority of aberrant cells will be lethal. However, some cells with nonlethal changes, such as reciprocal translocations, inversions, or small deletions of chromosomes, can survive if the cells are washed and recultured in fresh medium. We have established a fibroblast cell line with a normal diploid karyotype from Chinese hamster thymus and have treated them with different doses of MNNG for 48 h. Chromosomal aberrations were induced with a good dose response. After reculture, a number of colonies survived and were tested for their karyotype using the G-banding technique. Finally, we found that ca. 50% of the colonies showed karyotypes differing from their original cells (Sawada et al. 1978, unpubl. data). These findings suggest that the cells, which could survive after exposure to a clastogen, will proliferate as a new cell population with a stable chromosomal mutation. These more subtle changes may have important consequences for both somatic and germ cells. Further studies are needed on the specificity of chromosomal changes induced by chemicals, and on the significance of chromosomal mutation in relation to carcinogenicity and other genotoxic effects.

5.2 Variable Results from Different Protocols

From the USA-Japan collaborative studies on the comparison of the CHO and CHL cell systems, it was suggested that the discrepancy in the results depends in part on the difference in their sensitivity but mostly on the difference in the experimental protocols. These differences may be overcome by testing the same chemical with CHO cells using the CHL protocol and vice versa. For example, N,N'-diphenyl-p-phenylenediamine is negative in the CHO system and positive in the CHL system without S9 mix, according to the standard protocols. However, if the CHO cells are treated for a longer period, 24 instead 12.5 h, the results become positive (Fig. 5a). Similarly, triallyl isocyanurate becomes positive if the CHO cells are treated with S9 mix for a longer period, 6 instead 2 h, although the concentration of S9 in the medium is higher in the CHL than in the CHO protocol (Fig. 5b). Further studies are now in progress in this area at an international level.

5.3 Studies on the Significance of Induction of Aneuploid and Polyploid Cells

Some carcinogens, such as benzene, DES, and asbestos, are usually negative in bacterial mutation assays, but are clastogenic and interfere with chromosome seg-

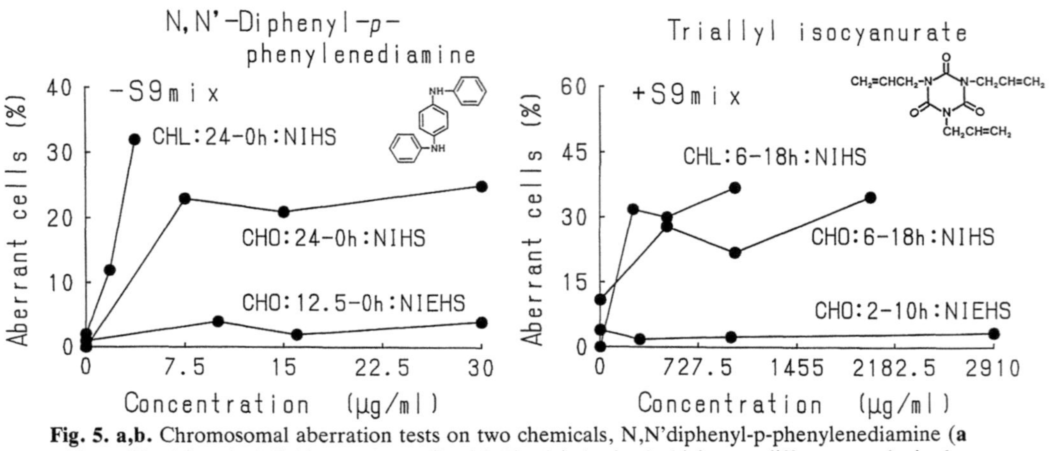

Fig. 5. a,b. Chromosomal aberration tests on two chemicals, N,N'diphenyl-p-phenylenediamine (**a** without S9 mix) and triallyl isocyanurate (**b** with S9 mix), both of which gave different results in the CHO and CHL systems; 6–18 h: The cells were treated for 6 h and cultured for 18 h in fresh medium. *NIHS:* Tests were performed at the National Institute of Hygience Science, Tokyo. *NIEHS:* Tests were performed in the USA under the project of NTP.

regation during mitosis and result in aneuploidy. In routine work, however, the counting of individual chromosomes in metaphase plates and identifying the frequency of cells differing from the diploid or normal number seems to be very time-consuming. It is almost impossible to observe the exact numbers on the air-dried preparations, especially when an established cell line is used.

The significance of such effects is still not well understood but there is a possibility that aneuploidy is associated with spontaneous abortions, congenital malformations, mental retardation, and carcinogenesis. In regulatory guidelines, however, it is difficult to require counting of aneuploid cells. On the other hand, counting of polyploid cells can be required, since the counting is easily performed during the process of metaphase analysis for structural aberrations.

References

Ashby J, de Serres FJ, Draper M, Ishidate M Jr, Margolin BH, Matter BE, Shelby MD (1985) Evaluation of short-term tests for carcinogens. Progress in mutation research vol 5. Elsevier, Amsterdam, pp 117–174

Barry DJ, Litchfield MH (1985) A review of the current regulatory requirements for mutagenicity testing. In: Ashby J, de Serres JF et al. (eds) Evaluation of short-term tests for carcinogens. Mutation Research vol 5. Elsevier, Amsterdam, pp 727–740

Cifone MA, Myhr B, Eiche A, Bolcsfoldi G (1987) Effect of pH shifts on the mutant frequency at the thymidine kinase locus in mouse lymphoma L5178Y TK+/- cells. Mutat Res 189: 39–46

Evans H, O'Riordan ML (1975) Human peripheral blood lymphocytes for the analysis of chromosome aberrations in mutagen tests. Mutat Res 31: 135–146

Evans HJ, Ishidate M Jr., Leng M, Miller CT, Mitelman F, Vogel E (1980) Cytogenetic damage as an endpoint in short-term assay systems for detecting environmental carcinogens, Reprt 8, in long-term and short-term screening assays for carcinogens: a critical appraisal. IARC/WHO Monograph, Suppl 2, IARC, Lyon, pp 227–244

Galloway SM, Bloom AD, Resnich M, Margolin BH, Nakamura F, Archer P, Zeiger E (1985) Development of a standard protocol for in vitro cytogenetic testing with Chinese hamster overy cells: comparison of results from 22 compounds in two laboratories. Environ Mutagen 7: 1–51

Galloway SM, Armstrong MT, Reuben S et al. (1987) Chromosome aberrations and sister chromatid exchanges in Chinese hamster ovary cells: evaluations of 108 chemicals. Environ Mol Mutagen 10: 1–175

ISCN (1985) An International System for Human Cytogenetic Nomenclature. Birth Defeats: Original Article Series, Vol 21, No 1, National Foundation –Merch of Dimes, New York

Ishidate M Jr (ed) (1988) Data book of chromosomal aberration tests in vitro. LIC/Elsevier, Amsterdam

Ishidate M Jr, Sofuni T (1985) The in vitro chromosomal aberration test using Chinese hamster lung (CHL) fibroblast cells in culture. In: Ashby J, de Serres FJ et al. (eds) Evaluation of short-term tests for carcinogens. Progress in Mutation Research, vol 5. Elsevier, Amsterdam, pp 427–432

Ishidate M Jr, Sofuni T, Yoshikawa K, Hayashi M, Nohmi T, Sawada M, Matsuoka A (1984) Primaray mutagenicity screening of food additives currently used in Japan. Fd Chem Toxic 22: 623–636

Ishidate M Jr, Harnois MC, Sofuni T (1988) A comparative analysis of data on the clastogenicity of 951 chemical substances tested in mammalian cell cultures. Mutat Res 195: 151–213

Koyama H, Utakoji T, Ono T (1970) A new cell line derived from newborn Chinese hamster lung tissue. Gann 61: 161–167

Ministry of Agriculture, Forestry and Fisheries of Japan (1985) Requirements for safety evaluation of agricultural chemical. Plant Protection Division, Agricultural Production Bureau

Ministry of Health and Welfare of Japan (1986) Guidelines for screening, toxicity testings of chemicals, law concerning the examination and regulation of manufacture, etc. of chemical substances.

Ministry of International Trade and Industry, Ministry of Environment, Ministry of Health and Welfare
Ministry of Health and Welfare of Japan (1984, 1989) Information on the guidelines of toxicity studies required for applications for approval to manufacture (import) drugs (part 1). Notification No. 118 of Pharmaceutical Affairs Bureau
Ministry of Labour of Japan (1979) On the standards of the mutagenicity test using micro-organisms in the labour safety and hygiene law. Labour Standard Bureau
Natarajan AT, Tates AD, Buul PPW van, Meijers M, de Vogel N (1976) Cytogenetic effects of mutagens/carcinogens after activation in a microsomal system in vitro. Mutat Res 37: 83–90
OECD (1988) Introduction to the OECD guidelines on genetic toxicology testing and guidance on the selection and application of assays, genetic toxicology. OECD
Sawada M, Ishidate M Jr (1978) Colchicine-like effect of diethylstilbestrol (DES) on mammalian cells in vitro. Mutat Res 57: 175–182
Sofuni T, Matsuoka A, Sawada M, Ishidate M Jr, Zeiger E, Shelby MD (1989) A comparison of chromosome aberration induction in the CHO and CHL cell systems by 25 compounds, Mutat Res (in press)
Thust R, Mendel J, Schwarz H, Warzok R (1980) Nitrosated urea pesticide metabolites and other nitrosamides. Activity in clastogenicity and SCE assays, and aberration kinetics in Chinese hamster V79-E cells. Mutat Res 79: 239–248
UKEMS Guideline (1983, 1989) Metaphase chromosome aberration assays in vitro. In: Scott D, Danford N, Dean B, Kirklan D, Richardson C (eds) UKEMS

Clastogenesis in Vitro Under Extreme Culture Conditions[1]

D. Scott[2,3]

Contents

1 Introduction

Approximately 50% of chemicals that are clastogenic in mammalian cells in vitro do not induce chromosome aberrations or micronuclei in the bone marrow of rodents (Thompson 1986; Ishidate et al. 1988). There are many possible reasons for this discrepancy (see Waters et al. 1988) including:

1. The absence of detoxification or excretion processes in vitro;
2. Metabolic processes unique to the in vitro bioassay that create active clastogens;
3. The insensitivity of the bone marrow test;
4. Extreme culture conditions producing clastogenesis in vitro that could not be achieved in vivo.

The International Commission for Protection against Environmental Mutagens and Carcinogens (ICPEMC) has commissioned a report on the latter topic covering various genotoxic end-points. This chapter presents a summary of that part of the ICPEMC report which relates to clastogenesis.

Four in vitro conditions have been considered as possibly generating "false positive" results; excessively high concentrations, high levels of cytotoxicity, extremes

[1] A working paper from ICPEMC Task Group 9.

[2] Chairman, ICPEMC Task Group 9. Other Task Group members: S.M. Galloway, R.R. Marshall, M. Ishidate Jr., D. Brusick, J. Ashby, and B.C. Myhr

[3] Paterson Institute for Cancer Research, Christie Hospital and Holt Radium Institute, Manchester, M20 9BX, Great Britain

of pH and the use of metabolic activation systems which in themselves may be clastogenic.

2 High Concentrations

2.1 High Osmolality

There have been a number of reports of clastogenesis at high concentrations of highly soluble non-DNA-reactive compounds which have been attributed to the increased osmolality of the culture medium rather than to the test compounds themselves. Some examples are sodium chloride (Ishidate et al. 1984; Galloway et al. 1987), potassium chloride (Galloway et al. 1987), sucrose (Ishidate et al. 1984; Galloway et al. 1987), urea (Oppenheim and Fishbein 1965; Ishidate 1988) and ethylene glycol (Ishidate 1988). Few of these studies have been sufficiently extensive to establish accurate dose/response relationships but a threshold response is suggested by some investigations (e.g. Ishidate et al. 1984; Ashby and Ishidate 1986). Typically, clastogenesis is observed at concentrations of > 50 mM but some salts of saccharin are clastogenic at approximately 20 mM (Ashby and Ishidate 1986) possibly as a result of high osmolality. In an attempt to avoid this problem the Japanese Guidelines (1986) for clastogenicity testing in vitro recommend the use of a maximum concentration level of 10 mM. If clastogenesis at these high concentrations is a consequence of the osmolality of the culture medium it may be better to define upper limits in terms of increase in osmolality. Clastogenesis associated with high osmolality has been observed when the increase is greater than 50 mOsmol kg^{-1}.

The mechanism(s) whereby high osmolality levels can result in chromosome aberrations is not understood but induction of intracellular ionic imbalance, leading to changes in chromatin structure and/or enzyme activity, has been suggested (Galloway et al. 1987). Not all chemicals are clastogenic at high osmolality (Ishidate et al. 1984, 1988; Galloway et al. 1987). For example, glycerol is non-clastogenic in CHO (Chinese hamster ovary cells) cells even with an increase of osmolality of more than 1100 mOsmol kg^{-1}, at a concentration of 1000 mM, probably because rapid equilibration across the cell membrane precludes osmotic stress (Galloway et al. 1987).

Clastogenesis occurring only at high levels of osmolality in vitro is unlikely to be relevant to human exposure. Although sodium saccharin is weakly mutagenic in rodents (Ashby 1985), this is at doses of about 10 g kg^{-1} because the chemical is tolerated at high levels in experimental animals (LD_{50} = 17 g kg^{-1} in mice). The relevance of such observations to human risk has been seriously questioned (Ashby 1985).

2.2 Lowest Effective Concentrations (LEC's) in Vitro for Chemicals Which Are Clastogenic in Vivo

The validity of recommending an upper concentration limit for in vitro testing in mammalian cells, to avoid osmolality effects, can be investigated by examining dose-reponse data in vitro for those chemicals which are clastogenic in vivo. Ishidate et al. (1988) have listed the lowest effective concentrations (LEC's) for in vitro clastogenicity for those chemicals which induce micronuclei in rodent bone marrow tests. (Designation of a lowest effective concentration does not necessarily imply the existence of a threshold. Extensive dose-response data are required for this purpose.) Thompson (1986) listed those chemicals which are clastogenic in vitro and induce either micronuclei or metaphase aberrations in bone marrow. From these reviews we present LEC's in vitro for 62 chemicals which are clastogenic in rodent bone marrow and are clastogenic in mammalian cells in vitro (Fig. 1, cross-hatched columns). We see that LEC's range over almost 10 orders of magnitude from 4.3×10^{-8} mM to as high as 100 mM. Thus, even for chemicals which are clastogenic in vivo, some require very high exposure concentrations to be detected in vitro. In part, this probably reflects the inadequacies of metabolic activation systems. Of the 62 in vivo clastogens, 7 (i.e. 11%) had LEC's of > 10 mM or were non-clastogenic at > 10 mM in at least one cell type (Table 1). However, the testing of these seven chemicals did not

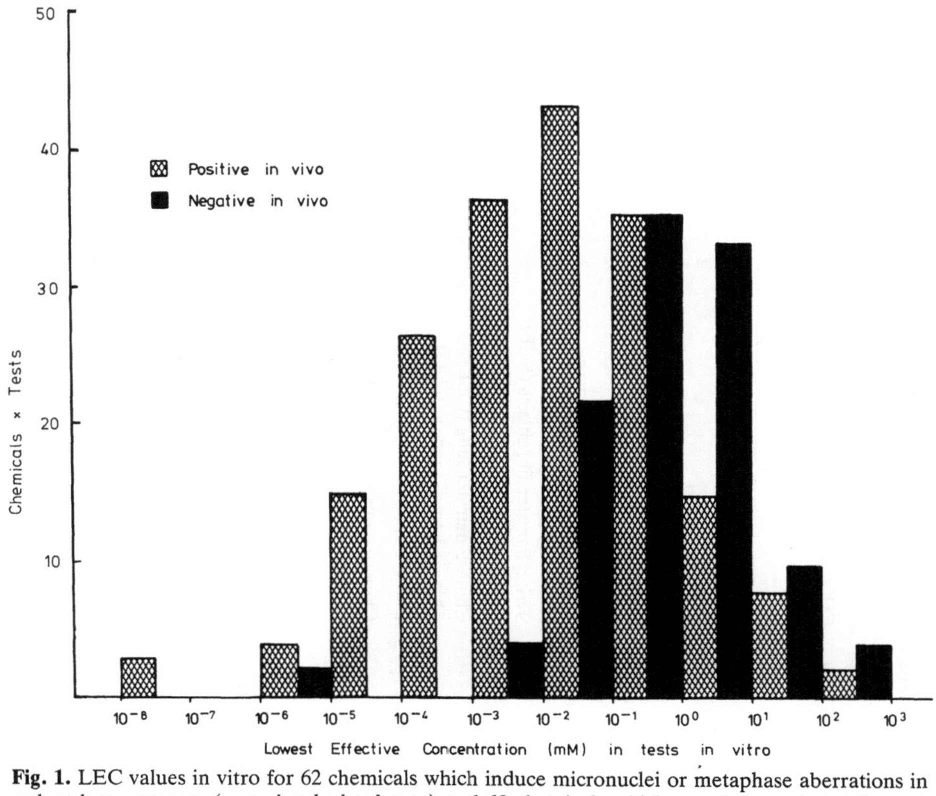

Fig. 1. LEC values in vitro for 62 chemicals which induce micronuclei or metaphase aberrations in rodent bone marrow (*cross-hatched columns*) and 68 chemicals which are clastogenic in vitro but not in vivo (*solid columns*). The height of each column is the product of the number of chemicals tested and the number of reported tests on each chemical, often in different cell types. Data from Ishidate et al. (1988) and Thompson (1986)

Table 1. Chemicals whose LEC values for clastogenicity in vitro in particular cell types are > 10 mM and which induce micronuclei or metaphase aberrations in rodent bone marrow (data from Ishidate et al. 1988)

Chemical	Cell type	With exogenous metabolic activation?	LEC (mM)	LED[a] in vivo ($mg\ kg^{-1}$)	Comments
Barbital	Chinese hamster lung (CHL)	–	11	330	
Benzene	Rat liver (RL4)	–	> 12.8	0.125	LEC < 10 mM in other cell types
Dimethylaminoazobenzene	Chinese hamster ovary (CHO)	+	> 55	185	LEC = 0.11 mM in Chinese hamster liver (CHI-L)
Dimethylnitrosamine	Chinese hamster ovary (CHO)	+	27		Difficult to activate?
	Human lymphocytes	+	50	18.5	
	Syrian hamster fibroblasts (SHF)	+	100		
Hexamethylphosphoramide	Rat liver (RL4)	–	> 11.2	0.4	LEC = 0.56 mM in human lymphocytes and 0.01 mM in CHI-L
	Chinese hamster lung (CHL)	–	33.5		
	Chinese hamster ovary (CHO)	+	> 56		
Tetramethylbenzidine	Chinese hamster ovary (CHO)	+	> 21	113	
Urethane	Chinese hamster fibroblasts (DON)	–	> 80	180	
	Chinese hamster lung (CHL)	–	90		
		+	225[b]		

[a] Lowest effective dose in vivo.
[b] Personal communication, M. Ishidate Jr. Not included in Fig. 1.

necessarily utilize modern testing protocols, particularly with respect to adequate metabolic activation and sampling times. Indeed many of the studies were not designed to establish LEC values. In carefully evaluating the data on these seven chemicals we conclude that the majority would probably have given positive results at < 10 mM in the cell type used if a rigorous testing protocol had been adopted. A striking exception is urethane which, even when thoroughly tested, gave very high LEC values (e.g. 90 mM in CHL cells, Table 1).

In conclusion, it is likely that with current protocols almost all in vivo clastogens would be detected if an upper concentration limit of 10 mM was adopted for in vitro testing. Certainly a significant number would be "missed" if the cut off concentration was lowered, whereas raising the level to 20 mM would begin to pick up effects associated with high levels of osmolality. If tests are performed at concentrations above 10 mM the osmolality of the culture medium should be measured. If there is a substantial increase (> 50 mOsmol kg^{-1}) and the chemical nature of the test agent does not suggest DNA reactivity, clastogenesis as a consequence of the high osmolality should be suspected.

2.3 LEC's in Vitro for Chemicals Which Are Non-Clastogenic in Vivo

Figure 1 gives LEC's of 68 chemicals which are clastogenic in vitro but not in vivo (solid columns). The distribution of LEC's differs from that for in vivo clastogens (cross-hatched columns) attesting to the importance of concentration dependency of in vitro clastogenesis in determining the probability of in vivo activity. Chemicals which are clastogenic in vitro at < 10^{-2} mM (more so at 10^{-3} mM) have a high probability of being clastogenic in vivo. However, as we have seen (Sect. 2.2) the converse is not the case; chemicals with high in vitro LEC's are not necessasrily non-clastogenic in vivo.

Only 10 of the 68 chemicals (i.e. 14%) had LEC's of > 10 mM. This, therefore, represents the maximum percentage of chemicals whose clastogenicity might be attributable to high osmolality. Consideration of the nature of these chemicals suggests that the proportion is likely to be less, at about 5%. Reasons other than osmolality effects must be sought for the majority of false positive results. One possibility which relates to concentration dependency is when clastogenesis is detected in vitro at concentrations (which may be quite low) which cannot be achieved in vivo because they exceed the tolerance of the test animals. Fluoride clastogenicity is an example. Of the various in vitro tests with fluoride, the lowest LEC (4.5 μg ml^{-1} for 48 h) has been found in human fibroblasts (Scott and Roberts 1987) where there was a fairly clearcut threshold response at this concentration. This concentration/duration in the plasma would be considerably higher than the LD_{50} in rodents (DeLopez et al. 1976; Lim et al. 1978) so it is not surprising that bone marrow clastogenesis tests have been negative (e.g. Martin et al. 1979). There would also appear to be a large safety margin for human exposure because even in areas with fluoridated water supplies the steady-state plasma level is only around 0.05 μg ml^{-1}.

However, there are likely to be relatively few chemicals, other than those producing osmolality effects at high concentrations, whose in vitro clastogenicity can be regarded as irrelevant to the situation in vivo by virtue of dose dependency. This is because any such chemicals would have to show a true threshold response at a concentration well above that which could be achieved in vivo. For DNA reactive agents there are very few examples of concentration thresholds for genotoxicity (Ehling et al. 1983) although there are some mechanisms of clastogenesis not involving direct DNA interaction which might be expected to be of the threshold type. For example, some chemicals which inhibit the enzymes involved in DNA synthesis and repair (e.g. hydroxyurea, cytosine arabinoside, aphidicolin; see Kihlman and Natarajan 1984) are clastogenic, and if the inhibition is not rate-limiting at low concentrations a true threshold response will be seen. Other possible examples are chemicals which are clastogenic through cellular energy depletion (e.g. cyanide; Kihlman 1957) and the production of active oxygen species (e.g. paraquat, Nicotera et al. 1985). The extent of dose-response studies required to demonstrate a threshold response for a particular chemical is very considerable (see Scott and Roberts 1987). Confidence in the existence of a threshold must come primarily from an understanding of the mechanisms involved.

A further difficulty in trying to predict in vivo activity even for chemicals which show threshold responses in vitro is that the sensitivity of cultured mammalian cells may not adequately reflect the sensitivity of cells in vivo. The insensitivity of some in vitro tests relative to in vivo responses is seen in Table 1, and Ishidate et al. (1988) cite some striking examples of differences in sensitivity to particular chemicals between different types of cultured cells and between different protocols using the same cell type.

3 High Levels of Cytotoxicity

Various guidelines for the in vitro clastogenicity testing recommend inclusion of cytotoxic dose levels, with the intention of demonstrating a biological response in the system. Since cytotoxicity may be less well tolerated in vivo than in vitro, the question arises whether upper cytotoxicity levels should be recommended for in vitro testing as has been the case for concentration levels (Sect. 2.1).

3.1 Direct and Indirect Clastogenesis and Cytotoxicity

The relationships between clastogenicity and cytotoxicity are complex because there are many clastogenic and cytotoxic mechanisms and many end points used in measuring cytotoxicity, some of which are measurements of cell death whereas others are not necessarily associated with lethality (see Sect. 3.2).

Clastogenesis can result from *direct* interaction of chemicals with DNA through covalent binding or intercalation, or through a wide variety of *indirect* mechanisms including interference with DNA replication and repair (see Kihlman and Natarajan 1984), interaction with specific chromosomal non-histone proteins (topoisomerase II and peripheral proteins; Gaulden 1987), nuclease release from lysosomes (Bradley et al. 1987), protein denaturation (Ashby et al. 1987), production of active oxygen species (Cerutti et al. 1983), cellular energy depletion (Kihlman 1957) and, of course, hypertonicity (Sect. 2.1) and pH changes (Sect. 4).

Agents which are clastogenic via direct or indirect mechanisms are also cytotoxic, either as a result of damage to the DNA itself (e.g. DNA strand breaks, chromosome aberrations) or to other cellular targets (e.g. enzymes, membranes, structural proteins). For some clastogens the cytotoxicity will be expressed mainly through effects on DNA, whether induced directly (e.g. alkylating agents) or indirectly (e.g. DNA synthesis inhibition) whereas for others the toxicity will be mediated mainly through damage to non-DNA targets (e.g. agents inducing lipid peroxidation, energy depletion, protein denaturation or ionic imbalance). A simplified scheme of the relationships between direct and indirect clastogenesis and cytotoxicity is shown in Fig. 2.

DNA degradation in association with necrosis of cultured mammalian cells has been reported after treatment with a variety of agents both genotoxic and non-genotoxic (Williams et al. 1974; Afanas'ev et al. 1986). The highly fragmented chromosomes sometimes seen in mitotic cells in clastogenicity assays at doses which induce significant cell death may be a manifestation of DNA degradation. Such a phenomenon, taking place whilst cells are dying, would not, of course, constitute a genetic hazard. This possibiliy is incorporated into Fig. 2.

3.2 Assays of Cytotoxicity

In clastogenesis assays, various methods are used to assess the associated cytotoxicity. Those which measure cell death include loss of colony-forming ability or loss of membrane integrity detected by using vital dyes, or enzyme assays which detect membrane leakage. The most popular methods are those which measure reduction in cell numbers or mitotic index; these detect cell death and/or a decreased proliferation rate. Other assays measure reduction in metabolic acitivty.

3.3 Quantitative Relationships Between Clastogenicity and Cytotoxicity

In attempting to establish quantitative relationships between clastogenicity and cytoxicity there are two major problems. For most cytotoxicity assays the extent of toxicity is markedly dependent upon the interval between treatment and sampling. This is illustrated for mitotic index (MI) determinations in Fig. 3 in which the MI has been measured in cultured human fibroblasts treated with various clastogens. Similar difficulties arise when membrane damage, cell counts or metabolic activity are used as end-points. In most clastogenesis studies, cytotoxicity is measured at only one sampling time. The most relevant assay from this point of view is colony-forming efficiency (CFE) because, not only is this a true measure of cell death, provided sufficient time elapses between treatment and observation to allow colony formation by surviving cells (usually 1 week or more), the surviving fraction remains constant, thus giving a single numerical value. Unfortunately, CFE is used infrequently as a measure of cytotoxicity in clastogenesis studies. The second problem is that when asynchronous populations are treated, the chromosome aberration yields also vary markedly with sampling time and usually only one or two harvesting times are used in clastogenicity tests.

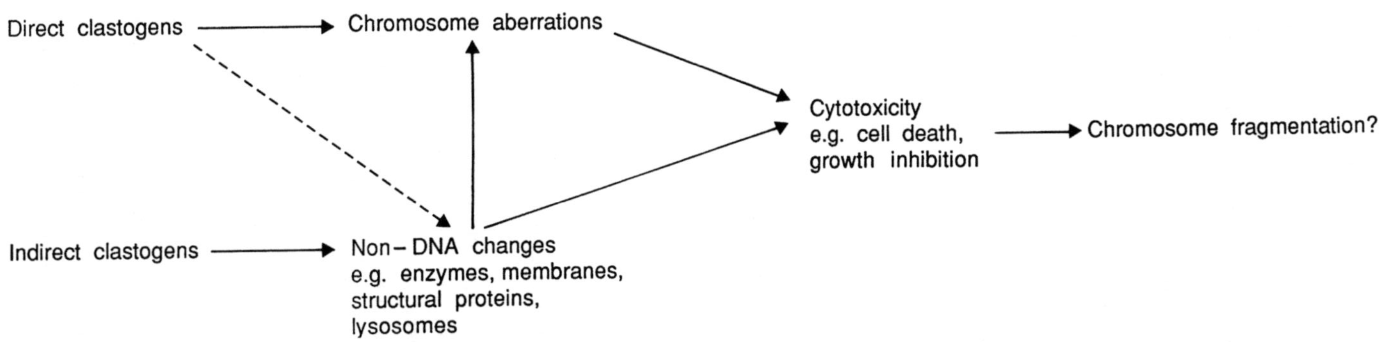

Fig. 2. Relationships between direct and indirect clastogenicity and cytotoxicity

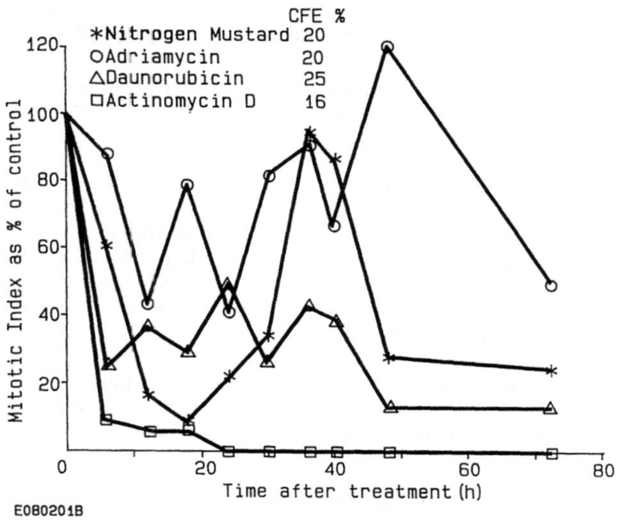

Fig. 3. Mitotic index as a percent of controls after treatment of asynchronous human fibroblasts with the clastogens, nitrogen mustard ($0.2\ \mu g\ ml^{-1}$), adriamycin ($0.25\ \mu g\ ml^{-1}$), daunorubicin ($0.1\ \mu g\ ml^{-1}$) or actinomycin D ($0.2\ \mu g\ ml^{-1}$) for 1 h. These doses induced a similar degree of loss of colony-forming efficiency (CFE). Corresponding aberration data summarized in Table 2

In view of these two problems it is not surprising that there are few useful quantitative data relating clastogenesis to cytotoxicity in vitro from which to make recommendations for in vitro testing. Nevertheless, it is clear from the limited data available that even chemicals giving a low clastogenesis/cytotoxicity ratio in vitro may be clastogenic in vivo, even in man. For example, Parkes and Scott (1982) examined in detail the relationship between chromosome aberration yields and CFE in cultured human fibroblasts treated with anti-tumour agents which are clastogenic in human lymphocytes or bone marrow in vivo (Table 2). Aberration yields were determined at 6-h intervals from 6 to 48 h after treatment. Although cell-killing levels ranged from 55–87% the average aberration yield over the entire sampling period did not exceed 15 aberrations per 100 cells. Peak aberration yields ranged from 8.5–100%, but clastogenicity tests are not designed to detect peak frequencies; an unacceptably large number of sampling times would be required. At 24 h, a time commonly used in testing, the yields were only 0–3% at 75–80% cell killing; at lower levels of killing these chemicals would not have been detectable as clastogens. Using MI as the end-point of cytotoxicity, the value was reduced by > 50% at 24 h (Fig. 3) in spite of the low aberration frequencies. Thus, even for chemicals which are clastogenic in vivo in man, testing must be done at high levels of cytotoxicity, at least in human fibroblasts. In a similar study in human fibroblasts with the potent in vivo clastogen mitomycin C, only a 20% aberration yield was observed at 80–90% cell killing and a 50% reduction in MI after a continuous 48-h exposure (Scott and Roberts 1987). In rodent bone marrow, mitomycin C clastogenesis appears to be detectable with minimal cell killing (Hayashi et al. 1984) suggesting that clastogenesis/cytotoxicity ratios can differ markedly between different cell types. Therefore, determination of such ratios in vitro may be of little value in predicting in vivo response.

It is conceivable that clastogenesis/cytotoxicity ratios are atypically low in human fibroblasts, which are not commonly used in clastogenicity tests. Unfortunately, for Chinese hamster cells, which *are* commonly used, there are few such detailed data

Table 2. Relationships between chromosome aberration yields and cell killing (loss of CFE) in human skin fibroblasts treated in vitro with anti-tumour agents[a] (Parkes and Scott 1982)

Chemical	Concentration[b] (μg ml^{-1})	Aberrations per 100 cells			Cell killing (%)
		Peak[c]	24 h[d]	Mean[e]	
Nitrogen mustard	0.2	10.2 (36 h)	0	4.6	80
Adriamycin	0.25	11.0 (6 h)	3.0	3.9	80
Daunorubicin	0.075	8.5 (6 h)	2.0	4.4	55
Daunorubicin	0.10	44.0 (6 h)	2.0	11.8	75
Daunorubicin	0.15	100.0 (6 h)	16.0	15.2	87

[a] Mitotic index data given in Fig. 3.
[b] Treatments were for 1 h.
[c] The time of maximum aberration frequency is given in parentheses.
[d] Yield at 24 h given because this is a sampling time regularly used in clastogenicity tests.
[e] The mean yield between 6 and 48 h post-treatment.

available, and for human lymphocytes CFE assays are not used. However, an indication that CHO cells may sustain more chromosome damage than human fibroblasts at a given level of cytotoxicity comes from a recent study by Armstrong, Galloway et al. (pers. comm.) in which CHO cells were treated with adriamycin or mitomycin C and sampled at 10 and 24 h after treatment (Table 3). Cytotoxicity was measured by reduction in CFE and MI. In contast to human fibroblasts, significant increases in aberration yields were detected at concentrations which produced little or no reduction in CFE and no reduction in MI (except 10 h after mitomycin C). However, not all in vivo clastogens have high clastogenesis/cytotoxicity ratios in CHO cells. Malathion is a relatively weak clastogen from this viewpoint when reduction in cell number is used to measure toxicity (Galloway et al. 1987). More such studies with known in vivo clastogens are required before clastogenicity/cytotoxicity ratios could be considered for use in predicting in vivo activity.

The possibility that one might be able to recommend upper cytotoxicity limits for testing would require that, for some chemicals, there be a threshold relationship between clastogenicity and cytotoxicity (i.e. that clastogenesis would only occur above a certain level of cytotoxicity) in the same way that in the relationship between clastogenicity and concentration a threshold response is believed to occur when the osmolality of the culture medium exceeds a certain level (Sect. 2.1). From the limited data available no such chemicals have been found and, indeed, if cell death is taken as the cytotoxic end-point, it is difficult to envisage mechanisms leading to cell death which are accompanied by clastogenicity only at high levels of killing and not at lower levels. For direct clastogens (Fig. 2) this would require that at lower concentrations the induced lesions produce lethal but no clastogenic events and that the latter are only produced when the burden of DNA lesions is above a certain level. For indirect clastogens the non-DNA changes (Fig. 1) at lower concentrations would have to lead exclusively to cell death and, only at higher doses, to both cell death and chromosome aberrations.

At our present state of knowledge it is not possible to define upper cytotoxicity limits for in vitro clastogenesis testing, and clastogenicity/cytotoxicity ratios in vitro cannot be used to predict in vivo response.

Table 3. Relationship between chromosome aberration, cell killing (loss of CFE) and mitotic index (MI) in CHO cells (Armstrong, Galloway et al., pers. comm.)

Chemical	Concentration[a] (μM)	Aberrations per 100 cells		MI		Cell killing (%)
		10 h	24 h	10 h	24 h	
Adriamycin	0	4	3	9.8	8.5	0
	0.10	4	2	12.2	10.8	0
	0.25	29	7	12.7	11.3	0
	0.50	85	13	8.8	11.6	11
	0.75	154	34	7.4	11.1	39
	1.00	460	85	2.0	12.5	68
Mitomycin C	0	3	4	15.9	10.5	0
	0.25	13	15	7.4	13.8	4
	0.75	20	140	7.5	16.2	25
	1.00	28	340	5.2	16.1	52
	2.00	38	530	4.2	10.0	90
	4.00	58	Toxic	0.9	6.4	98

[a] Treatments were for 3 h.

4 Extremes of pH

Thilager et al. (1984) first reported the clastogenic effects of low pH in mammalian cells after treatment of CHO cells with hydrochloric or acetic acid. However, aberrations were only found in cultures to which S9 mix had been added (details in Brusick 1986) suggesting that low pH may cause decomposition of S9 to generate clastogens (see Sect. 5). It is now clear, however, that low pH per se can be clastogenic. Morita et al. (1989) found that acidifying the culture medium of CHO cells with HCl or H_2SO_4 produced aberrations whether or not S9 was present. Without metabolic activation, initial pH levels of 5.5 and 4.5, respectively, were required for aberration induction by HCl and H_2SO_4. If an organic buffering system was used to maintain the acidity of the medium for the 24-h treatment period a pH level of 6.2 was found to be clastogenic. The chromosome damage did not result from decomposition products of the culture medium plus serum because phosphate-buffered saline acidified with HCl was also clastogenic at pH 5.2 (3-h treatment). High pH levels of up to 10.6 or 10.9, respectively, produced by addition of NaOH or KOH were not clastogenic in the absence of S9 (see Sect. 5).

Clearly, if test chemicals cause a reduction in pH of > 1 unit any observed chromosome damage could be an indirect consequence of acidification of the medium, at least in CHO cells. Neutralization of the medium and the use of enhanced buffering capacity may be required. Certainly, the effect of the test chemical on the pH of the treatment medium always be noted.

Not all cell types may suffer from this problem. Sinha et al. (1989) exposed rat lymphocytes to pH levels as low as 2.73 for 4 h, using HCl, with no evidence of clastogenesis.

Mechanisms of clastogenesis at low pH are not known.

5 The Clastogenicity of S9 Mix

There have been several reports of clastogenicity of S9 mix, particularly at non-physiological pH. However, clastogenesis at normal pH has been reported by Kirkland et al. (1989) who found that incubation of CHO cells with certain batches of Aroclor-1254 induced rat S9 mix gave exceptionally high aberration yields. These could be dramatically reduced by co-incubation with catalase or vitamin E implicating the involvement of active oxygen species (AOS). It is known that under certain conditions the cytochrome P-450 enzyme system which is induced in rat liver microsomes can generate AOS and that these are clastogenic (Phillips et al. 1984). The fact that the batches of S9 mix which were clastogenic in CHO cells were ineffective in whole blood cultures of human lymphocytes suggests that the AOS were inactivated by blood components.

Myhr and Mayo (1987) found that prolonged exposure of cultured L5178Y mouse lymphoma cells to S9 mix, or prolonged preincubation of S9 mix with culture medium prior to treatment of cells, induced small, trifluorothymidine-resistant mutant colonies which are thought to arise from clastogenic events (Clive et al. 1983).

Apart from the above studies of Kirkland et al. (1989) and Myhr and Mayo (1987) there have been reports of S9 clastogenesis under conditions of low or high pH. Thilager et al. (1984) found that S9 was clastogenic in CHO cells if the pH of the treatment medium was reduced to 5.5 (see Sect. 4). Cifone et al. (1987) observed an increased level of chromosome damage in mouse lymphoma cells exposed to S9 for 4 h at pH 6.2, and Morita et al. (1989) found that S9-containing medium was clastogenic in CHO cells at pH 6.0 (HCl) and 6.2 (H_2SO_4) for 6 h. However, in the latter study, low pH alone was clastogenic (Sect. 4) although there was an indication that a combination of low pH and S9 was more effective than low pH alone, suggesting the possibility of two different clastogenic mechanisms; one via low pH and the second via acidified S9. Morita et al. (1989) also found S9 to be clastogenic at high pH levels of 10.8 (with NaOH) or 10.4 (with KOH); high pH alone was not clastogenic.

If the production of active oxygen species is the mechanism whereby S9 is clastogenic the process may be acentuated at extremes of pH. Myhr and Mayo (1987) have warned that if a test chemical is found to be clastogenic only in the presence of S9 mix it may not be that the chemical requires metabolic activation but that it causes "decomposition" of S9 mix, generating clastogenic species. Clearly, a better understanding of the mechanism(s) of S9 clastogenesis is required.

The implications of the observation that S9 mix may be clastogenic are that batches should be screened for clastogenic activity and that extremes of pH should be avoided in tests with metabolic activation, as well as without activation (Sect. 4) by use of neutralization and enhanced buffering capacity. The use of lymphocytes in whole blood cultures may avoid or reduce these problems.

6 Concluding Remarks

It is now clear that false positive results in clastogenicity tests in vitro can result from the extreme culture conditions which sometimes arise in such tests, i.e. high osmolality or low pH produced by some test chemicals and by sub-optimal use of S9 mix. A better understanding of the underlying mechanisms is required. Recognition of these clastogenic conditions will help in the interpretation of in vitro clastogenicity data and improve the credibility of such tests in predicting in vivo response.

Acknowledgements. D. Scott is supported by the Cancer Research Campaign.

References

Afana's ev VN, Korol BA, Mantsygin YA, Nelipovich PA, Pechatnikov VA and Umansky SR (1986) Flow cytometry and biochemical analysis of DNA degradation characteristic of two types of cell death. Febs Lett 194: 347–350

Ashby J (1985) The genotoxicity of sodium saccharin and sodium chloride in relation to their cancer-promoting properties. Fd Chem Toxic 23: 507–519

Ashby J, Ishidate M Jr (1986) Clastogenicity in vitro of the Na, K, Ca and Mg salts of saccharin; and of magnesium chloride; consideration of significance. Mutat Res 163: 63–73

Ashby J, Mohammed R, Collander RD (1987) N-chloropiperidine and calcium hypochlorite: Possible examples of toxicity-dependent clastogenicity in vitro. Mutat Res 189: 59–68

Bradley MO, Taylor VI, Armstrong MJ, Galloway SM (1987) Relationships among cytotoxicity lysosomal breakdown, chromosome aberrations and DNA double strand breaks. Mutat Res 189: 69–79

Brusick D (1986) Genotoxic effects in cultured mamalian cells produced by low pH treatment and increased ion concentrations. Environ Mutagen 8: 879–886

Cerutti PA, Amstrad P Emerit I (1983) Tumour promoter phorbol-mysistate-acetate induces membrane mediated chromosome damage. In: Nygaard OF, Simic MG (eds) Radioprotectors and anticarcinogens. Academic Press, London, pp 527–538

Cifone MA, Myhr B, Eiche A, Bolcsfoldi G (1987) Effect of pH shifts on the mutant frequency at the thymidine kinase locus in mouse lymphoma L5178Y $TK^{+/-}$ cells. Mutat Res 189: 39–46

Clive D, Hozier J, Moore M (1983) "Single gene" and viable chromosome aberrations affecting the TK locus in L5178Y mouse lymphoma cells. Ann NY Acad Sci 407: 420–422

Ehling UH, Averbeck D, Cerutti PA, Friedman J, Greim H, Kolbye AC, Mendelsohn ML (1983) Review of the evidence for the presence or absence of thresholds in the induction of genetic effects by genotoxic chemicals. Mutat Res 123- 281–341

Galloway SM, Deasy DA, Bean CL, Kraynack AR, Armstrong MJ, Bradley MO (1987) Effects of high osmotic strength on chromosome aberrations, sister-chromatid exchanges and DNA strand breaks, and the relation to toxicity. Mutat Res 189: 15–26

Gaulden ME (1987) Hypothesis: some mutagens directly alter specific chromosomal proteins (DNA topoisomerase II and peripheral proteins) to produce chromosome stickness, which causes chromosome aberrations. Mutagenesis 2: 357–365

Hayashi M, Sofuni t, Ishidate M Jr (1984) Kinetics of micronucleus formation in relation to chromosomal aberrations in mouse bone marrow. Mutat Res 127: 129–137

Ishidate M Jr (1988) Data book of chromosomal aberration tests in vitro. Elsevier, Amsterdam

Ishidate M Jr, Sofuni T, Yoshikawa K, Hayashi M, Nohmi T, Sawada M, Matsuoka A (1984) Primary mutagenicity screening of food additives currently used in Japan. Fd Chem Toxic 22: 623–636

Ishidate M Jr, Harnois MC, Sofuni T (1988) A comparative analysis of data on the clastogenicity of 951 chemical substances tested in mammalian cell cultures. Mutat Res 195: 151–213

Japan Agency of Environment, Ministry of International Trade and Industry and Ministry of Health and Welfare (1986) The Guidelines for Toxicity Testing of Chemicals.

Kihlman BA (1957) Experimentally induced chromosome aberrations in plants. J Biophys Biochem Cytol 3: 363–380

Kihlman BA, Natarajan AT (1984) Potentiation of chromosomal aberrations by inhibitors of DNA repair. In: Collin SA, Downes CS, Johnson RT (eds). DNA repair and its inhibition. pp 319–339

Kirkland DJ, Marshall RR, Mc Enaney S, Bidgood J, Rutter A, Mullineux (1989) Aroclor-1254 induced rat liver S-9 causes chromosomal aberrations in CHO cells but not human lymphocytes: a role for active oxygen? Mutat Res 214: 115–122

Lin JK, Renaldo GJ, Chapman P (1978) LD_{50} of SaF_2, NaF and Na_2PO_3F in mouse compared with rat. Cancer Res 12: 177–179

Lopez OH de, Smith FA, Hodge HC (1976) Plasma fluoride concentrations in rats acutely poisoned with sodium flouride. Toxicol appl Pharmacol 37: 75–83

Martin GR, Brown KS, Matheson DW, Lebowitz H, Singer L, Ophaug R (1979) Lack of cytogenetic effects in mice or mutation in *Salmonella* receiving sodium flouride. Mutat Res 66: 159–167

Morita T, Watanabe Y, Takeda K, Ukumura K (1989) Effects of pH in the in vitro chromosomal aberration test. Mutat Res 225: 55–60

Myhr BC, Mayo JK (1987) Mutagenicity of rat-liver S9 to L5178Y mouse lymphoma cells. Mutat Res 189: 27–38

Nicotera TM, Block AW, Gibas Z, Sandberg AA (1985) Induction of superoxide dismutase, chromosomal aberrations and sister-chromatid exchanges by paraquat in Chinese hamster fibroblasts. Mutat Res 151: 263–268

Oppenheim JJ, Fishbein WN (1965) Induction of chromosome breaks in cultured normal human leukocytes by potassium arsenate, hydroxyurea and related compounds. Cancer Res 25: 980–983

Parkes DJG, Scott D (1982) A quantitative comparison of cytogenetic effects of anti-tumour agents. Cytogenet Cell Genet 33: 27–34

Phillips BJ, James TEB, Anderson D (1984) Genetic damage in CHO cells exposed to enzymically generated active oxygen species. Mutat Res 126: 265–271

Scott D, Roberts SA (1987) Extrapolation from in vitro tests to human risk; experience with sodium fluoride clastogenicity. Mutat Res 189: 47–58

Sinha AK, Gallapudi BB, Linscombe VA, McClintock ML (1989) Utilisation of rat lymphocytes for the in vitro chromosomal aberration assay. Mutagenesis 4: 147–153

Thilager AK, Kumaroo PV, Kott S (1984) Effect of low pH caused by glacial acetic acid and hydrochloric acid on chromosome aberrations in CHO cells (abstract). Toxicologist 4: 51

Thompson ED (1986) Comparison of in vivo and in vitro cytogenetic assay results. Environ Mutagen 8: 753–767

Waters MD, Bergman HB, Nesnow S (1988) The genetic toxicology of Gene-Tox non-carcinogens. Mutat Res 205: 139–182

Williams JR, Little JB, Shipley WU (1974) Association of mammalian cell death with a specific endonucleolytic degradation of DNA. Nature 252: 754–755

Life-Style and Genetic Factors that Determine the Susceptibility to the Production of Chromosome Damage

K. Morimoto[1]

Contents

1 Introduction

Chromosome alterations, which are directly visible changes in the DNA, have close associations to cancer development (German 1973; Carins 1981; Klein 1981). Lymphocytes in the circulating blood are human cell samples that can most easily be obtained, carrying information on both exposure doses and genetic effects of the mutagenic and carcinogenic factors. Thus, short-term cultures of phytohemagglutinin (PHA)-stimulated human lymphocyte are widely used to detect chromosome-damaging agents and possible human exposure to mutagens/carcinogens, and to study the immune response of blood.

Among various types of chromosomal changes, chromosome-structural aberrations and sister-chromatid exchanges (SCEs) are widely used indicators of chromosomal DNA damage; the former are efficiently induced by chemical adducts and modifications of DNA while the latter are produced directly by DNA-strand(s) breaks. Because the formation mechanisms of these two types of chromosome alterations are totally different, both end points are necessary to fully assess human genetic risks.

[1] Department of Hygiene and Preventive Medicine, Osaka University, School of Medicine, Nakanoshima, Kita-ku, Osaka 530, Japan

2 Cell Cycle Kinetics in Peripheral Lymphocyte Cultures

Although the lymphocytes are in the resting G_0 stage in the body, the cultures soon contain cells that have divided different times after phytohemagglutinin (PHA) stimulation. In our experiments, a simple combination of sister-chromatid differential staining, to determine how many times a cell has divided, and autoradiography, to determine when a cell synthesized DNA, was used to analyze the proliferating fate of human lymphocytes in culture. Cells were incubated contiuously with 5-bromodeoxyuridine (BrdU) and pulse-labelled with 0.1 μCi/ml ^{3}H-thymidine at various times after PHA stimulation. The cells were then harvested at 4-h intervals up to 72 h, and the percentage of labelled mitoses was determined separately in first, second, or third division cells. The data showed that cycling cells, whether they began cycling at earlier or later times after stimulation, had about the same generation times of 12–14 h (Figs. 1–3). This confirms that the heterogeneity of cell generations seen in short-term lymphocyte cultures is to a large part due to the difference in the times when cells began cycling in response to PHA (Morimoto and Wolff 1980; Morimoto et al. 1983).

Experiments have been carried out using human whole-blood cultures to determine the effects of sampling times and the duration of BrdU treatment before fixation on SCE frequencies following exposure to mitomycin-C (MMC). Cells were pulse-

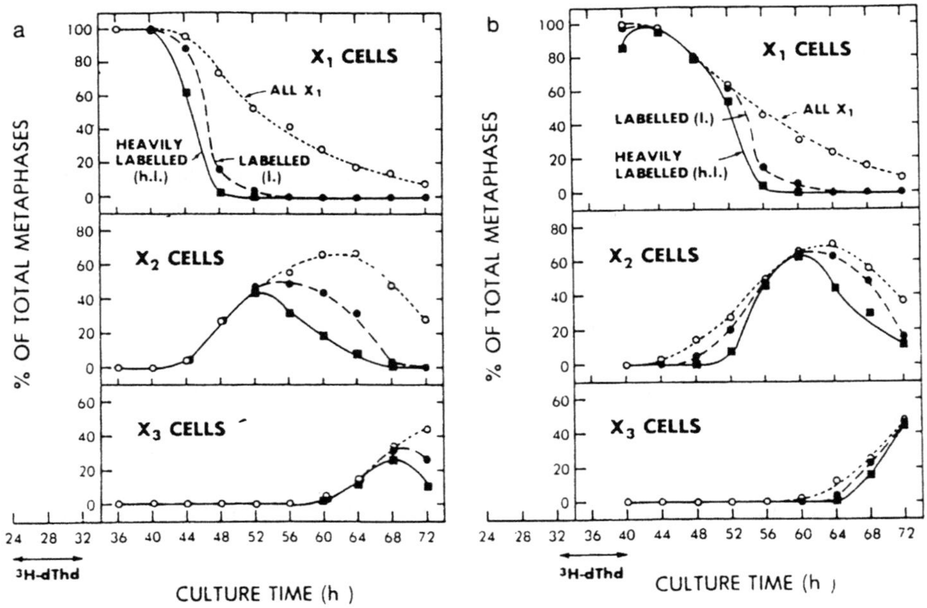

Fig. 1. a, b. Fate of proliferating lymphocytes after PHA stimulation. Cells were incubated continuously in medium containing BrdU and labelled with 0.1 μCi/ml^{3}H thymidine at **a** 24–32 h; or **b** 32–40 h following culture initiation. The relative percentages of ○ total; ● labelled (l.); ■ heavily labelled (h.l.) metaphases in the first (X_1), second (X_2), and third (X_3) generation cycle were plotted as a function of cultue time. For criteria used for the degree of labelling, see section 2

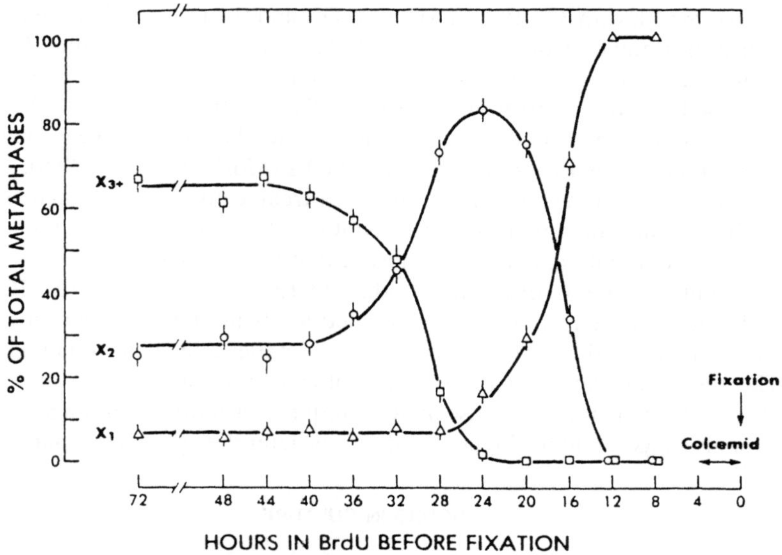

Fig. 2. Proportions of △, first (X_1); ○, second (X_2); and □, third or subsequent division (X_{3+}) cells in cultures exposed to BrdU for various times before fixation at 72 h after PHA stimulation. *Bars* = SE

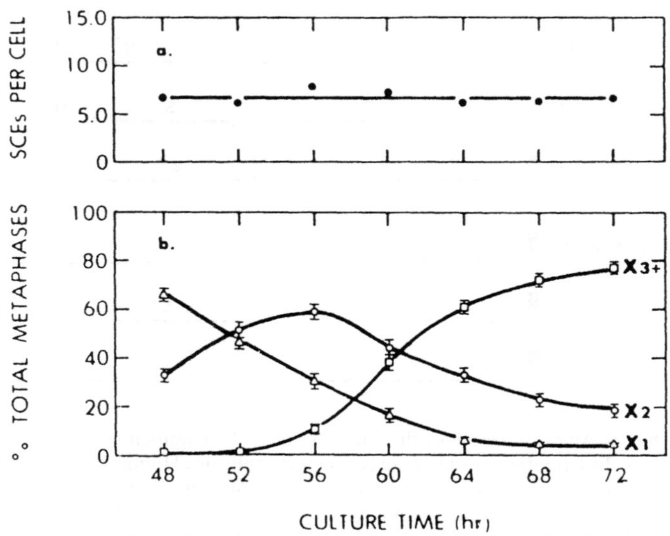

Fig. 3. a The frequency of SCEs and **b** the percentage of X_1, X_2, and X_{3+} cells after various times in culture. *Bars* = SE

treated for 1 h with 3 × 10^{-6} M MMC at G_1, and then sampled at 4-h intervals up to 88 h after stimulation of cultures with PHA. Results showed that this MMC treatment induced a 5–6-h proliferation delay per cell cycle, and that SCE frequencies first increased with time of fixation, peaking at 68 h, and then decreased. The data indicate that, in mutagen-treated cultures, lymphocytes having relatively longer cell-cycle times show a higher mean frequency of SCEs (Morimoto et al. 1985).

In a following experiment, cells were treated for 1 h with increasing doses of MMC or 4-nitroquinoline-1-oxide (4NQO) at 0, 24, or 48 h, and then fixed at 72 h after PHA stimulation. Results showed that the optimal treatment times at which the agents could most efficiently produce SCEs were different for MMC and 4NQO, and that the dose-response curves tended to "bend down" at very high doses; that is, treatments with very high doses induced smaller than expected numbers of SCEs. The results indicate that there is an optimal time for sampling at which one can observe the maximum increase in SCE frequencies following mutagen exposure, and strongly suggest that the higher the dose, the later the optimal sampling time (Fig. 4).

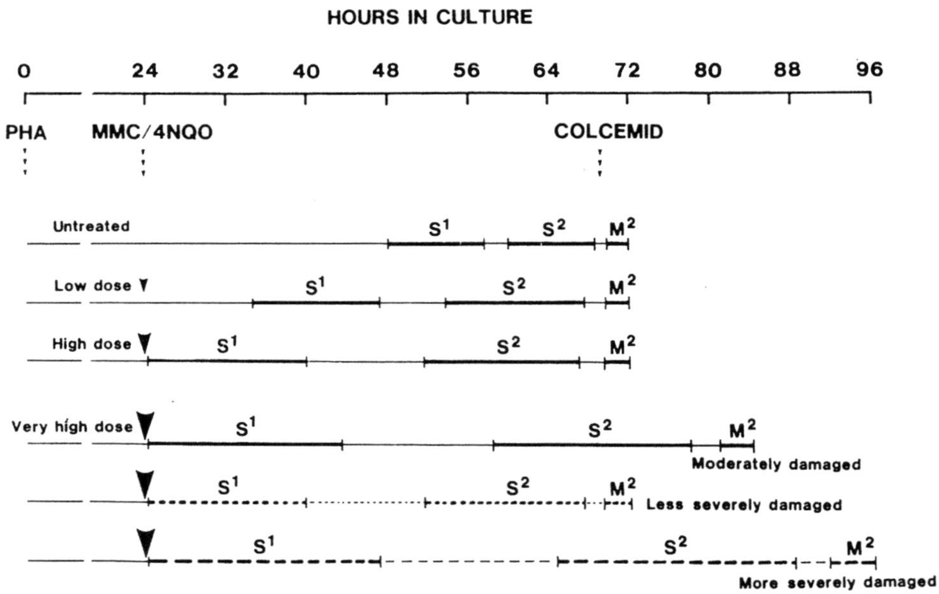

Fig. 4. An explanation for the "bending-down" phenomenon of dose-response curves seen after very high dose treatments of human lymphocyte cultures. Human lymphocytes commence their first DNA synthesis at different times, starting about 24 h after PHA stimulation of cultures, and continue thereafter to proliferate with a mean generation time of 12–14 h, whether they have begun cycling earlier or later. Treatments with increasing doses induce accordingly longer cell-cycle delays, and consequently lead to collections of subpopulations of X_2 metaphases that have started their first DNA synthesis at earlier times, when cultures were fixed at 72 h after initiation. When cells were exposed to very high doses that could induce longer than 24-h delays of the cell cycle as a whole, a subpopulation of cells that were moderately damaged would reach their second mitoses later than 72 h; 72-h cultures would contain only a subpopulation of cells that were less-severely damaged, thus, having relatively shorter cell-cycle delays. Such a differential sampling of less severely damaged cells having a lower than expected mean frequency of SCEs results in a bending-down of dose curves at very high doses

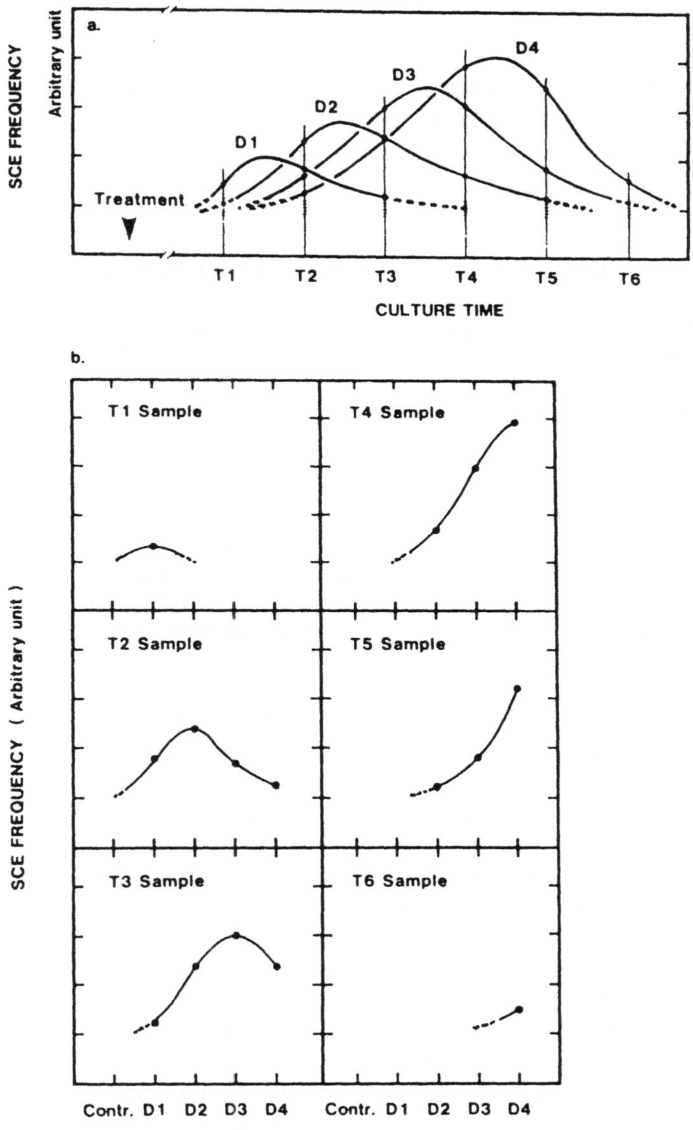

Fig. 5. a, b. Deformity of dose-response curves seen with sampling at suboptimal culture times after treatments. **a** A typical set of SCE curves with increasing doses (*D1-D4*) are shown as a function of cultue times (*T1-T6*). When cells are sampled for SCE analysis at various times after exposure to DNA-damaging agents, SCE frequencies usually show a peak when plotted as a function of fixation times. In general, the higher the dose, the later the optimal time of sampling for SCE analysis, because treatments with the higher doses that induce greater numbers of SCEs also induce longer cell-cycle delays. **b** Short-term SCE tests are usually carried out with only one sampling time. When earlier than optimal times are adopted for sampling, the dose-response curves may have a peak and show a "bending-down" phenomenon (*T1, T2, and T3*). On the other hand, with sampling at later than optimal times, the increases in SCE frequencies are smaller than expected and less marked (*T5* and *T6*). Only with adequate sampling times do the dose-response curves show a clear dose dependence (*T4*). It should also be noted here that the shapes of the dose-response curves readily change depending on the balance between the lesions that lead to SCE formation, those that lead to cell-cycle delays and those that result in cell lethality

Because of the apparent deformity of dose-response curves obtained after various treatments and sampling times, it seems necessary that extra fixation-time points be included in test protocols so as to avoid false negatives or confirm possible positives (Fig. 5).

3 Human Health Situations and Chromosome Alterations

3.1 Proliferative Kinetics and Mitomycin C-Induced Chromosome Damage in Fanconi's Anemia (FA) Lymphocytes

To investigate effects of hereditary factors on the sensitivity of the induction of chromosome alterations, peripheral blood samples were drawn from patients with FA (sisters, aged 7, 9, and 11 years), their father (34 years), and their mother (36 years) (Miura et al. 1983). Healthy, unrelated adults were examined as controls. Whole blood from each donor (0.2 ml) was added to 5 ml RPMI 1640 medium containing 15% fetal bovine serum and 20 μM BrdU. After the stimulation by PHA-M the cultures were incubated at 37 °C for 96 h.Colcemid (2×10^{-7} M) was added to each culture before fixation for 3 h. The cells were then collected by centrifugation, and air-dried chromosome preparations were made. A modification of the fluorescence-plus-Giemsa method (Goto et al. 1978) was used to obtain harlequin chromosomes. Cells dividing for the first (X_1), second (X_2), third or more (X_{3+}) time in culture can be determined in such preparations (Tice et al. 1976; Morimoto and Wolff 1980).

For chemical treatments, MMC was first dissolved in distilled water and after necessary dilutions in phosphate-buffered saline (PBS) aliquots of this freshly made solution were added to each culture to give the appropriate final concentration for the final 48 h of the 96-h culture.

Two hundred metaphase cells were scored for chromosomal aberrations and cell-cycle distribution, and 25–35 second-division metaphase cells were scored for SCEs per point per person.

Lymphocytes from FA patients showed consistently higher frequencies of SCEs than normal cells in MMC-treated and nontreated cultures (Fig. 6a). The ratio of the average SCE frequency in FA cells to that in normal cells was approximately 1.4 (Fig. 6b). Chromosomal aberrations were examined separately in X_1, X_2, and X_{3+} metaphase cells. In each division metaphase, FA cells showed several times higher frequencies of chromosomal aberrations than normal cells (Fig. 7a,b). When lymphocytes from untreated cultures fixed at 96 h were examined, FA cells showed much slower proliferation kinetics than normal cells (Fig. 8a,b), manifested as higher frequencies of X_1 and X_2 metaphases.

3.2 Genotoxicity of Diesel Exhaust Emissions

We have performed a series of experiments to investigate how genetic factors determine the sensitivity to the induction of chromosome alterations (Morimoto et al. 1986 a, b). The chromosomes in cells from patients with cancer-prone hereditary

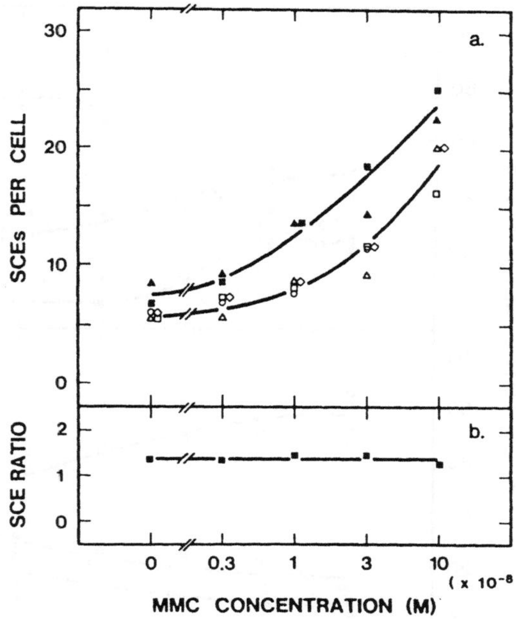

Fig. 6. **a** Frequencies of SCEs in FA (*solid symbols*) and normal (*open symbols*) X_2 cells exposed to various concentrations of MMC. **b** Ratio of SCEs in FA cells to SCEs in normal cells

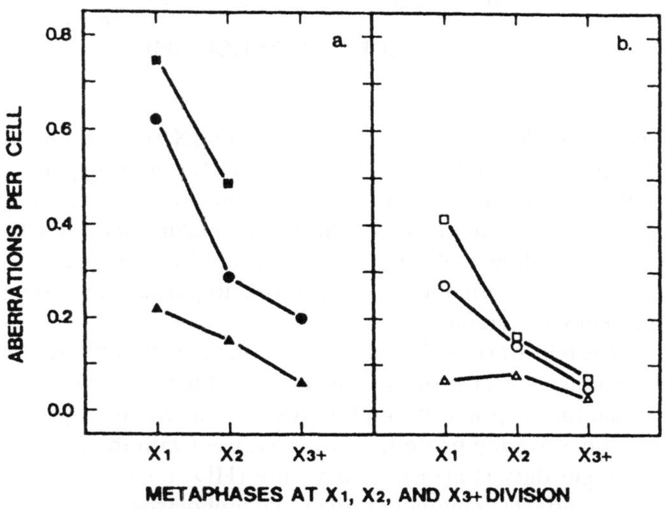

Fig. 7. **a, b** Frequencies of chromosomal aberrations in **a** FA cells and **b** normal cells in their 1st (X_1), 2nd (X_2), and 3rd or subsequent (X_{3+}) divisions in cultures exposed to various concentrations of MMC: ▲ △, OM; ●○, 3×10^{-9} M; ■ □, 3×10^{-8} M

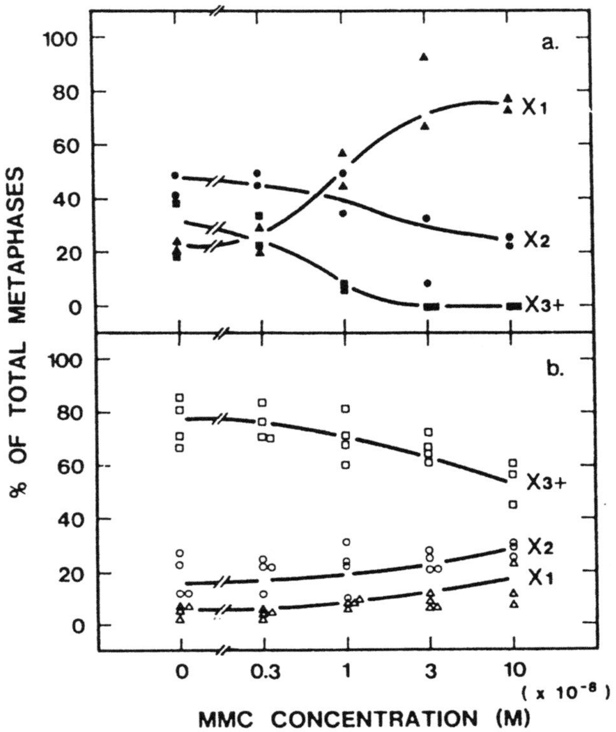

Fig. 8. a, b. Percentages of 1st (X_1) (▲△), 2nd (X_2) (●○), and 3rd or subsequent (X_{3+}) (■□) division metaphase cells from **a** FA patients and **b** control subjects in cultures exposed to various concentrations of MMC

diseases such as xeroderma pigmentosum (XP), ataxia telangiectasia (AT), Fanconi anemia (FA), and Bloom syndrome (BL) show hypersensitivity to some typical mutagens/carcinogens. Little information, however, is available on the susceptibilities of these cells to induction of chromosome damage by actual environmental genotoxicants. We show that XP and BL cells are hypersensitive, but FA cells are hyposensitive to SCE formation when exposed to particulates from diesel-engine emissions, a mixture of carcinogens.

Epstein-Barr (EB) virus-transformed lymphoblastoid cells from patients with hereditary cancer-prone diseases were cultured in a medium of the same composition as that for the primary lymphocytes for two cell cycles (40 h). They were treated with the same fixation procedures as those used with the primary lymphocytes.

Light-duty (LD) and heavy-duty (HD) tar samples were investigated for cytogenetic effects in human peripheral lymphocytes and lymphoblastoid cells. Both LD and HD samples induced SCEs in PHA-stimulated human lymphocyte cultures in a dose-dependent fashion (Fig. 9). LD samples seemed to be more potent in inducing chromosome damage than HD samples.

Experiments using hypersensitive lymphoblastoid cell lines from XP, FA, BL, and AT revealed that the diesel particulate extracts produced a great variability in the cytogenetic responses of these human cells from persons with hereditary diseases (Fig. 10).

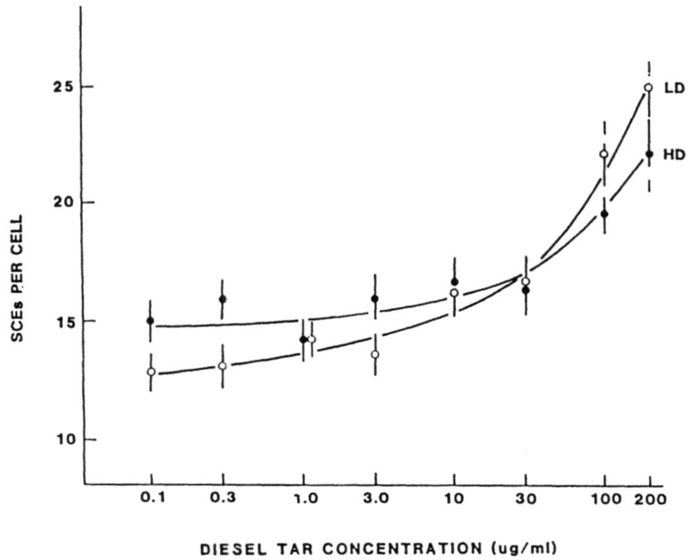

Fig. 9. Induction of sister-chromatid exchanges (SCEs) in human lymphocytes treated with LD and HD tar samples. *Bars* = SE

3.3 Studies of Chromosome Sensitivities in Twin Children

To elucidate the contribution of genetic factors in determining chromosomal sensitivity to genotoxicant exposure in healthy persons, we have investigated chromosome damage in lymphocytes from mono- and dizygotic twins (Miura et al. 1987). Cells from monozygotic twins are expected to have virtually identical genetic constitutions, whereas those from dizygotic twins may show genetic variations as observed between normal siblings. We investigate the frequencies of chemically induced SCEs or radiation-induced chromosome aberrations in lymphocytes drawn from twin children, and compared the intraclass correlation coefficients between monozygotic (MZ) and dizygotic (DZ) twins. Analyses of baseline and mitomycin-C-induced SCEs and radiation-induced chromosome aberrations revealed that intraclass correlation coefficients were significantly higher for MMC-induced SCE frequencies among monozygotic twins (Table 1).

4 Effects of Life-Styles on Chromosome Alterations

4.1 Effects of Life-Styles on SCE Frequencies in Peripheral Lymphocytes

A great deal of epidemiological and clinical studies have shown that ways of daily living as well as exposure to environmental carcinogens determine the risks for cancer development (Berkman and Breslow 1983; Morimoto 1986). Chromosome altera-

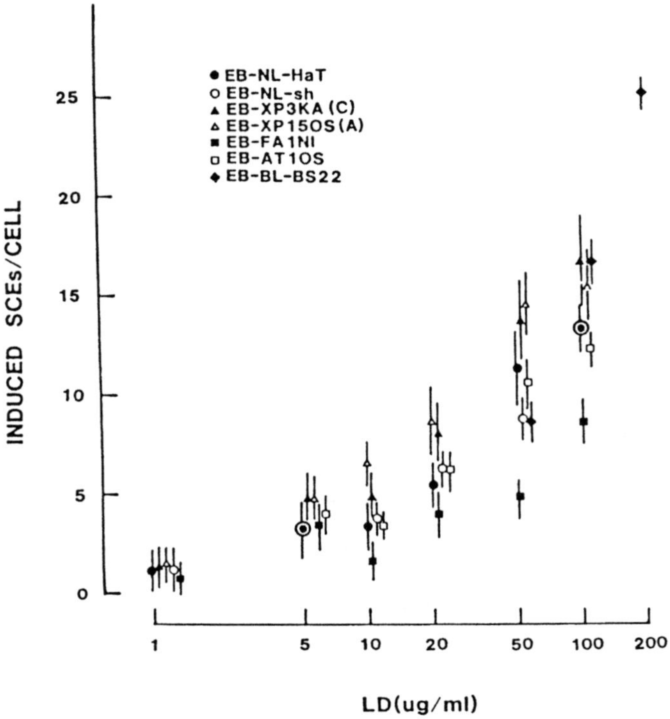

Fig. 10. Induction of SCEs by treatments with LD tar samples in EB-virus transformed human lymphoblastoid cells from normal persons (*NL*) and from patients with xeroderma pigmentosum (*XP*), Bloom syndrome (*BL*), Fanconi anemial (*FA*), and ataxia teleangiectasia (*AT*) showing increased risks of cancer development

Table 1. Frequencies of SCE and D+R in lymphocytes from monozygotic (MZ) and dizygotic (DZ) twins

	MZ	DZ
SCE (Baseline)	7.30 ± 0.12[a]	7.23 ± 0.17
SCE (MMC-induced)	26.59 ± 0.65	25.30 ± 0.89
D+R (γ-irradiation)	0.212 ± 0.015	0.274 ± 0.019
D+R (γ-irradiation + ara-C)	1.147 ± 0.228	1.602 ± 0.104

[a] Mean ± SE.

tions have close correlations with the initiation of cancer development. It is interesting to see the relationships between the level of chromosome damage and daily life styles such as (1) sleeping pattern; (2) working pattern; (3) cigarette smoking; (4) alcohol consumption; (5) leisure-time physical activity; (6) nutritional balance; and (7) subjective mental stress.

We have, therefore, conducted a series of experiments to quantitatively analyze the relationship between chromosome sensitivities and life styles. Blood samples were obtained from randomly selected persons who were examined as part of the multiphasic health check-ups that took place in the Katsushika Ward, Tokyo, Japan.

To obtain life style information, a self-recording questionnaire was used. The questionnaire covered the above-mentioned health-releated practices, medical history, and standard demographic variables. Answers were checked and if necessary completed by a brief interview by the investigators. The Health Practic Index (HPI), i.e., number of good life styles practiced, was calculated. The Health Practice Index value (mean ± SD) was 4.62 ± 1.50 for males and 5.51 ± 1.31 for females.

When baseline SCE frequencies were correlated with the HPI scores, a clear tendency was observed, i.e., cells from those persons who have low HPI scores show higher SCE frequencies than cells from those who have high HPI scores. When SCE frequencies were analyzed by one-way analysis of variance, the tendency of decreasing SCEs with increasing HPI score was statistically significant ($p < 0.05$) for baseline SCEs in male subjects.

4.2 Life-Styles Determining Susceptibility to the Induction of SCE by Mutagens

Experiments have also been conducted on blood samples from the same subjects to examine whether life style affects the cytogenetic sensitivities to mutagen/carcinogen exposure. For treatment, mitomycin-C (MMC) was first dissolved in distilled water and diluted by phosphate-buffered saline (PBS). Aliquots of this freshly made solution were added to the cultures to give a final concentration of 3×10^{-8} M. Cells were treated with MMC for the entire culturetime, i.e., 72 h. The results show that cells from male subjects who have good life styles show lower SCE frequencies than cells from those who have poor life styles. This suggests that life styles influence chromosome susceptibility to mutagen exposure.

4.3 Life-Styles Determining Sensitivity to Chemical Inhibition of Chromosome Repair

In our daily environment, we are exposed to a mixture of suspected hazardous agents containing substances which inhibit repair of DNA damage. To investigate the effects of life styles on repair kinetics of radiation-induced chromosomal damage, we analyzed the inhibitory effects of cytosine arabinoside (ara-C) on the rejoining of DNA-breaks. It has been shown that ara-C inhibits the repair of radiation-induced chromosomal aberrations via incorporation into DNA (Preston 1980; Ohno et al. 1988).

For irradiation, freshly drawn blood (0.3 ml) mixed with 4.7 ml culture medium was irradiated with ^{137}Cs-γ-rays or tritiated water (HTO)-β-rays at 37 °C. Ara-C dissolved in distilled water was added to the culture medium during irradiation to give a final concentration of 50 μM. When the irradiation was completed, cells were washed with PBS three to five times and mixed with complete medium containing PHA to start the culture.

The enhancement ratio (E.R.) was calculated for each subject: $E.R. = (D+R)_{radiation+ara\text{-}C}/[(D+R)_{radiation} + (D+R)_{ara\text{-}C}]$, where $(D+R)_{radiation+ara\text{-}C}$ denotes the frequencies of dicentric and ring chromosomes induced by irradiation and ara-C treatment, etc. When the mean ara-C enhancement ratio was calculated for three HPI categories, a clear, but not statistically significant, tendency was observed.

Our results strongly suggest that it is necessary to take into consideration not only hereditary predisposition but comprehensive life styles to assess the repair inhibition in the human population.

5 Adaptation-Like Response to the Induction of SCE by Mutagens

Experiments have been performed to determine whether human lymphocytes in primary cultures can show an adaptation-like response to the induction of cellular lesions (manifested as SCEs) as previously found in bacteria and established in human and mammalian cell lines (Morimoto et al. 1986a). Human lymphocytes were pretreated with various subtoxic concentrations (5–50 ng/ml) of N-methyl-N'-nitro-N-nitrosoguanidine (MNNG) once every 6 h for 72 h, and subsequently challenged by a high dose (4μg/ml) of MNNG (Fig. 11). The lymphocytes in MNNG-challenged cultures had the lowest frequency of SCEs when pretreated with 10 ng/ml MNNG (Fig. 12). Further studies on cross-resistance revealed that repeated pretreatments of lymphocytes with 10 ng/ml MNNG for 72 h can render the cells resistant to the induction of SCEs by the following challenge with a high dose of MNNG, but not of MMC or ENU. The data also indicate an interindividual variation in the degree of the adaptation-like response (Fig. 13).

6 New in Vivo Test for Extraporating Animal Data to Human Risk Assessment

An improved host-mediated assay system for evaluating the mutagenic potential of environmental agents, and for directly comparing the difference in sensitivities between humans and mice to the induction of SCEs by cyclophosphamide (CP), or chromosome aberrations by γ-rays was investigated. Isolated human peripheral lymphocytes were injected into mice peritoneal cavities, and exposed to CP or γ-rays in

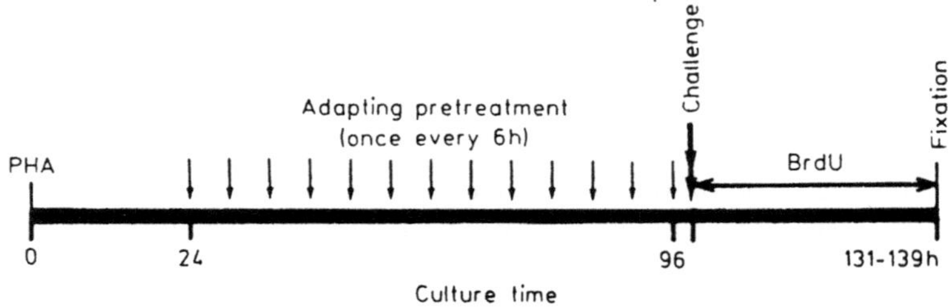

Fig. 11. Treatment protocol of the experiments on adaptive response. Human lymphocytes were repeatedly pretreated with subtoxic doses of MNNG once every 6 h for 72 h starting 24 h after PHA stimulation, and exposed to a challenge dose of MNNG, ENU, or MMC for 1 h, and then cultured in medium containing BrdU for two rounds of DNA replication (32–40 h before fixation)

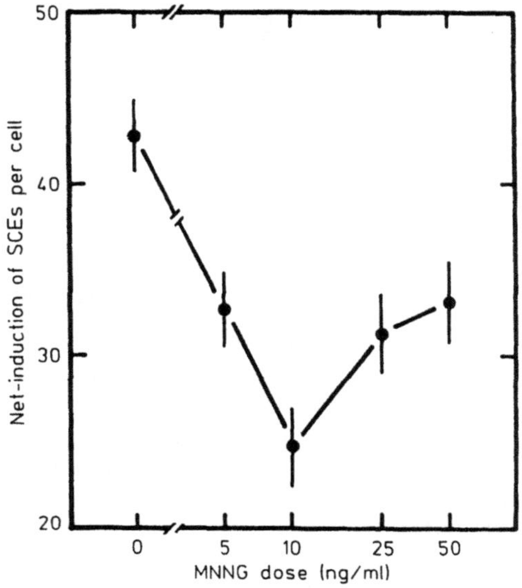

Fig. 12. Adaptive effect of repeated pretreatments with low doses of MNNG (0–50 ng/ml) on the induction of SCEs by the following challenge treatment with MNNG (4μg/ml) in cultured human lymphocytes. *Bars* indicate SE of four blood donors

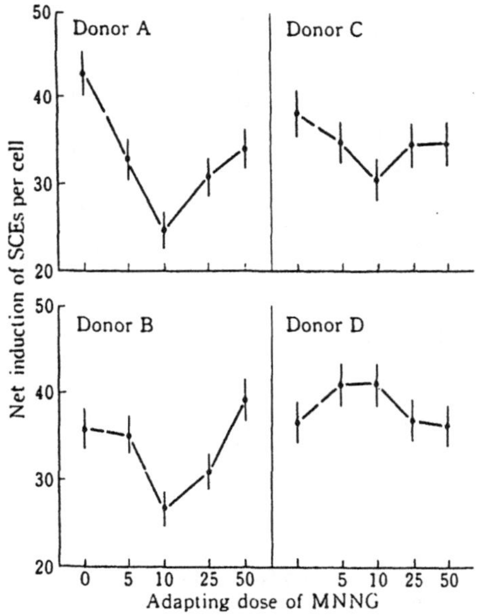

Fig. 13. Intraindividual variation in adaptive response of peripheral lymphocytes

vivo and/or in vitro. Cells recollected from the peritoneal cavities of mice were stimulated with PHA (human lymphocytes) or lipopolysaccaride (mouse lymphocytes) (Fig. 14). The results show almost the same sensitivities between human and mouse lymphocytes with respect to chemical induced SCEs and radiation-induced chromosome aberrations. This assay system seems to be a highly useful tool for human risk assessment by environmental hazardous agents using data obtained in animal experiments.

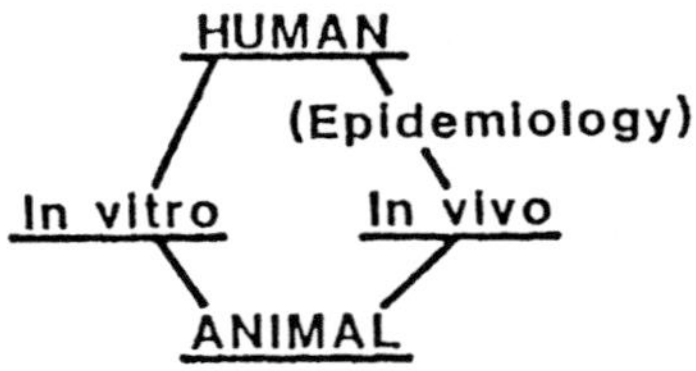

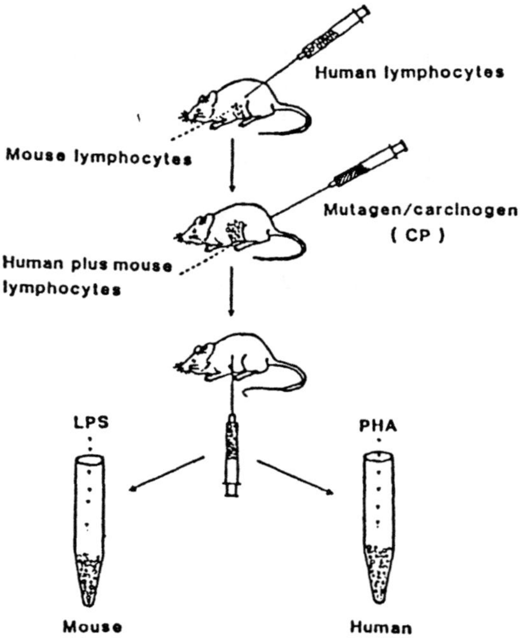

Fig. 14. Experimental procedure for a host-mediated assay with human peripheral lymphocytes and mice

References

Berkman LF, Breslow L (1983) Health and ways of living. Oxford University Press, New York

Carins J (1981) The origin of human cancers Nature 289: 353–357

German J (1973) Oncogenic implications of chromosomal instability. Hosp Pract February: 93–104

Goto K, Maeda S, Kano Y, Sugiyama T (1978) Factors involved in differential Giemsa-staining of sister chromatids. Chromosoma (Berl) 66: 351–359

Klein G (1981) The role of gene dosage and genetic transpositions in carcinogenesis. Nature 294: 313–318

Miura K, Morimoto K, Koizumi A (1983) Proliferative kinetics and mitomycin C-induced chromosome damage in Fanconi's anemia lymphocytes. Hum Genet 63: 19–23

Miura K, Morimoto K, Koizumi A, Yamada K, Asaka A (1987) The frequencies of the chromosomal aberration and the sister chromatid exchange in lymphocytes from twins. Jpn J Hum Genet 32: 187 (Meeting abstract)

Morimoto K (1986) Life-styles and health situations: importance of genetic health and quality of life. Position papr for the workshop on social learning and coping options. WHO Document of the International Conference on Health Promotion, WHO/EURO Office, Copenhagen

Morimoto K, Wolff S (1980) Cell cycle kinetics in human lymphocyte cultures. Nature 288: 604–606

Morimoto K, Sato M, Koizumi A (1983) Proliferative kinetics of human lymphocytes in culture measured by autoradiography and sister chromatid differential staining. Exp Cell Res 145: 349–356

Morimoto K, Sato Mizuno M, Koizumi A (1985) Sister-chromatid exchanges and cell cycle kinetics in human lymphocyte cultures exposed to alkylating mutagens: apparent deformity in dose-response relationships. Mutat Res 152: 187–196
Morimoto K, Sato-Mizuno M, Koizumi A (1986 a) Adaptation-like response to the chemical induction of sister chromatid exchanges in human lymphocytes. Hum Genet 73: 81–85
Morimoto K, Kitamura M, Kondo H, Koizumi A (1986b) Genotoxicity of diesel exhaust emissions in a battery of in-vitro short-term and in-vivo bioassays. In: Ishinishi N, Koizumi A, McClellan RO, Stöber E (eds) Carcinogenic and mutagenic effects of diesel engine exhaust. Elsevier, Amsterdam
Ohno Y, Spriggs D, Matsukage A, Ohno T, Kufe D (1988) Effects of 1-β-D-arabinofuranosylcytosine incorporation on elongation of specific DNA sequences by DNA polymerase β. Cancer Res 48: 1494–1498
Preston RJ (1980) The effect of cytosine arabinoside on the frequency of X-ray-induced chromosome aberrations in normal human lymphocytes. Mutat Res 69: 71–79
Tice R, Schneider EL, Rary JM (1976) The utilization of bromodeoxyuridine incorporation into DNA for the analysis of cellular kinetics. Exp Cell Res 102– 232–236

Subject Index

MIX
Papier aus verantwortungsvollen Quellen
Paper from responsible sources
FSC® C105338

If you have any concerns about our products,
you can contact us on
ProductSafety@springernature.com

In case Publisher is established outside the EU,
the EU authorized representative is:
Springer Nature Customer Service Center GmbH
Europaplatz 3, 69115 Heidelberg, Germany

Printed by Libri Plureos GmbH
in Hamburg, Germany